THE ROCHE PROBLEM

ASTROPHYSICS AND SPACE SCIENCE LIBRARY

A SERIES OF BOOKS ON THE RECENT DEVELOPMENTS
OF SPACE SCIENCE AND OF GENERAL GEOPHYSICS AND ASTROPHYSICS
PUBLISHED IN CONNECTION WITH THE JOURNAL
SPACE SCIENCE REVIEWS

VOLUME 152
CURRENT RESEARCH

THE ROCHE PROBLEM

AND ITS SIGNIFICANCE FOR DOUBLE-STAR ASTRONOMY

by

ZDENĚK KOPAL

Department of Astronomy,
University of Manchester, England

KLUWER ACADEMIC PUBLISHERS

DORDRECHT / BOSTON / LONDON

Library of Congress Cataloging in Publication Data

Kopal, Zdeněk, 1914-
The Roche problem and its significance for double-star astronomy / Zdeněk Kopal.
p. cm. -- (Astrophysics and space science library)
Bibliography: p.
Includes index.
ISBN 0-7923-0129-3
1. Stars, Double. 2. Roche equipotentials. I. Title. II. Series.
QB421.K66 1989
523.8'41--dc19 88-37515

ISBN 0-7923-0129-3

Published by Kluwer Academic Publishers,
P.O. Box 17, 3300 AA Dordrecht, The Netherlands.

Kluwer Academic Publishers incorporates
the publishing programmes of
D. Reidel, Martinus Nijhoff, Dr W. Junk and MTP Press.

Sold and distributed in the U.S.A. and Canada
by Kluwer Academic Publishers,
101 Philip Drive, Norwell, MA 02061, U.S.A.

In all other countries, sold and distributed
by Kluwer Academic Publishers Group,
P.O. Box 322, 3300 AH Dordrecht, The Netherlands.

Printed in The Netherlands

LIST OF CONTENTS

EDOUARD ALBERT ROCHE
(1820 - 1883)

PREFACE

The words of this preface were written when the book was ready to go to the press; and are limited to only a few points which are best made in this place.

As is intimated by the sub-title, the whole volume was written with applications in mind to double-star astronomy. The latter is, however, not the only branch of our science which could benefit from its contents. The same is true of certain aspects of the dynamics of stellar systems or galaxies (the stellar populations of which are also characterized by the fact that the mean-free-path of their constituent stars are long in comparison with the dimensions of the respective systems); the central condensations of which are high enough to approximate the gravitational action of a "mass-point". This fact did not, to be sure, escape the attention of previous investigators (in the case of globular clusters, in particular, the Roche model was introduced in their studies under the guise of polytropic models characterized by the index $n = 5$); though no particular attention will be paid to these in this book.

But possible applications of the Roche model are not limited to problems arising in stellar astrophysics. With Coulomb forces replacing gravitation, the equilibrium model finds a close analogy in the field of electrostatics—as was pointed out already at the beginning of this century by (then young) J. H. Jeans (cf. his *Mathematical Theory of Electricity and Magnetism*, Cambridge University Press, 1907, pp. 50-56). Jeans knew the astronomical problem well, but did not seem to have followed up the analogy between gravitational and electrostatical equipotential surfaces to the extent which the respective phenomena would have deserved. Naturally we know much more on the underlying problems now than the future Sir James did 80 years ago; and it is only to be hoped that many results presented in this book under the name of Roche Problem may become of interest also in domains beyond the limits of more classical astronomy!

Last, but not least, it is for the fourth time in the past half a century that the present writer addressed himself to the problems arising in connection with the Roche model. This preface may perhaps give him the last opportunity to apologize for misprints which crept into some of his earlier publications at the time when these were printed by letterpress offering but limited opportunities to correct them. It is to avoid such a situation as far as it is humanly possible that the author accepted the invitation of his Publishers to have the contents of this book set up at the University of Manchester in camera-ready form, suitable for direct reproduction. This task was expertly performed by Mrs. Ellen B. Carling—the first reader of all my books written since 1955—and her proximity to the author enabled proofreading to become an almost continuous process. If, as a result, this has led to an elimination (or, at least, drastic reduction) of misprints which could have distracted the reader's attention, Mrs. Carling's skill in composition of complicated mathematics has been primarily responsible for this outcome.

November, 1988 Z.K.

Chapter I

INTRODUCTION

Long before the problems discussed in this book emerged to become almost a byword in double-star astronomy in the second half of the present century, astronomers were well aware of their existence which can be traced back—and hence its name—to the French mathematician Edouard Albert Roche (1820–1883). Their roots, to be sure, go much further back in the history of our subject; for—through their connection with the energy integral of the problem of three bodies—many of their properties were known to the German mathematician C. G. J. Jacobi (1804–1851) with whose name they are still sometimes associated. In reality, however, its properties go back to the 18th century to the work of J. L. Lagrange (1736–1813) and, in fact, to Leonard Euler (1707–1789)—and thus to the early applications of the infinitesimal calculus to dynamical astronomy in the days of Isaac Newton (1642–1727). Yet all these grand men of our science were concerned with certain problems raised in this book solely from the viewpoint of particle mechanics (i.e., the motion of individual particles in the course of time); their relevance to the problems of double-star astronomy was glossed over mainly on account of the fact that no binary stars (or, at least, close binaries) were known to exist in the sky at that time. However, a study of the particular case of *equipotential surfaces* associated with gravitational dipoles (to which the energy associated with close gravitational dipoles—and to which the energy integral reduces to surfaces of zero velocity (or, more precisely, to surfaces over which the squares of the velocity components are small enough to be ignorable)—has remained the historical task of E. A. Roche; and it is, therefore, eminently appropriate that these be permanently associated with his name.

By his roots, Edouard Albert Roche was in a very true sense the son of Languedoc; having been born at Montpellier on 17 October 1820 of a family associated with the University of Montpellier for several generations. To scientists of more recent times, this University may have seemed to be somewhat of a provincial institution; but this certainly was not true of its past. Founded in the second half of the 12th century, very soon it became (with Bologna and Paris) one of the most famous universities of Europe; and (politically a part of the Kingdom of Aragon) it remained so until at least the middle of the 14th century (when it relinquished its academic pre-eminence to Paris); though it recovered some of its prestige from the 17th century on to the present time.

It was this institution from which Roche received his doctorate in 1844; and (with the exception of a brief spell between 1845–1848 in Paris under the influence of Augustin Cauchy) he spent his whole life on its academic staff (from 1849 as *chargé de cours*, and from 1852 as professor of mathematics) until the time of his death.

Viewed from the point of the last decades of this millenium, the fame of E. A. Roche rests on two achievements, which were too far ahead of their time to be appreciated fully by his contemporaries (and, perhaps, even by himself): namely, his studies on the stability of rotating homogeneous masses which are, moreover, distorted by tidal action of an external mass-point ("*Mémoirs divers sur l'equilibre d'une mass fluide*"; Roche, 1850); and of the geometry of the equipotential surfaces which surround a rotating gravitational dipole (in his "Essai sur la constitution et l'origine du système solaire"; Roche, 1873). The first will be of only marginal interest to us in this tract; and for subsequent developments of this problem first formulated by Roche see, e.g., Darwin (1906), Jeans (1919); and (in more recent times) by Chandrasekhar (1963) or Kopal and Song (1983). It is to the efflorescence of the second problem (Roche, 1873), to which we shall devote our principal attention in this book.

It is very unlikely that, in this latter work, Roche received much encouragement of his contemporaries (who reciprocated with disinterest of their own); but this fact could have been to the advantage of a good cause. Relative isolation in Montpellier seems, in this particular case, to have only stimulated the emergence of original ideas (which could have been "pooh-poohed" in more closely-knit academic communities) to the benefit of our science; though not necessarily to the benefit of the originator. It is true that, in December 1873, the Parisiens elected Roche to corresponding membership of the Académie des Sciences; but when, ten years later, his name was put forward for election to membership of the astronomical section of the Academy (to fill the vacancy created by the death of Joseph Liouville the year before), the outcome of the election held on 16 April 1883 (cf. Compt. Rend. Acad. Paris, **96**, 1116; 1883) was devastating; for out of 56 votes cast, Edouard Roche received only 1 (one)—no doubt that of his proposer—while the rest of the academicians (trying no doubt to avoid the injustice of electing him as one of their equals) voted for others whose names are all forgotten, and their work as dead as mutton. Fate anesthesized the candidate for the consequences at least somewhat; for only two days after that memorable election E. A. Roche died (18 April 1883) at Montpellier (of pneumonia), hardly aware of the iniquity which the Parisian establishment had just perpetrated on our science.

Should Roche have survived the debacle, he could have derived at least some satisfaction from the fact that the one vote he received was cast by François Tisserand (1845–1896), successor (since 1878) to Le Verrier both in the Academy and at the Sorbonne, who also paid marked posthumous attention to Roche and his work in the second volume of his *Traité de la Mécanique Céleste* (Paris, 1891). It was, moreover, regrettable that Henri Poincaré (1854–1912) was too young in 1883 to double the favourable vote for Roche (Poincaré was not elected to the Academy till 1887); though vote for him he would assuredly have, judging from the respect with which he lectured on Roche's work later at the Sorbonne (cf.

Poincaré, 1902, 1911). Moreover, in subsequent years the work by Roche was kept in good memory by G. H. Darwin (cf. his *Scientific Papers*, Vol. 2; Cambridge 1908) and his pupil J. H. Jeans (1919, 1928).

Perhaps the main cause of delay to recognize the fundamental nature of Roche's contributions to modern astrophysics was the extent of the ignorance of his age of the internal structure of the stars; for even by the time when Roche died, scholars of the calibre of Henri Poincaré (not to speak of "*minores gentium*") did not find it unreasonable to regard the stars as homogeneous (and sometimes incompressible!) self-gravitating globes for the sake of mathematical simplifications permissible on such a basis. That the actual structure of the stars must be very far from such a model did not begin to transpire until the beginning of the 20th century: the roots were indigenous, but they centred around the life work of A. S. Eddington (1882–1944) who (together with his contemporaries) established that the internal constitution of the stars—far from being homogeneous—must be characterized by a high degree of central condensation of the star's mass. And if so, it quickly transpired that the Roche mass-point model of 1873 can represent a much closer approximation to reality than any type of homogeneous configuration.

The first investigator to establish this on a quantitative basis was S. Chandrasekhar (1933) who demonstrated (albeit for one particular class of polytropic models) that—with an increasing degree of central condensation—the geometrical properties of such configurations approach asymptotically those of the Roche model of 1873, which are known to us in a closed and relatively simple form. It is true that complete polytropes can represent the actual structure of the stars as rarely as homogeneous configurations did of the 18th or 19th centuries. However, it will be shown later in this book (Section IV.4) that asymptotic properties of the Roche model hold good for a much wider class of stellar models than the polytropes; and that, therefore, the closed properties of the Roche model will continue to offer an invaluable tool for the studies of many observed aspects of close binary systems.

It is, moreover, very important to note that a generalization of the Roche equipotentials to self-gravitating configurations of *finite* central condensation antedate Roche (though this was not evident in his lifetime)—and, in fact, the surface of zero-velocity of the restricted problem of three bodies—through the work of another Frenchman of immortal memory: namely, Alexis-Claude Clairaut (1713–1765)—perhaps less well-known today than Leonard Euler or Jean-Louis Lagrange (possibly because of his shorter lifetime); but of equally remarkable personality, and a worthy confrère of other great mathematicians in that "century of genius". Born in Paris on 7 May 1713, Clairaut presented to the French Academy his first paper (on "*Quatre problèmes sur deux nouvelles courbes*") at the prodigious age of 12—an event whose like we were not to see again till after 1777 with the birth of C. F. Gauss. At the age of 16, Clairaut completed his "*Recherches sur la courbe à double courbure*", and was elected member of the Academy. His *Théorie de la Figure de la Terre, tirée des principes d'Hydrostat-*

ique—a fundamental work which, according to Mach (1901) inaugurated the rise of mathematical hydrostatics as a scientific discipline—did not appear in Paris until 1743, when Clairaut reached the relatively advanced age of 30 (a German translation of this work appeared in 1903 in Leipzig); though he is known to have harboured the essential ideas incorporated there already at the time of his election to the Academy 12 years before.

It was in this book that Clairaut took the first steps towards the development of a mathematical theory of remarkable generality—potentially valid for an *arbitrary* distribution of density $\rho(r)$ in the interior of the respective configuration—not necessarily continuous (provided only that its singularities are integrable; cf. Liapounov, 1904); and one such solution turned out to be identical with the model constructed (heuristically) by Roche more than a hundred years later. However, this was not established till long after both Clairaut and Roche departed from this world. It should be stressed, in this connection, that the applications Clairaut had in mind were geophysical rather than astrophysical; and as such, directed to the other extreme end of homogeneous configurations (or such in which the density varied but slowly throughout their interiors).

At any rate, Clairaut's death on 17 May 1765 brought to a premature end the first epoch of the development of our subject; and subsequent development of his ideas (characterized by a complete freedom from the need to adopt any particular "equation of state") remained in French hands till, in effect, the end of the 19th century (cf. Legendre, 1793; Laplace, 1825; up to Poincaré, 1902). Geophysical (or planetary) applications in the minds of these investigators made it, moreover, superfluous to push a construction of the solutions of Clairaut's problem beyond the terms of first order in surficial distortion caused by rotation or tides. Its systematic extension beyond this order of accuracy to terms of second order in centrifugal force commenced with the work of Darwin (1900) and de Sitter (1924); and was extended by the present writer (Kopal, 1960) to include the effects of the tides as well as those due to interaction between rotation and tides.

The need to extend Clairaut's theory of the figures of equilibrium of self-gravitating fluid bodies to terms of the orders higher than the second became in the past 30 years irresistible, because of observational discoveries that many of the most interesting phenomena (in the domain of Be-stars, explosive variables, pulsars, etc.) are caused by binary nature of the respective object, requiring the construction of models capable of accounting for the observed facts. This led, in turn to renewed efforts to extend the Clairaut theory to third-order accuracy (cf. Lanzano, 1962; or Kopal, 1973) in terms caused by axial rotation, or (by Rahimi Ardabili, 1979) in terms caused by the tides—and accuracy extended to fourth-order rotational terms investigated by Kopal and Kamala-Mahanta (1974).

The essential features of these and subsequent investigations will be duly described in this book. From the physical point of view, the contents of the present volume can be divided in three parts. The first—consisting of Chapters II–III—will be devoted to a discussion of the Roche model within the framework of *particle mechanics*—or, if applied to continua—to cases in which the mean free path of the constituent particles is large in comparison with the scale of the respective system. In Chapters IV–VI we then propose to deal with *continuous* systems, in which the pressure P and density ρ are related with an "equation of state" of the form $P = f(\rho)$; and shall set out to extract the maximum amount of information on the system without having to specify the actual form of the function $f(\rho)$. The concluding Chapter VII will then be devoted to *applications* of the theory developed in the first two parts of this book to the observations of close binary systems or (to a more limited extent) to the systems of stars at large.

The principal aim of the theory developed in the first two parts of this volume will, however, be not only to compare the theory with observations, but also to provide the methods for *regularizing* the equilibrium form of the respective bodies to facilitate investigations of their internal constitution or atmospheric structure. Spherical forms of equilibrium can, strictly speaking, be characteristic only of single stars of zero rotational momentum. In the case of our Sun (or of the terrestrial planets which attend our central luminary) the centrifugal force created by axial rotation is weak enough to permit us to treat its consequences as small perturbations. This is, however, no longer the case for major planets like Jupiter or Saturn—let alone for close binaries whose components may be nearly (or actually) in contact.

As is well known, the principal obstacle to our fuller understanding of the internal structure as well as of external manifestations of interacting components in close binary systems has been the fact that the boundary conditions of problems arising in this connection must be formed over equipotential surfaces whose departures from spherical form are too large to be treated as ordinary perturbations. An effective treatment of interaction phenomena in such systems will, therefore, necessitate a resort to new coordinate systems, in which the total potential of all forces acting upon the components—constant, by definition over their level surfaces of equilibrium—will replace the radial coordinates of spherical polars; while the angular coordinates may (but need not) be orthogonal to the equipotentials.

Such coordinates are not entirely new. Their orthogonal form ("Roche coordinates") was developed by the present writer more than ten years ago (cf. Kopal, 1969, 1970, 1971; Kopal and Ali, 1971); while their non-orthogonal version has been described (as "Clairaut coordinates") in Kopal (1980). Each system possess merits (and drawbacks) of its own; and the present book is the first source in which they are presented (see Chapters II and V) in systematic form. Moreover, in its applications to the problem of double-star astronomy the Roche model is rapidly being transformed into a *Roche problem* of many facets, of which only some have been touched upon in the text but which are already now sufficiently numerous for their developments to bestow their name on this book.

I.1 Bibliographical Notes

The concept of the Roche model as introduced in this chapter owes its origin to an investigation by Edouard Roche (1873) more than a hundred years ago; but through its relation to the energy integral of the problem of three bodies (as its limit for zero kinetic energy) goes back much further to the past—certainly to Leonard Euler (1772) and Jean-Louis Lagrange (1772) to whom all its essential features of interest to us in this book were well known. These giants of mathematical analysis discovered, however, so much during their long lifetimes that the respective energy integral became subsequently better known under the name of Jacobi (1836). Incidentally, the term "*problème restreint*" of three bodies, referred to frequently in this book, was introduced only in Vol. 1 of Poincaré's celebrated *Méthodes Nouvelles de Mécanique Céleste* in 1892.

For a more complete account of the history of the subject in the 20th century (insofar as it concerns the Roche problem) the reader should be referred to Bibliographical Notes to Chapter IV. Two summarizing publications of more recent date should, however, be mentioned in this place: namely, Chapter VI of *The Analytical Foundations of Celestial Mechanics* by Aurel Wintner (1941), and the *Theory of Orbits: the Restricted Problem of Three Bodies*, by V. G. Szebehely (1967). Chapters 4 and 7 of this latter volume underline multiple connections between the energy integral of the "*problème restreint*" and the problem of Roche, to which individual references will subsequently be made.

Chapter II

THE ROCHE MODEL

One of the fundamental problems of the astronomy of close binary systems is to investigate the equilibrium forms of their components, of arbitrary structure, distorted by rotation and tides, defined as the surfaces over which the potential of all forces acting within the system remain constant. Should we insist that such results be applicable to stars of any structure, the problem of the equilibrium forms has so far been solved only to a limited degree of accuracy—insufficient for an interpretation of the observed phenomena of very close systems. The aim of the present chapter will, however, be to demonstrate that if this latter requirement is given up, and the density concentration of the stars constituting our binary is allowed to approach infinity, their shape can be described in a closed *algebraic* form, which is *exact* for any such configuration *irrespective of the proximity of its components or their mass ratio*. Such a model is generally known in the literature, under the name of its originator, as the *Roche Model*; and the aim of the present chapter will be to summarize its most important geometrical and other properties which should be of interest for the students of close binary systems.

II.1 Roche Model: A Definition

In order to introduce to the reader such a model, let $m_{1,2}$ denote the masses of the two components of a close binary system; and R, the separation of their centres of mass. Suppose, moreover, that the positions of these centres are referred to a rectangular system of Cartesian coordinates, with the origin at the centre of gravity of mass m_1—the x-axis of which coincides with the line joining the centres of the two stars (i.e., the radius-vector of the relative orbit of the two masses which will—in this chapter—be regarded as constant); while the z-axis is perpendicular to the plane of the orbit. If so, the coordinates of the centre of gravity of the system are

$$\frac{m_2 R}{m_1 + m_2},\ 0,\ 0; \tag{1.1}$$

and the total potential Ψ of all forces acting at an arbitrary point $P(x, y, z)$ becomes expressible as

$$\Psi = G\frac{m_1}{r} + G\frac{m_2}{r'} + \frac{\omega^2}{2}\left\{\left(x - \frac{m_2 R}{m_1 + m_2}\right)^2 + y^2\right\} \tag{1.2}$$

where

$$r^2 = x^2 + y^2 + z^2; \quad r'^2 = (R - x)^2 + y^2 + z^2, \tag{1.3}$$

represent squares of the distance of P from the centres of gravity of the two components, and ω denotes the angular velocity of rotation of the system about an axis perpendicular to the orbital plane and passing through the centre of gravity of the system whose coordinates are given by (1.1). The first term on the right-hand side of Equation (1.2) represents the potential arising from the mass m_1; the second, the disturbing potential of its companion of mass m_2; and the third, the potential arising from the centrifugal force.

Let, furthermore, the angular velocity ω on the right-hand side of Equation (1.2) be identified with the Keplerian angular velocity

$$\omega^2 = \frac{G(m_1 + m_2)}{R^3} \tag{1.4}$$

of the system. If we insert (1.4) in (1.2) and, moreover, adopt m_1 as our unit of mass; R, as the unit of length while the unit of time is chosen so that $G = 1$, Equation (1.2) may be expressed in terms of spherical polar coordinates

$$\left.\begin{aligned} x &= r\cos\phi\ \sin\theta = r\lambda\,, \\ y &= r\sin\phi\ \sin\theta = r\mu\,, \\ z &= r\cos\theta = r\nu\,, \end{aligned}\right\} \tag{1.5}$$

as

$$\xi = \frac{1}{r} + q\left\{\frac{1}{\sqrt{1 - 2\lambda r + r^2}} - \lambda r\right\} + \frac{q+1}{2}r^2(1 - \nu^2)\,, \tag{1.6}$$

where

$$\xi \equiv \frac{R\Psi}{Gm_1} - \frac{m_2^2}{2m_1(m_1 + m_2)} \tag{1.7}$$

and

$$q \equiv \frac{m_2}{m_1} \tag{1.8}$$

the two masses $m_{1,2}$ (and, therefore, their ratio q being likewise regarded as constant.

Within the scheme of definitions adopted, none of the terms constituting the normalized potential ξ depends on the time; and Equation (1.6) defining it will generate surfaces described by the polar coordinates r, λ, ν; the forms of which are governed by the non-dimensional values of ξ and q. If ξ is large, the corresponding surfaces—hereafter referred to as the *Roche Equipotentials*—will consist of two separate ovals (see Figure II.1) closed around each of the two mass-points; for the right-hand side of (1.6) can be large only if r (or $r' = \sqrt{1 - 2\lambda r + r^2}$) becomes small; and if the left-hand side of (1.6) is to be constant, so must be (very nearly) r or r'. Large values of ξ correspond, therefore, to equipotentials differing but little from spheres—the less so, the greater ξ becomes. With diminishing value

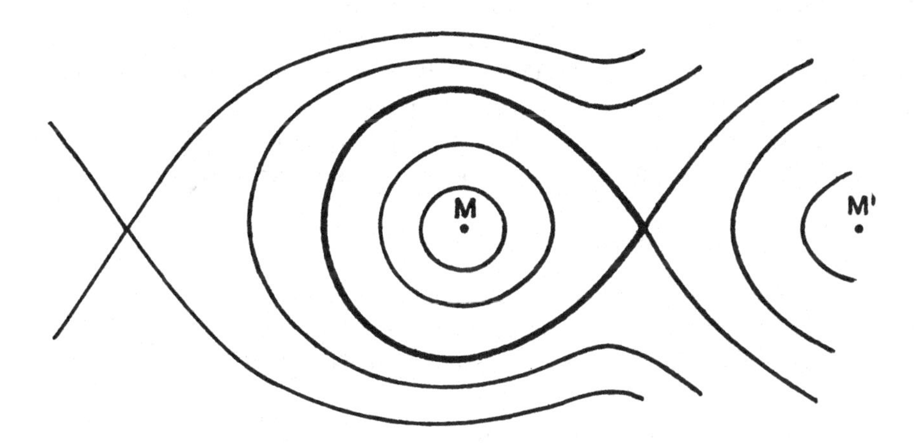

Figure II.1: A cross-section of the Roche Model of a binary system of masses m and m'; the heavy line representing the Roche limit (after Jeans, 1919).

of ξ the ovals defined by (1.6) become increasingly elongated in the direction of the centre of gravity of the system — until, for a certain critical value of ξ_1 characteristic of each mass-ratio, both ovals will unite in a single point on the x-axis to form a dumb-bell-like configuration (cf. again Figure II.1) which we propose to call the *Roche Limit*.[1] For still smaller values of ξ the connecting part of the dumb-bell opens up and the corresponding equipotential surfaces would envelop both bodies. This latter case is, however, of no direct interest to us in this connection; as for $\xi < \xi_1$ the two initially distinct bodies would coalesce in one and we should no longer have the right to speak of a binary system. In what follows we shall, therefore, limit ourselves to a study of the geometry of surfaces characterized by $\xi \geq \xi_1$.

II.2 Geometry of Roche Equipotentials

Equation (1.6) of the Roche equipotentials represents an implicit function defining, for given values of ξ and q, r as a function of λ and ν. When it has been rationalized and cleared of fractions, the result is an algebraic equation of *eighth* degree in r, whose analytical solution presents unsurmountable difficulties. In

[1]Not to be confused with a concept used, under the same name, in other literature to signify the minimum distance to which a fluid satellite of infinitesimal mass can approach with impunity an oblate planet. This latter term, coined in the latter half of the 19th century by G. H. Darwin, has nothing to do with the "Roche Limit" as defined in this section (and introduced under this name by Kopal (1955) to describe the fractional size of "contact" components in semi-detached eclipsing systems).

the case of pure rotational distortion (obtaining if $q = 0$) Equation (1.6) can be reduced to a cubic solvable in terms of circular functions. In the case of a purely tidal distortion ($\omega = 0$), Equation (1.6) becomes a quartic, which could also be solved for r in a closed form (though its solution would be very much more involved). However, in the general case of rotational *and* tidal distortion interacting, any attempt at an *exact* solution of (1.6) for r becomes virtually hopeless; and approximate solutions must be sought by successive approximations.

A: Radius and Volume

In order to obtain them, let us begin by expanding the radical $(1 - 2\lambda r + r^2)^{-1/2}$ on the right-hand side of (1.6) in terms of the Legendre polynomials $P_j(\lambda)$. Doing so and removing fractions we find it possible to replace (1.6) by

$$(\xi - q)r = 1 + q\sum_{j=2}^{\infty} r^{j+1}P_j(\lambda) + nr^3(1 - \nu^2)\,, \tag{2.1}$$

where we have abbreviated

$$n = \frac{q+1}{2} \tag{2.2}$$

If r is small in comparison with unity (i.e., if the linear dimensions of the equipotential surfaces are small in comparison with our unit of length R), the second and third terms on the right-hand side of (2.1) may be neglected in comparison with unity—in which case

$$r_0 = \frac{1}{\xi - q}\,. \tag{2.3}$$

This result asserts that if ξ is large, the corresponding Roche equipotential will differ but little from a sphere of radius r_0.

Suppose now that

$$r_1 = r_0 + \Delta' r = r_0\left(1 + \frac{\Delta' r}{r_0}\right) \tag{2.4}$$

should represent our next approximation to r. Inserting it in (2.1) we find that

$$1 + \frac{\Delta' r}{r_0} = 1 + q\sum_{j=2}^{\infty} r_0^{j+1}P_j(\lambda) + nr_0^3(1 - \nu^2) \tag{2.5}$$

where, in small terms on the right-hand side, r was legitimately replaced by r_0. The foregoing equation then yields

$$\frac{\Delta' r}{r_0} = q\sum_{j=2}^{4} r_0^j P_j(\lambda) + nr_0^3(1 - \nu^2) \tag{2.6}$$

correctly to quantities of the order of r_0^5 (i.e., as far as squares and higher terms of first-order distortion remain negligible).

In order to improve upon this approximation let us set, successively,

$$r_2 = r_1 + \Delta'' r = r_0 \left\{ 1 + \frac{\Delta' r}{r_0} + \frac{\Delta'' r}{r_0} \right\} , \tag{2.7}$$

$$\begin{aligned} r_3 &= r_2 + \Delta''' r &= r_0 \left\{ 1 + \frac{\Delta' r}{r_0} + \frac{\Delta'' r}{r_0} + \frac{\Delta''' r}{r_0} \right\} \\ &\vdots & \vdots \end{aligned} \tag{2.8}$$

$$r_{j+1} = r_j + \Delta^{(j+1)} r = r_0 \left\{ 1 + \sum_{i=0}^{j} \frac{\Delta^{(i+1)} r}{r_0} \right\} , \tag{2.9}$$

where

$$\frac{\Delta^{(i+1)} r}{r_0} = q \sum_{k=3}^{3(N-i)} (r_i^k - r_{i-1}^k) P_{k-1}(\lambda) + n(r_i^3 - r_{i-1}^3)(1-\nu^2) ; \tag{2.10}$$

$3N$ denoting the highest power of r_0 to which Equation (2.9) represents a correct solution for r. We may note that, in general, the leading terms of the expression (2.10) for $\Delta^{(i+1)} r)/r_0$ will be of $3(i+1)$st degree in r_0; and, similarly, the difference $r_i^k - r_{i-1}^k$ in higher terms on the right-hand side of (2.10) will be of the order of r_0^{3i+k}.

Suppose that, in what follows, we wish to construct the explicit form of an approximate solution of Equation (2.1), in the form of (2.8), correctly to (say) quantities of the order of $\Delta''' r/r_0$—which should, therefore, differ from the exact solution of (2.1) at most in quantities of the order of r_0^{12}. By use of the expression already established for $\Delta' r/r_0$ the explicit forms of $\Delta'' r/r_0$ and $\Delta''' r/r_0$ can successively be found[2] and their insertion in (2.8) leads to the equation

$$\begin{aligned} \frac{r - r_0}{r_0} = \; & r_0^3\{qP_2 + n(1-\nu^2)\} + r_0^4\{qP_3\} + r_0^5\{qP_4\} + \\ & + r_0^6\{qP_5 + 3[qP_2 + n(1-\nu^2)]^2\} + \\ & + r_0^7\{qP_6 + 7q[qP_2 + n(1-\nu^2)]P_3\} + \\ & + r_0^8\{qP_7 + 8q[qP_2 + n(1-\nu^2)]P_4 + 4q^2P_3^2\} + \\ & + r_0^9\{qP_8 + 9q[qP_2 + n(1-\nu^2)]P_5 + 9q^2P_3P_4 + \\ & \quad + 6[qP_2 + n(1-\nu^2)]^3 + 6[q^3P_2^3 + n^3(1-\nu^2)^3]\} + \\ & + r_0^{10}\{qP_9 + 10q[qP_2 + n(1-\nu^2)]P_6 + 5q^2[P_4^2 + 2P_3P_5] + \\ & \quad + 45q[qP_2 + n(1-\nu^2)]^2P_3\} + \\ & + r_0^{11}\{qP_{10} + 11q([qP_2 + n(1-\nu^2)]P_7 + 11q^2[P_3P_6 + P_4P_5] + \\ & \quad + 55q[qP_2 + n(1-\nu^2)]^2P_4 + \\ & \quad + 55q^2[qP_2 + n(1-\nu^2)]P_3^2\} + \ldots , \end{aligned} \tag{2.11}$$

[2] For fuller details of this process, cf. Kopal (1954, 1959).

where we have abbreviated $P_j \equiv P_j(\lambda)$, and which represents the desired approximate solution of Equation (2.1) for r as a function of λ and ν in the form of an expansion in ascending powers of r_0 (as defined by Equation (2.3)).

The volume V of a configuration whose radius-vector r is given by the foregoing Equation (2.12) will be specified by

$$V = \frac{2}{3}\int_{-1}^{1}\int_{-\sqrt{1-\lambda^2}}^{\sqrt{1-\lambda^2}} \frac{r^3\, d\lambda\, d\nu}{\mu}, \tag{2.12}$$

where $\mu^2 = 1 - \lambda^2 - \nu^2$. By virtue of the algebraic identity

$$r^3 = r_0^3\left\{1 + \frac{r - r_0}{r_0}\right\}^3 \tag{2.13}$$

we find it convenient to express the integrand in (2.19) in terms of (2.19) as a function of λ and ν. This integrand will, in general, consist of a series of terms of the form $\lambda^m\nu^n/\mu$, factored by constant coefficients; therefore, the entire volume V will be given by an appropriate sum of partial expressions V_n^m of the form

$$V_n^m = \int_{-1}^{1}\int_{-\sqrt{1-\lambda^2}}^{\sqrt{1-\lambda^2}} \frac{\lambda^m\nu^n}{\mu}\, d\lambda\, d\nu\,. \tag{2.14}$$

These expressions vanish (on grounds of symmetry) if either m or n is an odd integer. If, however, both happen to be even and such that $m = 2a$ and $n = 2b$, an evaluation of the foregoing integrals readily reveals that

$$V_{2b}^{2a} = \frac{\sqrt{\pi}\,\Gamma(a+\frac{1}{2})\Gamma(b+\frac{1}{2})}{\Gamma(a+b+\frac{3}{2})}, \tag{2.15}$$

where Γ denotes the ordinary gamma function. As, accordingly,

$$\int_{-1}^{1}\int_{-\sqrt{1-\lambda^2}}^{\sqrt{1-\lambda^2}} \frac{P_j(\lambda)\, d\lambda\, d\nu}{\sqrt{1-\lambda^2-\nu^2}} = \begin{cases} 2\pi & \text{if } j = 0 \\ 0 & \text{if } j > 0 \end{cases} \tag{2.16}$$

and

$$\int_{-1}^{1}\int_{-\sqrt{1-\lambda^2}}^{\sqrt{1-\lambda^2}} \frac{\nu^{2j}\, d\lambda\, d\nu}{\sqrt{1-\lambda^2-\nu^2}} = \frac{2\pi}{j+1}, \tag{2.17}$$

we eventually find that the volume of a configuration whose surface is a Roche equipotential will be given by

$$\begin{aligned} V \;=\; & \frac{4}{3}\pi r_0^3\left\{1 + \frac{12}{5}q^2 r_0^6 + \frac{15}{7}q^2 r_0^8 + \frac{18}{9}q^2 r_0^{10} + \ldots \right. \\ & + \frac{22}{7}q^3 r_0^9 + \frac{157}{7}q^3 r_0^{11} + 2nr_0^3 + \frac{32}{5}n^2 r_0^6 + \frac{176}{7}n^3 r_0^9 + \ldots \\ & \left. + \frac{8}{5}nqr_0^6 + \frac{296}{35}nq(2q+n)r_0^9 + \frac{26}{35}nq(q+3n)r_0^{11} + \ldots\right\}, \end{aligned} \tag{2.18}$$

correctly to quantities of the order up to and including r_0^{11}. With n and r_0 as given by Equations (2.2) and (2.3) the volume V becomes an explicit function of ξ and q alone and can be tabulated in terms of these parameters.

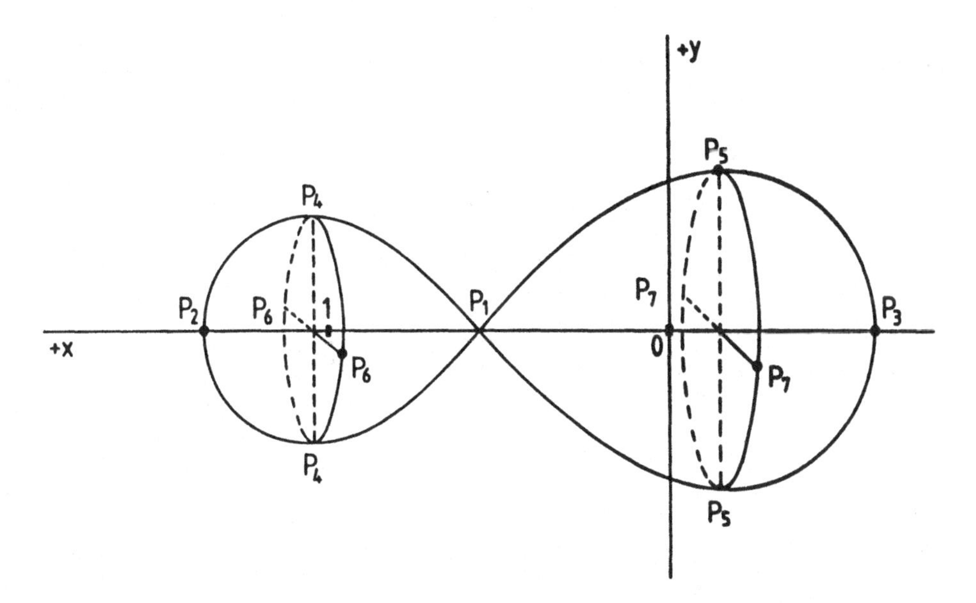

Figure II.2: Schematic view of a contact binary at the Roche Limit. In order to exhibit the essential features of the geometry of this model, the diagram has not been drawn to scale for any particular mass-ratio; and certain features of it (such as the distance of the P_5P_7-plane from the origin) have been exaggerated.

B: Roche Limit

It was pointed out already in Section II.1 that a diminution of the value of the constant ξ on the left-hand side of Equation (1.6) will cause the respective Roche equipotentials to expand from nearly spherical configurations to ovals of increased elongation in the direction of the attracting centre until, for a certain critical value of ξ characteristic of each mass-ratio, these ovals unite in a single point on the line joining their centres. Such configurations represent the largest *closed* equipotentials capable of containing the whole mass of the respective components, and will hereafter be referred to as their *Roche limits*. Any star filling its Roche limit will therefore be termed a *contact component*; and a binary system consisting of a pair of such components, a *contact system*. The fact that close binaries in which one, or both, components have attained their Roche limits actually exist in considerable numbers in the sky underlines the importance of a study of the geometry of Roche limits in binary systems of different mass-ratios.

In order to do so, our first task should be to specify the values of ξ for which the two loops of the critical equipotential (cf. Figure II.2) develop a common point of contact at P_1; but its determination presupposes a knowledge of the position of P_1 on the x-axis. The location of this point is characterized by the vanishing of the gravity due to all forces - which means that, at that point,

$$\xi_x = \xi_y = 0 . \tag{2.19}$$

Now a differentiation of (1.6), rewritten in terms of rectangular coordinates, with respect to x and y yields

$$\xi_x = -xr^{-3} + q\{(1-x)(r')^{-3} - 1\} + 2nx , \tag{2.20}$$

$$\xi_y = -y\{r^{-3} + q(r')^{-3} - 2n\} , \tag{2.21}$$

$$\xi_z = -zr^{-3} - qz(r')^{-3} , \tag{2.22}$$

where $r^2 = x^2 + y^2 + z^2$ and $r'^2 = (1-x)^2 + y^2 + z^2$ continue to be given by Equations (1.3) and $2n = q + 1$ in accordance with (2.2).

The partial derivative ξ_y vanishes evidently everywhere along the x-axis; but the vanishing of ξ_x renders the x-coordinate of P_1 to be a root of the equation

$$x^{-2} - x = q\{(1-x)^{-2} - (1-x)\} \tag{2.23}$$

which, after removal of the fractions, assumes the form

$$(1+q)x^5 - (2+3q)x^4 + (1+3q)x^3 - x^2 + 2x - 1 = 0 . \tag{2.24}$$

For $q = 0$ the foregoing equation would evidently reduce to

$$(1-x)^3(x^2 + x + 1) = 0 , \tag{2.25}$$

the value $x = 1$ becoming a triple root. Therefore, for small values of q, the root x_1 of Equation (2.23) which is interior to the interval $0 < x < 0$ can be approximated by the expansion

$$x_1 = 1 - w + \frac{1}{3}w^2 + \frac{1}{9}w^3 + \dots \tag{2.26}$$

in terms of the auxiliary parameter

$$w^3 = \frac{1}{3(1+q)} ; \tag{2.27}$$

and more accurate values of x_1 can further be obtained by the method of differential corrections.

Once a sufficiently accurate value of x_1 has thus been obtained, the actual value of ξ corresponding to our critical equipotential follows as

$$\xi_1 \equiv \xi(x_1, 0, 0) . \tag{2.28}$$

Moreover, the points $P_{4,5}$ in the xy-plane (see again Figure II.2) are evidently characterized by the vanishing of the derivative dy/dx at the Roche limit. Their coordinates $x_{4,5}$ and $y_{4,5}$ can, therefore, be evaluated by solving the simultaneous system

$$\left.\begin{aligned} \xi(x, y, 0) &= \xi_1 , \\ \xi_x(x, y, 0) &= 0 ; \end{aligned}\right\} \tag{2.29}$$

q	x_1	ξ_1	x_4	$\pm y_4$	x_5	$\pm y_5$	$\pm z_6$	$\pm z_7$
1.0	0.50000	3.75000	1.01134	0.37420	-0.01134	0.37420	0.35621	0.35621
0.8	0.52295	3.41697	1.01092	0.35388	-0.01168	0.39501	0.33770	0.37491
0.6	0.55234	3.06344	1.01029	0.32853	-0.01198	0.42244	0.31431	0.39909
0.4	0.59295	2.67810	1.00926	0.29465	-0.01213	0.46189	0.28260	0.43278
0.3	0.62087	2.46622	1.00847	0.27204	-0.01204	0.49015	0.26123	0.45599
0.2	0.65856	2.23273	1.00735	0.24233	-0.01163	0.52989	0.23294	0.48714
0.15	0.68392	2.10309	1.00656	0.22280	-0.01117	0.55774	0.21425	0.50781
0.1	0.71751	1.95910	1.00552	0.19746	-0.01034	0.59609	0.18991	0.53451
0.05	0.76875	1.78886	1.00397	0.15979	-0.00859	0.65804	0.15366	0.57291
0.02	0.82456	1.65702	1.00245	0.11992	-0.00618	0.73070	0.11522	0.61434
0.01	0.85853	1.59911	1.00165	0.09613	-0.00457	0.77779	0.09231	0.62867
0.005	0.88635	1.56256	1.00110	0.07689	-0.00327	0.81807	0.07379	0.64170
0.001	0.93231	1.52148	1.00041	0.04550	-0.00137	0.88816	0.04361	0.65762
0.0002	0.96001	1.50737	1.00015	0.02678	-0.00052	0.93264	0.02566	0.66348
0	1.00000	1.50000	1.00000	0.00000	0.00000	1.00000	0.00000	0.66667

Table II.1:

and once the values of $x_{4,5}$ have thus been found, the z-coordinates of points $P_{6,7}$ in the xz-plane (cf. Figure II.2) follow as roots of a single equation

$$\xi(x_{4,5},\, 0,\, z) = \xi_1 \,. \tag{2.30}$$

The accompanying Table II.1 lists five-digit values of ξ_1, x_1; $x_{4,5}$, $y_{4,5}$; and $z_{6,7}$ for Roche limits appropriate for 15 discrete values of the mass-ratio.

It may further be noted that if, in place of ξ_1, we introduce a new constant C_1 as defined by the equation

$$C_1 = \frac{2\xi_1}{1+q} + \left(\frac{q}{1+q}\right)^2 = 2(1-\mu)\xi_1 + \mu^2 \,, \tag{2.31}$$

where we have abbreviated

$$\mu = \frac{q}{1+q} = \frac{m_2}{m_1+m_2},\; 1-\mu = \frac{1}{1+q} = \frac{m_1}{m_1+m_2} \,, \tag{2.32}$$

the values of C_1 remain largely invariant with respect to the mass-ratio, and sensibly equal to 4 provided that q does not depart greatly from unity. This is demonstrated by an inspection of the tabulation of C_1 as given in column (2) of the following Table II.2. In consequence, the simple expression

$$\xi_1 \doteq 2(1+q) - \frac{q^2}{2(1+q)} = \frac{4-\mu^2}{2(1-\mu)} \tag{2.33}$$

q	C_1	$(r_0)_1$	$(r_0)_2$	V_1	V_2	$(r*)_1$	$(r*)_2$	v_1	v_2
1	4.00000	0.36363	0.36363	0.22704	0.22704	0.37845	0.37845	0.072267	0.072267
0.8	3.99417	0.38212	0.34528	0.26459	0.19374	0.39825	0.35896	0.075799	0.069377
0.6	3.96993	0.40594	0.32199	0.31974	0.15665	0.42420	0.33441	0.081422	0.066485
0.4	3.90749	0.43896	0.29025	0.40923	0.11444	0.46057	0.30115	0.091184	0.063726
0.3	3.84744	0.46163	0.26876	0.48148	0.09089	0.48622	0.27892	0.099619	0.062683
0.2	3.74900	0.49195	0.24018	0.59399	0.06492	0.52147	0.24933	0.113443	0.061996
0.15	3.67456	0.51201	0.22121	0.68002	0.05079	0.54552	0.22973	0.124462	0.061967
0.1	3.57027	0.53789	0.19642	0.80715	0.03564	0.57760	0.20414	0.141308	0.062385
0.05	3.40962	0.57509	0.15931	1.0289	0.01910	0.62626	0.16584	0.17193	0.063854
0.02	3.24945	0.61087	0.11974	1.2700	0.007961	0.67179	0.12387	0.2062	0.06462
0.01	3.16665	0.62928	0.09606	1.4656	0.004042	0.70465	0.09882	0.2356	0.06497
0.005	3.10959	0.64203	0.07686	1.5950	0.002038	0.7248	0.07865	0.2551	0.06520
0.001	3.03992	0.65769	0.04549	1.868	0.0004114	0.764	0.04614	0.298	0.06554
0.0002	3.01414	0.66350	0.02679	2.067	0.0000826	0.790	0.02702	0.329	0.06575
0	3.00000	0.66667	0.00000	2.26663	0.0000000	0.81488	0.00000	0.36075	0.065843

Table II.2: The data collected in Tables II.1 and 2 are taken from Kopal (1959). More extensive tabulations of the same parameters for $q = 1(-0.02)\ 0.10$ to $5D$ have since been prepared by Plavec and Kratochvíl (1964). Cf. also ten Bruggencate (1934).

is found to approximate the exact values of ξ_1 within 1% if $1 \geq q \geq 0.5$, or within 10% for the wider range $1 \geq q \geq 0.1$. The mean radii $(r_0)_{1,2}$ of the two components of contact systems become (consistent with Equations (2.3) and (2.31)) equal to

$$(r_0)_{1,2} = \frac{2(1-\mu)}{C_1 - (1+\mu)^2 + 1} \tag{2.34}$$

where, for the primary component, $0 \leq \mu \leq 0.5$; while, for the secondary, $0.5 \leq \mu \leq 1$. Alternatively, we may fall back on Equation (2.3) and, by inserting for ξ_1 from (2.28) write

$$(r_0)_1 = \frac{2x_1}{2 + 2qx_1^3(1-x_1)^{-1} + (q+1)x_1^3}\,; \tag{2.35}$$

while $(r_0)_2$ is obtainable from the same expression if we replace x_1 by $1 - x_1$ and q by its reciprocal. The values of $(r_0)_{1,2}$ so determined are listed as functions of the mass-ratio in columns (3) and (4) of Table II.2.

Having evaluated them, we are in a position to invoke Equation (2.19) for expressing the volumes $V_{1,2}$ of contact components—the reader will find them tabulated in columns (5) and (6) of Table II.2—while columns (7) and (8) list the equivalent radii $(r*)_{1,2}$ of spheres having the same volume as the respective contact component. The penultimate and ultimate columns of Table II.2 then contain the quantities

$$v_{1,2} = \frac{\omega^2}{2\pi G\overline{\rho}_{1,2}} = \frac{2}{3}\left\{1+\frac{m_{2,1}}{m_{1,2}}\right\}(r*)^3_{1,2}\,, \tag{2.36}$$

where ω denotes the (Keplerian) angular velocity of axial rotation of each component and $\overline{\rho}_{1,2}$, their respective mean densities.

The series on the right-hand side of the volume equation (2.19)—which are at the basis of our numerical data as given in columns (5)–(10)—converge with satisfactory rapidity if the masses of the two components are not too unequal, but fail to do so if the mass of one component becomes very much larger than the other. In order to attain adequate representation of the radii and volumes in such cases, asymptotic solutions of Equation (2.1) must be sought as $\mu \to 0$ or 1.

In order to do so, we find it advantageous to rewrite (1.6) in the alternative form

$$(1-2\lambda r+r^2)\{(1-\nu^2)r^3-2\lambda\mu r^2+(\mu^2-C_1)r+2(1-\mu)\}^2 = 4\mu^2r^2\,, \tag{2.37}$$

where C_1 as well as μ are defined by Equations (2.31) and (2.32); and consider first the case of very small disturbing mass (when $\mu \to 0$). As long as quantities of the order of μ^2 remain ignorable, Equation (2.37) will admit of a real solution only if

$$(1-\nu^2)r^3-2\lambda\mu r^2-C_1r+2(1-\mu) = 0\,. \tag{2.38}$$

For small values of μ, the solution of this latter equation can be sought in the form

$$r = S_{10}+S_{11}\mu+\ldots\,, \tag{2.39}$$

where $S_{10}, S_{11}, \ldots$ are defined by the equations

$$(1-\nu^2)S_{10}^3 - C_1S_{10}+2 = 0\,, \tag{2.40}$$

$$3(1-\nu^2)S_{10}^2S_{11}-C_1S_{11}-2 = 2\lambda S_{10}^2\,, \tag{2.41}$$

etc., whose solutions become

$$S_{10} = 2\left\{\frac{C_1}{3(1-\nu^2)}\right\}^{1/2}\sin\left\{\frac{1}{3}\sin^{-1}\frac{3}{C_1}\sqrt{\frac{3(1-\nu^2)}{C_1}}\right\} \tag{2.42}$$

and

$$S_{11} = \frac{2(1+\lambda S_{10}^2)}{3(1-\nu^2)S_{10}^2-C_1}\,, \tag{2.43}$$

respectively

Equation (2.39) with its coefficients as given by (2.42) and (2.43) will closely approximate the form of the primary component of a contact system which is very much more massive than the secondary. Its first term S_{10} defines obviously the form of a Roche equipotential distorted by centrifugal force alone. If $\mu \to 0$, $\xi_1 \to$

1.5 and $C_1 \to 3$, in which case the parametric equation of the corresponding critical equipotential assumes the neat form

$$r = \frac{2}{\sqrt{1-\nu^2}}\left\{\sin\frac{1}{3}\cos^{-1}\nu\right\} = 2\,\frac{\sin\frac{1}{3}\theta}{\sin\theta}\,; \tag{2.44}$$

and its volume V_1, in accordance with Equation (2.12), becomes

$$\begin{aligned} V_1 &= \frac{32}{3}\pi\int_0^1 (1-\nu^2)^{-3/2}\sin^3\left(\frac{1}{3}\cos^{-1}\nu\right)d\nu = \\ &= \frac{4}{3}\pi\left\{3\sqrt{3}-4+3\log\frac{3(\sqrt{3}-1)}{\sqrt{3}+1}\right\} = 2.26662\ldots\,, \end{aligned} \tag{2.45}$$

so that, by (2.36), $v_1 \equiv V_1/2\pi = 0.36074$. It is the foregoing values, rather than those which would follow from a straightforward application of (2.19), which have been used to complete the last entries in columns (5) and (9) of Table II.2.

If the primary component accounts thus for most part of the total mass of our contact binary system, the volume of the secondary must clearly tend to zero. The form of its surface will, in turn, be given by an asymptotic solution of Equation (2.37) as $\mu \to 1$. Let us, therefore, expand this solution in a series of the form

$$r = S_{20}(1-\mu) + S_{21}(1-\mu^2) + \ldots\,; \tag{2.46}$$

inserting it in (2.37) we find the vanishing of the coefficients of equal powers of $(1-\mu)$ to require that

$$S_{20} = \frac{2}{C_1-3}\,,\quad S_{21} = -\left\{2+\frac{\lambda}{C_1-3}\right\}S_{20}^2\,, \tag{2.47}$$

etc.. An application of Equation (2.12) reveals, moreover, that the volume V_2 of the respective configuration should be approximable by

$$V_2 = \frac{4}{3}\pi\{1-\mu)^3S_{20}^3 - 6(1-\mu)^4S_{20}^4 - \ldots\}\,, \tag{2.48}$$

and the radius r_2^* of a sphere of equal volume becomes

$$r_2^* = (1-\mu)S_{20} - 2(1-\mu)^2S_{20}^2 + \ldots. \tag{2.49}$$

A glance at the second column of Table II.2 reveals that, as $q \to 0$, $C_1 \to 3$ and, as a result the product $(1-\mu)S_{20}$ tends to become indeterminate for $\mu = 1$. In order to ascertain its limiting value, let us depart from Equation (2.31) which, on insertion of ξ_1 from (2.28) assumes the form

$$C_1 = \frac{2(1-\mu)}{1-x_1} + \frac{2\mu}{x_1} + (1-\mu-x_1)^2\,, \tag{2.50}$$

with the root x_1 approximable by means of (2.26) where, by (2.27) and (2.32),

$$3w^3 = 1-\mu\,. \tag{2.51}$$

Inserting (2.26) in (2.50) we find that, within the scheme of our approximation,

$$C_1 = 3(1 + 3w^2 - 4w^3 + \dots); \tag{2.52}$$

so that

$$(1-\mu)S_{20} = \frac{2w}{3-4w} + \dots, \tag{2.53}$$

and therefore

$$r_2^* = \frac{2}{3}w - \frac{32}{27}w^3 + \dots. \tag{2.54}$$

In consequence, it follows from (2.36) that, for a secondary component of vanishing mass

$$v_2 = \frac{1}{3}\left(\frac{2}{3}\right)^4 \left\{1 - \frac{16}{3}w^2 + \dots\right\}. \tag{2.55}$$

An inspection of the last two columns of Table II.2 reveals that, for the primary (more massive) component, the value of v_1 increases monotonously with diminishing mass-ratio m_2/m_1 from 0.07227 for the case of equality of masses to 0.36075 for $m_2 = 0$, at which point the primary component becomes rotationally unstable and matter begins to be shed off along the equator if axial rotation is any faster. On the other hand, for the secondary (less massive) component the values of v_2 diminish with decreasing mass-ratio from 0.07227 until, as $m_2 \to 0$, the value of $2^4/3^5$ has been attained.

C: Geometry of the Eclipses

The data assembled in the foregoing section on the geometry of contact configurations lead to a number of specific conclusions regarding the eclipse phenomena to be exhibited by such systems. For suppose that a contact binary both of the components of which are at their Roche limits is viewed by a distant observer, whose line of sight does not deviate greatly from the x-axis of our model as shown on Figure II.2 If so, then in the neighbourhood of either conjunction one component is going to eclipse the other, and the system will exhibit a characteristic variation in brightness. if, in turn, the observed light variation is analyzed for the geometrical elements the fractional "radii" $r_{1,2}$ of the two components should (very approximately) be identical with the quantities $y_{4,5}$ as listed in columns (5) and (7) of Table II.2. In Table II.3 we have, accordingly, listed four-digit values of the sums $r_1 + r_2$ as well as the ratios r_2/r_1 of the "radii" of such contact components as functions of their mass-ratio.

An inspection of this tabulation reveals that, within the scheme of our approximation, *the sum* $r_1 = r_2$ *of fractional radii of both components in contact binary systems is very nearly constant* and equal to 0.75 ± 0.01 for a very wide range of the mass-ratios q; whereas the *ratio* r_2/r_1 *decreases monotonically with diminishing value of* q. Therefore, a photometric determination of the sum $r_1 + r_2$ — which, unfortunately, represents nearly all that can be deduced with any accuracy from an analysis of light curves due to shallow partial eclipses — cannot

q	$r_1 + r_2$	r_1/r_2
1.0	0.7474	1.0000
0.9	0.7486	0.9495
0.8	0.7489	0.8959
0.7	0.7496	0.8389
0.6	0.7510	0.7777
0.5	0.7529	0.7112
0.4	0.7565	0.6379
0.3	0.7622	0.5550
0.2	0.7722	0.4573
0.15	0.7805	0.3995
0.1	0.7935	0.3312

Table II.3:

be expected to tell us anything new about contact systems; or, in particular, about their mass-ratios. It is the ratio of the radii r_2/r_1 whose determination would provide a sensitive photometric clue to the mass-ratio of a contact system. This underlines the importance of photometric determination of the ratios of the radii of contact binary systems; but owing to purely geometrical difficulties this important task of light curve analysis is, unfortunately, not yet well in hand.

Suppose next that a contact binary system, consisting of two components at their Roche limits, is viewed by a distant observer from an arbitrary direction. What will be the range of such directions from which this observer will see both bodies mutually eclipse each other during their revolution? In order to answer this question, let us replace the actual form of the corresponding Roche limit by an *osculating cone* which is tangent to it at the point of contact P_1. The equation of this cone may readily be obtained if we expand the function $\xi(x, y, z)$ of Roche equipotentials in a Taylor series, in three variables, about P_1.

The first partial derivatives ξ_x, ξ_y and ξ_z have already been given by Equations (2.20)–(2.22) in the preceding part of this section. Differentiating these equations further we find that

$$\xi_{xx} = (3x^2 - r^2)r^{-5} + q\{3(1-x)^2 - r'^2\}(r')^{-5} + q + 1 , \tag{2.56}$$

$$\xi_{yy} = (3y^2 - r^2)r^{-5} + q\{3y^2 - r'^2\}(r')^{-5} + q + 1 , \tag{2.57}$$

$$\xi_{zz} = (3z^2 - r^2)r^{-5} + q\{3z^2 - r'^2\}(r')^{-5} , \tag{2.58}$$

$$\xi_{xy} = 3xyr^{-5} - 3q(1-x)y(r')^{-5} , \tag{2.59}$$

$$\xi_{xz} = 3xzr^{-5} - 3q(1-x)z(r')^{-5} , \tag{2.60}$$

$$\xi_{yz} = 3yzr^{-5} + 3qyz(r')^{-5} . \tag{2.61}$$

We note that all first (as well as mixed second) derivatives of ξ vanish at P_1. Hence, a requirement that the sum of nonvanishing second-order terms should add up to zero provides us with the desired equation of the osculating cone in the form

$$(x-x_1)^2(\xi_{xx})_1 + y^2(\xi_{yy})_1 + z^2(\xi_{zz})_1 = 0 , \tag{2.62}$$

where

$$\begin{aligned} (\xi_{xx})_1 &= \ \ 2p \ \ +q+1 , \\ (\xi_{yy})_1 &= -p \ \ +q+1 , \\ (\xi_{zz})_1 &= -p \qquad\quad , \end{aligned} \tag{2.63}$$

in which we have abbreviated

$$p \equiv x_1^{-3} + q(1-x_1)^{-3} . \tag{2.64}$$

The direction cosines l, m, n, of a line normal to the surface of this cone clearly are given by

$$l, m, n = \{f_\xi, f_y, f_z\} \div \{f_\xi^2 + f_y^2 + f_z^2\}^{1/2} , \tag{2.65}$$

where $f(\zeta, y, z)$ stands for the left-hand side of Equation (2.62) and $\zeta \equiv x - x_1$. Moreover, the direction cosines of the axis of this cone in the same coordinate system are (1, 0, 0). Consequently, the angle ϵ between any arbitrary line on the surface of the osculating cone and its axis will be defined by the equation

$$l = \cos(\frac{1}{2}\pi - \epsilon) = \sin\epsilon ; \tag{2.66}$$

so that

$$\tan^2\epsilon = \frac{l^2}{1-l^2} = \frac{f_\zeta^2}{f_y^2 + f_z^2} ; \tag{2.67}$$

where, by (2.62),

$$f \equiv \zeta^2(\xi_{xx})_1 + y^2(\xi_{yy})_1 + z^2(\xi_{zz})_1 = 0 . \tag{2.68}$$

Therefore,

$$\frac{f_\zeta^2}{f_y^2 + f_z^2} = \frac{[\zeta((\xi_{xx})_1 y]^2]}{[(\xi_{yy})_1]^2 + [(\xi_{zz})_1 z]^2} . \tag{2.69}$$

Since, moreover, it follows from Equation (2.62) that $\zeta^2(\xi_{xx})_1 = -y^2(\xi_{yy})_1 - z^2(\xi_{zz})_1$, it follows on insertion in (2.67) that

$$\begin{aligned} \tan^2\epsilon &= -\left\{\frac{(\xi_{yy})_1 y^2 + (\xi_{zz})_1 z^2}{(\xi_{yy})_1^2 y^2 + (\xi_{zz})_1^2 z^2}\right\}(\xi_{xx})_1 = \\ &= \left\{\frac{(p-q-1)y^2 + pz^2}{(p-q-1)^2 y^2 + p^2 z^2}\right\}(2p+q+1) , \end{aligned} \tag{2.70}$$

where the values of $\xi_{xx}, \xi_{yy}, \xi_{zz}$ at the conical point P of Figure II.2 are given by Equations (2.64).

All the foregoing results of this section have been based on the approximation of the Roche lobes in contact by an osculating cone at P_1. While this should always constitute a legitimate basis for computations of the limits of eclipses by the less massive component of a contact pair (in the sense that its surface is always interior to the common osculating cone), Chanan *et al* (1976) called attention recently to the fact that the surface of the more massive component can actually "overflow" this cone by amounts increasing with the disparity in masses of the two stars.

In order to demonstrate this, let us expand—in accordance with Equation (1.2)—the Roche equipotentials $\Psi(x, y, z)$ of a contact loop in the proximity of the point P_1 of coordinates x_1, 0, 0 in a Taylor series of the form

$$\begin{aligned}\Psi(x_1, y, z) &= \Psi(x_1,\ 0,\ 0) - \frac{1}{2}(2p+q+1)\zeta^2 + \\ &+ \frac{1}{2}(p-q-1)y^2 + \frac{1}{2}pz^2 + s\zeta^3 - \\ &- \frac{3}{2}s\zeta y^2 - \frac{3}{2}s\zeta z^2 + \ \ldots\ ,\end{aligned} \qquad (2.71)$$

correctly to terms of third order in x, y, z, where— as before—$\zeta \equiv x - x_1$ and where we have abbreviated

$$s = x_1^{-4} - q(1-x_1)^{-4}\ . \qquad (2.72)$$

Over an equipotential surface Ψ = constant and, therefore, $\Psi(x, y, z) = \Psi(x_1,\ 0,\ 0)$. If, moreover, we confine our attention to an intersection of these equipotentials with the plane $z = 0$, Equation (2.72) can be solved for y in terms of ξ in the form

$$\begin{aligned}y^2 &= \frac{(2p+q+1)\zeta^2 - 2s\zeta^3}{p-q-1-3s\zeta} = \\ &= \frac{2p+q+1}{p-q-1}\zeta^2 + \frac{4p+5q+5}{(p-q-1)^2}s\zeta^3 + \ \ldots\ .\end{aligned} \qquad (2.73)$$

If we truncate the expansion on the r.h.s. of the foregoing equation to its first term, we obtain the osculating cone identical with Equation (2.62) above. The next term, factored by an odd power of ζ, will change sign as $x \gtrless x_1$: for $x < x_1$ (i.e., in the direction of the less massive star) it will be negative and, hence, the actual value of y^2 will be less than that appropriate for the osculating cone. For $x > x_1$ the converse will, however, be the case; and the actual Roche surface will overflow the osculating cone.

The extent to which this is the case cannot, in general, be established analytically, and recourse must be had to numerical computation. This has recently

q	$\psi_{\rm max}$	$\psi_{\rm cone}$	$i_{\rm min}$	$i_{\rm cone}$
1.00	57.31	57.31	34.45	34.45
0.80	57.35	57.32	34.45	34.45
0.60	57.49	57.35	34.33	34.44
0.40	57.88	57.43	34.07	34.42
0.20	59.00	57.64	33.34	34.35
0.10	60.56	57.92	32.39	34.27

Table II.4: Eclipse and osculating cone angles (in degrees) as a function of mass-ratio for binary systems in which both components fill their Roche lobes (after Chanan *et al*, 1976).

been done by Chanan *et al* (1976), from whose paper the data given in columns (2) and (4) of the accompanying Table II.4 have been excerpted. In the *conical* approximation, *the maximum duration of eclipse* (i.e., the maximum value of the phase angle ψ_1 of first contact of the eclipse) can be expressed in a closed form in the following manner. Let xy stand for the orbital plane of the two components, inclined to our normal to the line of sight (i.e., tangent to the celestial sphere) at an angle i. If so, then obviously

$$\left.\begin{aligned} x &= \cos\psi_1 \ \sin i \equiv \cos\epsilon\ , \\ y &= \sin\psi_1 \ \sin i\ , \\ z &= \cos i\ ; \end{aligned}\right\} \tag{2.74}$$

and the apparent projected distance δ_1 between the centres of the two components at the phase angle ψ_1 will be given by

$$\delta_1^2 \equiv y^2 + z^2 = \sin^2\psi_1 \sin^2 i + \cos^2 i\ . \tag{2.75}$$

Since, moreover,

$$1 - \delta_1^2 = \cos^2\epsilon = \frac{1}{1+\tan^2\epsilon}\ , \tag{2.76}$$

a combination of (2.67) with (2.76) discloses that

$$\delta_1^2 = \frac{a\cos^2 i}{a-1}\left\{\frac{a+2-4\delta_1^2}{a+2-3\delta_1^2}\right\}\ , \tag{2.77}$$

where we have abbreviated

$$a \equiv \frac{q+1}{p} = \frac{q+1}{x_1^{-3}+q(1-x_1)^{-3}} \tag{2.78}$$

by (2.64).

The maximum duration of the eclipse will occur when $i = 90°$ and $Z \equiv \cos i = 0$. If so, however, Equation(2.77) can remain finite only if the denominator $a + 2 - 3\delta_1^2$ on the right-hand side of (2.77) will also vanish—which will be the case if

$$\delta_{\text{cone}}^2 \equiv \sin^2 \psi_{\text{cone}} = \frac{a+2}{3} \tag{2.79}$$

yielding

$$\cos^2 \psi_{\text{cone}} = \frac{1-a}{3} = \frac{p-q-1}{3p} \,. \tag{2.80}$$

Conversely, the eclipse becomes grazing (i.e., $\delta_0 \equiv \cos i_{min}$) if $\psi = 0°$. If so, however, Equation (2.77) will disclose that

$$\cos^2 i_{\text{cone}} = \frac{a+2}{a+3} = \frac{2p+q+1}{3p+q+1} \,. \tag{2.81}$$

The values of ψ and i_{cone} are then listed as functions of q in columns (3) and (5) of Table II.4. This is always bound to be true if the contact component of the Roche loop is the less massive one of the two; for the more massive component (i.e., the larger of the two, the values given in column (2) continue to apply.

A glance at these data discloses that *the values of ψ_{max} are remarkably insensitive to the mass-ratio*—a fact of considerable significance for the students of close binary systems, first noted by the present writer in 1954; for the variation of light exhibited by systems the components of which fill in the largest Roche lobes capable of containing their mass (in particular, eclipsing systems of the W UMa-type) in the course of an orbital cycle is so smooth and continuous that it is next to impossible to detect by a mere inspection of their respective light curves just where eclipses may set in. Our present analysis has now supplied a theoretical answer: namely, *the light changes of an eclipsing system will be unaffected by eclipses for all phase angles in excess of* $\pm 60°$ *even if both components are in actual contact*—at least as long as their mass-ratios do not become less than 1:10 (though for greater disparity in masses this limit will continue to increase; cf. Chanan *et al*, 1976). Therefore, the light changes exhibited at phases $\pm 30°$ (or more) around each quadrature should be due solely to the proximity effects associated with both stars (cf. Chapter VII), and may be analysed as such without fear of interference from eclipse phenomena. Moreover—and again almost regardless of the mass ratio q—Equation (2.81) makes it evident that *no binary system can exhibit eclipses if its orbit is inclined to the celestial sphere by less than* 33°–34°. For values of i greater than this limit eclipses may occur (and must occur for contact binary systems), of durations ψ_1 connected with i by Equation (2.77) in the conical approximation. For mass ratios $q \approx 1$ the relation between the two becomes again more involved; and for its tabulation the reader is referred to Chanan *et al* (1976).

D: External Envelopes

In the foregoing parts of this section we have been concerned with various properties of Roche equipotentials when $\xi > \xi_1$, and later we investigated the geometry of limiting double-star configurations for which $\xi = \xi_1$. The aim of the present section will be to complete our analysis of the geometrical properties of the Roche model by considering what happens when $\xi < \xi_1$. In the introductory part of this chapter we inferred on general grounds that, if $\xi < \xi_1$, the dumb-bell figure which originally surrounded the two components will open up at P_1 (cf. again Figure II.1 and the corresponding equipotentials will enclose *both* bodies.

When will these latter equipotentials containing the total mass of our binary system cease to form a *closed* surface? A quest for the answer will take us back to Equation (2.20) defining the partial derivative ξ_x. We may note that the right-hand side of this equation is positive when $x \to \infty$, but becomes negative when $x = 1 + \epsilon$, where ϵ denotes a small positive quantity. It becomes positive again as $x \to 0$, and changes sign once more for $x \to -\infty$. Since ξ_x is finite and continuous everywhere except at $x = \infty$ and for $r = 0$ or $r' = 0$, it follows that it changes sign *three* times by passing through zero at points x_1, x_2, x_3, whose values are such that

$$(a)\ 0 < x_1 < 1\,, \quad (b)\ x_2 > 1\,, \quad (c)\ x_3 < 0\,; \tag{2.82}$$

and of these, only the first one has been evaluated so far in this section, and its numerical values listed in column (2) of Table II.1.

An evaluation of the remaining roots $X_{2,3}$ offers, however, no greater difficulty. In embarking upon it we should merely keep in mind that, regardless of the sign of x, the distances r and r' as defined by Equations (1.3) are *positive* quantities. Thus, unlike in case (a)—when, by setting $r = x$ and $r' = 1 - x$, we were led to define x_1 as a root of Equation (2.24)—in case (b), when $x_2 > 1$, we must set $r = x$ but $r' = x - 1$; and in case (c), when $x_3 < 0, r = -x$ and $r' = 1 - x$. After doing so and clearing the fractions we may verify that the equation $\xi_x = 0$ in the case of (b) and (c) assumes the explicit form

$$(1+q)x^5 - (2+3q)x^4 + (1+3q)x^3 - (1+2q)x^2 + 2x - 1 = 0 \tag{2.83}$$

and

$$(1+q)x^5 - (2+3q)x^4 + (1+3q)x^3 + x^2 - 2x + 1 = 0\,, \tag{2.84}$$

respectively

For $q = 0$, the former Equation (2.83) becomes identical with (2.24) and reduces to (2.25) admitting of $x = 1$ as a triple root. Hence, for small values of q, the root $x_2 > 1$ of the complete Equation (2.83) should be expansible as

$$x_2 = 1 + \left(\frac{\mu}{3}\right)^{1/3} + \frac{1}{3}\left(\frac{\mu}{3}\right)^{2/3} + \frac{1}{9}\left(\frac{\mu}{3}\right) + \ldots \tag{2.85}$$

in terms of fractional powers of $\mu \equiv q/(q+1)$. Similarly, Equation (2.84) reduces for $q = 0$ to

$$(x-1)^2(x^3+1) \;=\; 0\,, \tag{2.86}$$

admitting of only one negative root (namely, -1). In consequence, the negative root x_3 of (2.84) should, for small values of μ, be approximable in terms of integral powers of μ by an expansion of the form

$$x_3 = -1 + \frac{7}{12}\mu - \frac{1127}{20736}\mu^3 + \dots . \tag{2.87}$$

The approximate values of x_2 and x_3 as obtained from (2.85)–(2.87) may, moreover, be subsequently refined to any degree of accuracy by differential corrections or any other standard method.

Once sufficiently accurate values of $x_{2,3}$ have thus been established, the values of ξ corresponding to equipotentials which pass through these points can be ascertained from the equation

$$\xi_{2,3} = \xi(x_{2,3}, 0, 0) ; \tag{2.88}$$

while the corresponding values of $C_{2,3}$ then can be found (cf. Equation (2.31) from

$$C_{2,3} = 2(1-\mu)\xi_{2,3} + \mu_2 . \tag{2.89}$$

A tabulation of five-digit values of $x_{2,3}$ and $C_{2,3}$ is given in columns (2)–(5) of the accompanying Table II.5. It may also be noticed that, to a high degree of approximation

$$\xi_3 \doteq \frac{3}{2} + 2q - \frac{q^2}{2(1+q)} \tag{2.90}$$

or

$$C_3 \doteq 3 + \mu ; \tag{2.91}$$

while, somewhat less accurately,

$$(x_2 - 1)^2 = 1 - x_3^2 . \tag{2.92}$$

A comparison of the values of $C_{2,3}$ as given in Table II.5 with those of C_1 from Table II.2 reveals that, for all values of $q > 0$,

$$C_1 > C_2 \geq C_3 . \tag{2.93}$$

For any value of C within the limits of the inequality $C_1 > C > C_2$ the corresponding equipotential will surround the whole mass of the system by a common *external envelope*, which may enclose the common atmosphere of the two stars. For $C = C_2$, this envelope will develop a conical point P_2 (at which $\xi_x = \xi_y = \xi_z = 0$) at $x = x_2$—i.e., behind the centre of gravity of the less massive component (see Figures II.1 or II.2); and if $C < C_2$, the respective equipotentials will open up at P_2. For $C = C_3$, a third conical point P_3 develops behind the centre of gravity of the more massive component at $x = x_3$; and if $C < C_3$, the equipotentials will open up at both ends. Their intersection with the xy-plane will then no longer represent a single closed curve, but will split up in two separate sections (symmetrical with respect to the x-axis), closing gradually around

q	x_2	C_2	$-x_3$	C_3	$C_{4,5}$
1.0	1.69841	3.45680	0.69841	3.45680	2.75000
0.8	1.66148	3.49368	0.73412	3.41509	2.75309
0.6	1.61304	3.53108	0.77751	3.35791	2.76563
0.4	1.54538	3.55894	0.83180	3.27822	2.79592
0.3	1.49917	3.55965	0.86461	3.22675	2.82249
0.2	1.43808	3.53634	0.90250	3.16506	2.86111
0.15	1.39813	3.50618	0.92372	3.12959	2.88658
0.1	1.34700	3.45153	0.94693	3.09058	2.91735
0.05	1.27320	3.34671	0.97222	3.04755	2.95465
0.02	1.19869	3.22339	0.98854	3.01961	2.98077
0.01	1.15614	3.15344	0.99422	3.00990	2.99020
0.005	1.12294	3.10301	0.99710	3.00498	2.99504
0.001	1.07089	3.03838	0.99942	3.00099	2.99990
0.0002	1.04108	3.01387	0.99988	3.00020	2.99980
0	1.00000	3.00000	1.00000	3.00000	3.00000

Table II.5: The data collected in this table are taken from Kopal (1959). For other tabulations of these quantities—in particular for very small values of the parameter $\mu = q/(q+1)$—cf., Rosenthal (1931), Kuiper and Johnson (1956), Szebehely (1967) or Kitamura (1970).

two points which make equilateral triangles with the centres of mass of the two components. The coordinates of such points are specified by the requirements that $r = r' = 1$; consequently, $x = 0.5$ and $y = \pm\sqrt{3}/2$. These triangular points represent also the loci at which our equipotentials vanish eventually from the real plane — if (consistent with Equations 1.6 and 2.31) their constants C reduce to

$$C_{4,5} = 3 - \mu + \mu^2 \ . \tag{2.94}$$

The values of $C_{4,5}$'s as given by this equation are listed in column (6) of Table II.5 for $1 > q > 0$, and represent the lower limits attainable by these constants; for if $C < C_{4,5}$, the equipotential curves ξ = constant in the xy-plane become imaginary, and thus devoid of any physical significance.

II.3 Time-Dependent Roche Equipotentials

The geometrical properties of equipotential surfaces surrounding the Roche gravitational dipoles discussed in the preceding section, are independent of the time. This is, however, true only within the framework of special assumptions underlying Equation (1.6): namely, for constant values of $m_{1,2}$ and R (i.e., if the two finite masses describe circular orbits with the Keplerian angular velocity $\omega \equiv \omega_K$), and if the equatorial plane of the rotating configuration coincides with that of their orbit. A breakdown of any of these assumptions is, however, bound to render the expression for the total potential Ψ as given by Equation (1.2)—and, consequently, the normalized potential ξ—a function of time; and the aim of the present section will be to develop the necessary consequences of this fact.

A: Inclined Axes of Rotation

In order to investigate such effects, consider first the case in which the component of mass m_1 (whose equipotential surfaces are distorted by tides raised by its companion of mass m_2) rotates about an arbitrarily oriented axis with an angular velocity ω_1 which may (but need not) be equal to the Keplerian angular velocity ω_K as given by Equation (1.4). If so, Equation (1.2) for the total potential Ψ should be replaced by

$$\begin{aligned}\Psi &= G\frac{m_1}{r} + G\frac{m_2}{r'} + \frac{\omega_K^2}{2}\left(\frac{m_2 R}{m_1+m_2}\right)^2 - \\ &\quad - \omega_K^2\left(\frac{m_2 R}{m_1+m_2}\right)x'' + \frac{\omega_1^2}{2}(x'^2+y'^2), \qquad (3.1)\end{aligned}$$

where the singly-primed rectangular coordinates x', y', z' *rotate* with the angular velocity ω_1 of the star of mass m_1, but x'' stands for the *revolving* coordinate the axis of which coincides with the radius-vector r of relative orbit of the two finite masses m_1 and m_2.

In order to relate the singly- and doubly-primed coordinates with each other, consider (also for future use) three systems of rectangular coordinates, defined as follows:

1) The (unprimed) axes XYZ will represent a system of *inertial* coordinates ("space axes") of direction fixed in space in such a way that the XY-plane coincides with the *invariable plane of the system*; while the Z-axis is perpendicular to it.

2) The singly-primed axes $X'Y'Z'$ will stand for a system of rectangular coordinates *rotating* with the body ("body-axes"), defined so that $X'Y'$-plane represents the (instantaneous) *equator* of the rotating star, inclined by an angle θ to the inertial XY-plane and intersecting it at the angle ϕ (see Figure II.3).

3) The doubly-primed axes $X''Y''Z''$ will hereafter represent a system of *revolving* rectangular coordinates, in which the X-axis is constantly coincident with the radius-vector between the origin and the centre of mass of the revolving star,

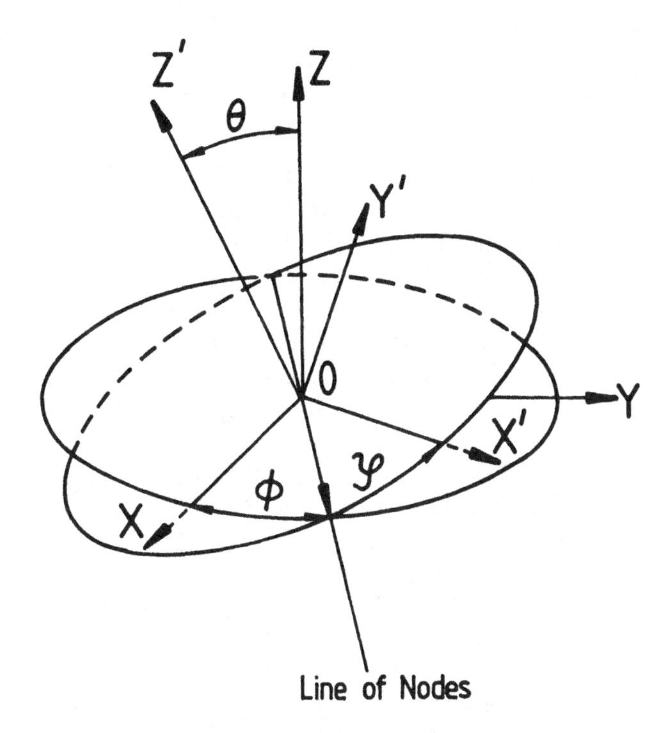

Figure II.3: Definition of Eulerian angles

and $Z'' = 0$ represents the (instantaneous) position of its orbital plane. As is well known, a transformation of coordinates from the inertial (space) to the rotating (body) axes is governed by the matrix equation

$$\left\{ \begin{array}{c} x \\ y \\ z \end{array} \right\} = \left\{ \begin{array}{ccc} a'_{11} & a'_{12} & a'_{13} \\ a'_{21} & a'_{22} & a'_{23} \\ a'_{31} & a'_{32} & a'_{33} \end{array} \right\} \left\{ \begin{array}{c} x' \\ y' \\ z' \end{array} \right\} , \tag{3.2}$$

where the direction cosines

$$\left. \begin{array}{rcl} a'_{11} & = & \cos\varphi\cos\phi - \sin\varphi\sin\phi\cos\theta \, , \\ a'_{12} & = & -\sin\varphi\cos\phi - \cos\varphi\sin\phi\cos\theta \, , \\ a'_{13} & = & \sin\phi\sin\theta \; ; \end{array} \right\} \tag{3.3}$$

$$\left. \begin{array}{rcl} a'_{21} & = & \cos\varphi\sin\phi + \sin\varphi\cos\phi\cos\theta \, , \\ a'_{22} & = & -\sin\varphi\sin\phi + \cos\varphi\cos\phi\cos\theta \, , \\ a'_{23} & = & -\cos\phi\sin\theta \; ; \end{array} \right\} \tag{3.4}$$

$$\left. \begin{array}{rcl} a'_{31} & = & \sin\varphi\sin\theta \, , \\ a'_{32} & = & \cos\varphi\sin\theta \, , \\ a'_{33} & = & \cos\theta \; ; \end{array} \right\} \tag{3.5}$$

where the Eulerian angles ϕ, θ, φ are defined by a scheme illustrated on Figure II.3.

A transformation of the inertial to revolving coordinates is similarly governed by the matrix equation

$$\begin{Bmatrix} x \\ y \\ z \end{Bmatrix} = \begin{Bmatrix} a''_{11} & a''_{12} & a''_{13} \\ a''_{21} & a''_{22} & a''_{23} \\ a''_{31} & a''_{32} & a''_{33} \end{Bmatrix} \begin{Bmatrix} x'' \\ y'' \\ z'' \end{Bmatrix}, \tag{3.6}$$

where the direction cosines

$$\left.\begin{aligned} a''_{11} &= \cos u \cos \Omega - \sin u \sin \Omega \cos i \,, \\ a''_{12} &= -\sin u \cos \Omega - \cos u \sin \Omega \cos i \,, \\ a''_{13} &= + \sin \Omega \sin i \,; \end{aligned}\right\} \tag{3.7}$$

$$\left.\begin{aligned} a''_{21} &= \cos u \sin \Omega + \sin u \cos \Omega \cos i \,, \\ a''_{22} &= -\sin u \sin \Omega + \cos u \cos \Omega \cos i \,, \\ a''_{23} &= - \cos \Omega \sin i \,; \end{aligned}\right\} \tag{3.8}$$

$$\left.\begin{aligned} a''_{31} &= \sin u \sin i \,, \\ a''_{32} &= \cos u \sin i \,, \\ a''_{33} &= \cos i \,; \end{aligned}\right\} \tag{3.9}$$

where Ω denotes the longitude of the nodes (i.e., of intersection of the $Z = 0$ and $Z'' = 0$ planes measured from the X-axis); i, the inclination of the orbital ($Z'' = 0$) to the invariable ($Z = 0$) plane of the system; and u, the angle between the line of the nodes and the instantaneous position of the radius-vector (if, in Equations (3.3)–(3.5) defining the singly-primed direction cosines a'_{ij}, we set $\phi = \Omega, \varphi = u$ and $\theta = i$, the a'_{ij}'s become identical with the doubly-primed direction cosines a''_{ij}).

Accordingly, a transformation from the rotating (singly-primed) to the revolving (doubly-primed) axes obeys the matrix equation

$$\begin{Bmatrix} x' \\ y' \\ z' \end{Bmatrix} = \begin{Bmatrix} \lambda''_1 & \lambda''_2 & \lambda''_3 \\ \mu''_1 & \mu''_2 & \mu''_3 \\ \nu''_1 & \nu''_2 & \nu''_3 \end{Bmatrix} \begin{Bmatrix} x'' \\ y'' \\ z'' \end{Bmatrix} \tag{3.10}$$

where (cf. p.155 of Kopal, 1978)

$$\begin{Bmatrix} \lambda''_j \\ \mu''_j \\ \nu''_j \end{Bmatrix} = \begin{Bmatrix} a'_{11} & a'_{21} & a'_{31} \\ a'_{12} & a'_{22} & a'_{32} \\ a'_{13} & a'_{23} & a'_{33} \end{Bmatrix} \begin{Bmatrix} a''_{1j} \\ a''_{2j} \\ a''_{3j} \end{Bmatrix} \tag{3.11}$$

for $j = 1, 2, 3$ — so that the direction cosines $\nu''_{1,2,3}$ of the axis Z' in the doubly-primed revolving system

$$\begin{aligned} \nu''_1 &= a'_{13}a''_{11} + a'_{23}a''_{21} + a'_{33}a''_{31} \\ &\equiv \mathcal{A} \sin u + \mathcal{B} \cos u \,, \end{aligned} \tag{3.12}$$

$$\nu_2'' = a_{13}'a_{12}'' + a_{23}'a_{22}'' + a_{33}'a_{32}'' \equiv \mathcal{A}\cos u - \mathcal{B}\sin u\,, \tag{3.13}$$

$$\nu_3'' = a_{13}'a_{13}'' + a_{23}'a_{23}'' + a_{33}'a_{33}'' \equiv (1 - \mathcal{A}^2 - \mathcal{B}^2)^{1/2}\,, \tag{3.14}$$

where we have abbreviated

$$\mathcal{A} = \cos\theta\sin i - \sin\theta\cos(\phi - \Omega)\cos i\,, \tag{3.15}$$

$$\mathcal{B} = + \sin\theta\sin(\phi - \Omega)\,; \tag{3.16}$$

such that

$$(1 - \mathcal{A}^2 - \mathcal{B}^2)^{1/2} = \cos\theta\sin i + \sin\theta\cos(\phi - \Omega)\sin i \equiv \nu_3''\,. \tag{3.17}$$

The foregoing Equations (3.16)–(3.17) for $\nu_{1,2,3}''$ are exact for any values of the Eulerian angle θ or the inclination i. If, however, we identify the Z''-axis with Z—i.e., set $i = 0$ rendering $Z'' = 0$ the invariable plane of the respective system (an identification permissible without any loss of generality for the Roche model, though *not* for one exhibiting a finite degree of central condensation; cf. Section VI.3A), Equations (3.12) – (3.14) will reduce to

$$\left.\begin{aligned} \nu_1'' &= -\sin\theta\sin(u + \Omega - \phi)\,, \\ \nu_2'' &= -\sin\theta\cos(u + \Omega - \phi)\,, \\ \nu_3'' &= +\cos\theta\,; \end{aligned}\right\} \tag{3.18}$$

in which the Eulerian angles θ and ϕ of Figure II.3 as well as the longitude Ω of the nodes at which the planes $Z = Z'' = 0$ intersect can be treated as constants.

After these preliminaries let us return to Equation (3.1) for the generalized potential with inclined axes of rotation, and rewrite the expression $x'^2 + y'^2$ on the r.h.s. of (3.1) in terms of the revolving coordinates x'', y'', z'' by means of the transformation (3.10): in doing so we find that

$$\begin{aligned} x'^2 + y'^2 &= r^2 - (\nu_1''x'')^2 - (\nu_2''y'')^2 - (\nu_3''z'')^2 - \\ &\quad - 2(\nu_1''\nu_2''x''y'' + \nu_1''\nu_3''x''z'' + \nu_2''\nu_3''y''z'')\,, \end{aligned} \tag{3.19}$$

$r^2 = x''^2 + y''^2 + z''^2$ exactly; or, alternatively,

$$x'^2 + y'^2 = r^2(1 - \cos^2\Theta) = \frac{2}{3}\,[1 - P_2(\cos\Theta)] \tag{3.20}$$

where

$$\cos\Theta = \lambda''\nu_1'' + \mu''\nu_2'' + \nu''\nu_3''\,; \tag{3.21}$$

and λ'', μ'', ν'' being the direction cosines of an arbitrary radius-vector r in the doubly-primed coordinates. Accordingly (by the addition theorem for the Legendre coefficients)

$$\begin{aligned} P_2(\cos\Theta) &= P_2(\cos\theta)P_2(\cos\theta'') + \\ &+ \frac{1}{3}P_2^1(\cos\theta)P_2^1(\cos\theta'')\sin(\phi-\Omega-u-\phi'') - \\ &- \frac{1}{12}P_2^2(\cos\theta)P_2^2(\cos\theta'')\cos 2(\phi-\Omega-u-\phi'')\,; \end{aligned} \tag{3.22}$$

and if, moreover, the Eulerian angle θ is small enough for its squares and higher powers to be ignorable, the foregoing expression simplifies to

$$P_2(\cos\Theta) = P_2(\nu'') + 3\lambda''\mu''\sin\theta\sin(\phi-\Omega-u) - 3\mu''\nu''\sin\theta\cos(\phi-\Omega-u) + \dots\,. \tag{3.23}$$

When $\theta = 0$ (i.e., if the equator and orbit are coplanar), $\cos\Theta$ becomes identical with ν'' as it was in Section II.1; but for $\theta \neq 0, \cos\Theta$ turns out to depend on the true anomaly u measured from the node. If, moreover, $\omega_1 = \omega_K$, we can set $\phi = \Omega$; but otherwise this need not be the case (cf. Chapter VI).

However, the presence of extra terms on the right-hand side of the Equation (3.1) for the Roche equipotentials—rendering the latter to depend on the time—will also influence the explicit form of the expansion (2.12) of their radius-vector r in ascending powers of r_0: in fact, proceeding in the same way as in Section II.2A we find that (correctly to terms of the order of r_0^5) the right-hand side of Equation (2.12) should be augmented by the term

$$+\, nr_0^3(\omega_1/\omega_K)^2(1-\cos^2\Theta)\,, \tag{3.24}$$

where (by Equation (3.21)),

$$\cos\Theta = \cos\theta\cos\theta'' + \sin\theta\sin\theta''\sin(\phi-\Omega-u-\phi'') \tag{3.25}$$

replacing $nr_0^2(1-\cos^2\theta'')$; but an evaluation of higher-order effects arising from inclination of the equator to the orbital plane should be left as an exercise for the interested reader.

B: Eccentric Orbits

In conclusion of our brief survey of different geometrical properties of the Roche Model, let us point out one additional case in which such properties are bound to depend on the time: namely, if the Keplerian orbit of the two finite masses m_1 and m_2 becomes eccentric. If so, the separation R of their mass centres will be bound to vary as

$$R = \frac{A(1-e^2)}{1+e\cos v}\,, \tag{3.26}$$

where A stands for the semi-major axis of relative orbit of the masses $M_{1,2}$; e, its eccentricity; and v is the true anomaly measured from the periastron passage. Within the framework of the Roche model (i.e., if $m_{1,2}$ can be regarded as mass-points) the parameters A and e on the right-hand side can be regarded as constants. If so, the right-hand side of the expression (1.2) for the total potential Ψ can be rewritten, more explicitly, as

$$\begin{aligned}\frac{\Psi(v)}{G(m_1+m_2)} &= \frac{1-\mu}{r}+\frac{\mu}{R}\left\{1+\left[1-\left(\frac{R}{A}\right)^3\right]\frac{x}{r}+\right. \\ &\left.+\sum_{j=2}^{\infty}\left(\frac{r}{R}\right)^j P_j\left(\frac{x}{r}\right)\right\}+\frac{x^2+y^2}{2A^3}+\frac{\mu^2R^2}{2A^3}, \qquad (3.27)\end{aligned}$$

where (in agreement with Equations (2.32)) $\mu \equiv m_2/(m_1+m_2)$; and for eccentric orbits, the Keplerian angular velocity

$$\omega_K^2 = \frac{G(m_1+m_2)}{A^3}. \qquad (3.28)$$

We wish to conclude this chapter by a proof that, in the case of elliptic orbits of masses $m_{1,2}$, the foregoing potential (3.27) is no longer identical with the zero-velocity surfaces of the restricted problem of three bodies. In order to do so, let us depart from the elliptic three-body problem in space, which in the revolving (doubly-primed) coordinates x'', y'', z'' assumes the form

$$\begin{aligned}\ddot{x}'' - 2\dot{y}'' &= (1+e\cos v)^{-1}\Omega_{x''}, \qquad (3.29)\\ \ddot{y}'' + 2\dot{x}'' &= (1+e\cos v)^{-1}\Omega_{y''}, \qquad (3.30)\\ \ddot{z}'' &= (1+e\cos v)^{-1}\Omega_{z''}; \qquad (3.31)\end{aligned}$$

in terms of the potential

$$\Omega = (1-\mu)(r_1^{-1}+\frac{1}{2}r_1^2)+\mu(r_2^{-1}+\frac{1}{2}r_2^2)-\frac{1}{2}z''^2(1+e\cos v); \qquad (3.32)$$

or, alternatively,

$$\begin{aligned}\ddot{x}'' - 2\dot{y}'' &= (1+e\cos v)^{-1}\Omega'_{x''}, \qquad (3.33)\\ \ddot{y}'' + 2\dot{x}'' &= (1+e\cos v)^{-1}\Omega'_{y''}, \qquad (3.34)\\ \ddot{z}'' + z'' &= (1+e\cos v)^{-1}\Omega'_{z''}, \qquad (3.35)\end{aligned}$$

where

$$\begin{aligned}\Omega' &= \Omega+\frac{1}{2}z''^2(1+e\cos v) = \\ &= (1-\mu)(r_1^{-1}+\frac{1}{2}r_1^2)+\mu(r_2^{-1}+\frac{1}{2}r_2^2), \qquad (3.36)\end{aligned}$$

in which

$$r_1^2 = (x'' + \mu)^2 + y''^2 + z''^2 \tag{3.37}$$

and

$$r_2^2 = (x'' + \mu - 1)^2 + y''^2 + z''^2 \; ; \tag{3.38}$$

differing from r and r' as defined by Equations (1.3) only insofar as the origin of coordinates has been shifted from the centre of mass m_1 to that of the system $m_1 m_2$ as a whole.

The dots on the left-hand sides of Equations (3.31)–(3.32) or (3.35)–(3.36) stand for ordinary differentiation with respect to the time t, and the terms factored by $\dot{x}$ and $\dot{y}$ represent the effects of the Coriolis force. It may also be noted that the potential Ω' as defined by Equation (3.36) does not depend explicitly on the eccentricity e of the orbit of the finite masses m_1 and m_2; nor does it depend explicitly on the time; the latter enters Equations (3.35)–(3.37) only through the cosine of the true anomaly v of the binary orbit.

In order to obtain the Jacobi energy integral of the systems (3.31)–(3.32) or (3.35)–(3.36), let us multiply these equations successively by $\dot{x}, \dot{y}, \dot{z}$ and add: the result will be (in each case)

$$\begin{aligned} \frac{1}{2}\frac{d}{dt}(\dot{x}''^2 + \dot{y}''^2 + \dot{z}''^2 + z'') &= (1 + e\cos v)^{-1}(\dot{x}''\Omega_{x''} + \dot{y}''\Omega_{y''} + \dot{z}''\Omega_{z''}) \\ &= (1 + e\cos v)^{-1}\frac{d\Omega}{dt} \end{aligned} \tag{3.39}$$

or

$$\frac{1}{2}\frac{d}{dt}\{(\dot{x}'')^2 + (\dot{y}'')^2 + (\dot{z}'')^2\} = (1 + e\cos v)^{-1}\frac{d\Omega'}{dt} \,. \tag{3.40}$$

If, moreover, we change over from the time t to the true anomaly v as the independent variable of our problem by means of Kepler's second law

$$r^2 \, dv = \sqrt{G(m_1 + m_2)} \, dt \tag{3.41}$$

valid for the orbit of the two finite masses, Equation (3.39) can be formally integrated to yield

$$(\dot{x}'')^2 + (\dot{y}'')^2 + (\dot{z}'')^2 + (z'')^2 = 2\int \frac{d\Omega}{1 + e\cos v} \tag{3.42}$$

or

$$(\dot{x}'')^2 + (\dot{y}''^2 + (\dot{z}'')^2 = 2\int \frac{d\Omega'}{1 + e\cos v} \,. \tag{3.43}$$

For $e = 0$ (corresponding to a "circular" restricted problem of three bodies), the foregoing Equations (3.42) and (3.43) can be readily integrated to yield

$$T = \Omega + C \,, \tag{3.44}$$

where

$$T = \frac{1}{2}\{(\dot{x}'')^2 + (\dot{y}'')^2 + (\dot{z}'')^2\} \tag{3.45}$$

stands for the (scaled) kinetic energy of the system; Ω, for its potential energy; and C is a constant of integration independent of the time.

Equation (3.44) as it stands represents the classical form of the Jacobi integral of the "circular" restricted problem of three bodies. If, however, $e \neq 0$, it is no longer admissible to integrate Equations (3.39) or (3.40) explicitly to the form (3.42) or (3.43): the latter become mere identities, satisfied by any solution of the Equations (3.31)–(3.32) or (3.35)–(3.36) of motion. The significance of the classical Jacobi integrals rests on the fact that, if $e = 0$, each side of Equations (3.42) or (3.43) constitutes a perfect differential; so that the values of the various terms depend only on the end-point positions of the infinitesimal mass particle. But if $e > 0$, in order to establish the instantaneous shape of the zero-velocity surface, we would need to know the value of $d\Omega$ at each value of (or v) all along any selected path; and to establish this from (3.42) or (3.43), an integration of Equations (3.31)–(3.33) or (3.35)–(3.37) would require the adoption of six arbitrary constants representing the initial conditions of our problem; and no closed formula can be invoked to establish the outcome.

Only in our particular case do the known solutions of the "circular" problem of three bodies transfer readily to the "elliptical" case: namely, the positions of the five Lagrangian points in the orbital plane of the two finite masses. As is well known, their locations are defined by the requirements that the velocities as well as accelerations vanish simultaneously at such points in the plane $z'' = 0$; and if so, the left-hand sides of Equations (3.31)–(3.33) or (3.35)–(3.37) vanish identically. However, in order that this be true, it is necessary that their right-hand sides must vanish—i.e., that

$$\Omega'_{x''} = \Omega'_{y''} = \Omega'_{z''} = 0 \tag{3.46}$$

as well; and the coordinates x'', y'' at which this occurs for $z'' = 0$ have already been established in Section 2C of this chapter. We stressed, however, before that the potential Ω' of the "elliptic" problem as given by Equation (3.36) is independent of the eccentricity of the Keplerian orbit of the finite masses m_1 and m_2—a fact which implies that the existence as well as relative positions of the five Lagrangian points—collinear as well as triangular—remains the same as given in Table II.5; and completely independent of e.

II.4 Bibliographical Notes

The contents of Sections 1 and 2 of this chapter follow largely the presentation of the subject in the first part of Chapter VI of the writer's *Dynamics of Close Binary Systems* (Kopal, 1978) and (partly) Chapter III of his previous treatise (Kopal, 1959); with corrections of misprints which crept into these previous sources. The geometry of the eclipses of contact systems represented by the Roche model can be traced to a previous source (Kopal, 1954); while for complications arising from a great disparity in mass-ratios of contact systems see Chanan, Middleditch and Nelson (1976).

It may be noted that the term "Roche Limit" in the sense used in this book (as well as in all other above-quoted sources) goes back to Kopal (1955). It is not to be mistaken for the concept signifying the minimum distance at which a fluid satellite of infinitesimal mass can approach an oblate planet (Roche, 1850), and referred to as "Roche Limit" by G. H. Darwin (1906, 1911) and Jeans (1919). In this sense, it likewise continues to be used (cf. Chandrasekhar, 1963; or Kopal and Song, 1983) up to the present time.

For previous literature concerning the time-dependent Roche equipotentials discussed in Section II.3A, cf. Plavec (1958); Limber (1963) or Kruszewski (1966).

Concerning the "elliptic" restricted problem of three bodies, the equations (3.35)–(3.37) used in Section II.3B were deduced first by Scheibner as far back as 1866. However, inasmuch as Scheibner's work appeared in a non-astronomical periodical and under a concealing title ("Satz aus der Störungstheorie"), it was generally overlooked by subsequent investigators, and remained unknown until it was re-discovered independently by Nechvíle (1926) and Rein (1940).

For the energy integral of the elliptic problem of three bodies cf. Ovenden and Roy (1961) or Kopal and Lyttleton (1963). A treatment of this subject in Section II.3B follows largely this latter source.

Chapter III

ROCHE COORDINATES

The preceding chapter of this monograph has been concerned primarily with the geometry of the Roche equipotential surfaces surrounding a gravitational dipole of two finite masses in Keplerian orbit. This geometry finds many applications to the phenomena exhibited by close binary systems—in particular, to an interpretation of photometric (and spectroscopic) observations of pairs whose components happen to eclipse each other in the course of their orbital cycle—phenomena to which we shall devote specific attention in the last part of this book. But it is not only to an analysis of descriptive observations that the subject matter of the last chapter can provide an appropriate basis. Much more important in this connection are the phenomena concerned with the stability (both vibrational and secular) of the components of close binary systems (and of their interaction) and gas streams which may revolve between them, that must be approached on the hydrodynamical basis, and in which the shape of the components is bound to play a crucial role.

All such phenomena—whether mathematically expressible in a closed form or described by differential equations of hydrodynamics—possess many features in common: in particular, the facts that the limits of expressions obtainable in integral forms, or the boundary conditions of systems of the respective differential equations, are specified over equipotential surfaces whose equilibrium forms have been described in the preceding chapter. Such surfaces are, however, quite complicated in shape—the more so, the closer the respective system—irrespective of the system of coordinates (rectangular, or spherical polar) we adopt for their description (see, e.g., the scaled Roche potential ξ as given by Equation (1.6) of the preceding chapter, or its parametric expansion of the form (2.11)). Yet a gravitational potential is known (by Dirichlet's theorem) to be a unique function of the coordinates—a fact which suggests the use of such a potential as one curvilinear coordinate of our problem in three-dimensional space; the other two being chosen so as to satisfy other prescribed conditions.

As the form of the equipotential surfaces associated with the Roche model has already been laid down in the preceding chapter, the aim of the present chapter should be to *regularize* their shape by choosing the Roche potential ξ = constant to play the role of the radial coordinate r (replacing the radius-vector of spherical polars); and, for the purpose of subsequent analysis, to adjoin to it two additional coordinates (say, η and ζ; playing the role of angular variables in spherical polars) to define together a new system of *curvilinear* coordinates which may (but need not) be *orthogonal* to each other. If, in particular, the equation ξ = const is to represent the Roche equipotentials as given by Equation (1.6) of Chapter II, while the surfaces η = const and ζ = const are made orthogonal to ξ as well as to each other, the respective system will hereafter be referred to as the *Roche Coordinates.* The geometry of such coordinates will be systematically developed in the first part of this chapter; and, in its second half, the fundamental equations of hydrodynamics will be rewritten in their terms for subsequent use; while an extension of the Roche coordinates to non-orthogonal systems—more suitable for other tasks—will be postponed for the second half of this book.

III.1 Metric Transformations

Consider a three-dimensional metric transformation of the general form

$$\begin{aligned}(ds)^2 &\equiv (dx)^2+(dy)^2+(dz)^2 = \\ &= h_1^2(d\xi)^2+h_2^2(d\eta)^2+h_3^2(d\zeta)^2+ \\ &\quad + 2g_{12}\,d\xi\,d\eta + 2g_{13}\,d\xi\,d\zeta + 2g_{23}\,d\eta\,d\zeta \end{aligned} \tag{1.1}$$

where ξ denotes the Roche potential defined by Equation (1.6) of Chapter II, and η, ζ are the remaining Roche coordinates, and

$$h_1^2 \equiv (x_\xi)^2+(y_\xi)^2+(z_\xi)^2\ , \tag{1.2}$$

$$h_2^2 \equiv (x_\eta)^2+(y_\eta)^2+(z_\eta)^2\ , \tag{1.3}$$

$$h_3^2 \equiv (x_\zeta)^2+(y_\zeta)^2+(z_\zeta)^2\ ; \tag{1.4}$$

and

$$g_{12} \equiv x_\xi x_\eta + y_\xi y_\eta + z_\xi z_\eta\ , \tag{1.5}$$

$$g_{13} \equiv x_\xi x_\zeta + y_\xi y_\zeta + z_\xi z_\zeta\ , \tag{1.6}$$

$$g_{23} = x_\eta x_\zeta + y_\eta y_\zeta + z_\eta z_\zeta\ ; \tag{1.7}$$

the subscripts denoting partial derivatives with respect to the Roche variables ξ, η, ζ.

From Equation (1.6) of Chapter II we know ξ as an explicit function of x, y and z; with $\eta(x,y,z)$ and $\zeta(x,y,z)$ still to be determined. Since, however,

$$\left.\begin{aligned} Jx_\xi &= \eta_y\zeta_z - \eta_z\zeta_y\ , \\ Jx_\eta &= \xi_z\zeta_y - \xi_y\zeta_z\ , \\ Jx_\zeta &= \xi_y\eta_z - \xi_z\eta_y\ ; \end{aligned}\right\} \tag{1.8}$$

$$\left.\begin{aligned} Jy_\xi &= \eta_z\zeta_x - \eta_x\zeta_z\ , \\ Jy_\eta &= \xi_x\zeta_z - \xi_z\zeta_x\ , \\ Jy_\zeta &= \xi_z\eta_x - \xi_x\eta_z\ ; \end{aligned}\right\} \tag{1.9}$$

$$\left.\begin{aligned} Jz_\xi &= \eta_x\zeta_y - \eta_y\zeta_x\ , \\ Jz_\eta &= \xi_y\zeta_x - \xi_x\zeta_y\ , \\ Jz_\zeta &= \xi_x\eta_y - \xi_y\eta_x\ ; \end{aligned}\right\} \tag{1.10}$$

where the Jacobian

$$J \equiv \frac{\partial(\xi, \eta, \zeta)}{\partial(x, y, z)} . \tag{1.11}$$

On insertion from (1.8)–(1.10) in (1.2)–(1.4), we establish that

$$\begin{aligned} J^2 h_1^2 &= (\eta_y \zeta_z - \eta_z \zeta_y)^2 + (\eta_z \zeta_x - \eta_x \zeta_z)^2 + (\eta_x \zeta_y - \eta_y \zeta_x)^2 = \\ &= (\eta_x^2 + \eta_y^2 + \eta_z^2)(\zeta_x^2 + \zeta_y^2 + \zeta_z^2) - \\ &\quad - (\eta_x \zeta_x + \eta_y \zeta_y + \eta_z \zeta_z)^2 , \end{aligned} \tag{1.12}$$

$$\begin{aligned} J^2 h_2^2 &= (\xi_z \zeta_y - \xi_y \zeta_z)^2 + (\xi_x \zeta_z - \xi_z \zeta_x)^2 + (\xi_y \zeta_x - \xi_x \zeta_y)^2 = \\ &= (\xi_x^2 + \xi_y^2 + \xi_z^2)(\zeta_x^2 + \zeta_y^2 + \zeta_z^2) - \\ &\quad - (\xi_x \zeta_x + \xi_y \zeta_y + \xi_z \zeta_z)^2 , \end{aligned} \tag{1.13}$$

$$\begin{aligned} J^2 h_3^2 &= (\xi_y \eta_z - \xi_z \eta_y)^2 + (\xi_z \eta_x - \xi_x \eta_z)^2 + (\xi_x \eta_y - \xi_y \eta_x)^2 = \\ &= (\xi_x^2 + \xi_y^2 + \xi_z^2)(\eta_x^2 + \eta_y^2 + \eta_z^2) - \\ &\quad - (\xi_x \eta_x + \xi_y \eta_y + \xi_z \eta_z)^2 . \end{aligned} \tag{1.14}$$

Moreover, from (1.5)–(1.7) it follows likewise that

$$\begin{aligned} J^2 g_{12} &= (\xi_x \zeta_x + \xi_y \zeta_y + \xi_z \zeta_z)(\eta_x \zeta_x + \eta_y \zeta_y + \eta_z \zeta_z) - \\ &\quad - (\xi_x \eta_x + \xi_y \eta_y + \xi_x \eta_z)(\zeta_x^2 + \zeta_y^2 + \zeta_z^2) , \end{aligned} \tag{1.15}$$

$$\begin{aligned} J^2 g_{13} &= (\xi_x \eta_x + \xi_y \eta_y + \xi_z \eta_z)(\eta_x \zeta_x + \eta_y \zeta_y + \eta_z \zeta_z) - \\ &\quad - (\xi_x \zeta_x + \xi_y \zeta_y + \xi_z \zeta_z)(\eta_x^2 + \eta_y^2 + \eta_z^2) , \end{aligned} \tag{1.16}$$

$$\begin{aligned} J^2 g_{23} &= (\xi_x \eta_x + \xi_y \eta_y + \xi_z \eta_z)(\xi_x \zeta_x + \xi_y \zeta_y + \xi_z \zeta_z) - \\ &\quad - (\eta_x \zeta_x + \eta_y \zeta_y + \eta_z \zeta_z)(\xi_x^2 + \xi_y^2 + \xi_z^2) ; \end{aligned} \tag{1.17}$$

and the angles $\omega_{1,2,3}$ of intersection of the surfaces $\xi = \eta =$ const, $\xi = \zeta =$ const and $\eta = \zeta =$ const are given by the equations

$$\cos \omega_1 = (h_1 h_2)^{-1} g_{12} , \tag{1.18}$$

$$\cos \omega_2 = (h_1 h_3)^{-1} g_{13} , \tag{1.19}$$

$$\cos \omega_3 = (h_2 h_3)^{-1} g_{23} , \tag{1.20}$$

respectively.

While the Roche coordinate ξ has already been identified with the potential defined by Equation (1.6) of Chapter II, the accompanying coordinates η and ζ

have so far been left unspecified. In order to remove this arbitrariness, let us assume that all three coordinates ξ, η, ζ are *orthogonal* — i.e., that the angles

$$\omega_1 = \omega_2 = \omega_3 = \frac{1}{2}\pi ; \tag{1.21}$$

so that, by (1.18)–(1.20), the mixed metric coefficients

$$g_{12} = g_{13} = g_{23} = 0 . \tag{1.22}$$

If true, Equations (1.14)–(1.16) disclose that, accordingly,

$$\xi_x\eta_x + \xi_y\eta_y + \xi_z\eta_z = 0 , \tag{1.23}$$

$$\xi_y\zeta_x + \xi_y\zeta_y + \xi_z\zeta_z = 0 , \tag{1.24}$$

$$\eta_x\zeta_x + \eta_y\zeta_y + \eta_z\zeta_z = 0 ; \tag{1.25}$$

which transform (1.2)–(1.4) to

$$h_1^{-2} = \xi_x^2 + \xi_y^2 + \xi_z^2 , \tag{1.26}$$

$$h_2^{-2} = \eta_x^2 + \eta_y^2 + \eta_z^2 , \tag{1.27}$$

$$h_3^{-2} = \zeta_x^2 + \zeta_y^2 + \zeta_z^2 ; \tag{1.28}$$

disclosing on insertion to (1.12)–(1.14) that

$$J\, h_1 h_2 h_3 = 1 . \tag{1.29}$$

Lastly, the direction cosines l_1, m_1, n_1 of a normal to the surface ξ = constant can be expressed alternatively as

$$\left.\begin{aligned} l_1 &= h_1\xi_x &&= h_1^{-1}x_\xi , \\ m_1 &= h_1\xi_y &&= h_1^{-1}y_\xi , \\ n_1 &= h_1\xi_z &&= h_1^{-1}z_\xi ; \end{aligned}\right\} \tag{1.30}$$

while the normals to the surfaces η = constant and ζ = constant (i.e., of the tangent and the binormal to the surface ξ = constant) are similarly characterized by the direction cosines

$$\left.\begin{aligned} l_2 &= h_2\eta_x &&= h_2^{-1}x_\eta , \\ m_2 &= h_2\eta_y &&= h_2^{-1}y_\eta , \\ n_2 &= h_2\eta_z &&= h_2^{-1}z_\eta ; \end{aligned}\right\} \tag{1.31}$$

and

$$\left.\begin{aligned} l_3 &= h_3\zeta_x &&= h_3^{-1}x_\zeta , \\ m_3 &= h_3\zeta_y &&= h_3^{-1}y_\zeta , \\ n_3 &= h_3\zeta_z &&= h_3^{-1}z_\zeta ; \end{aligned}\right\} \tag{1.32}$$

such that (for orthogonal coordinates)

$$\left.\begin{aligned}\cos\omega_1 &\equiv l_1l_2 + m_1m_2 + n_1n_2 = 0\,,\\ \cos\omega_2 &\equiv l_1l_3 + m_1m_3 + n_1n_3 = 0\,,\\ \cos\omega_3 &= l_2l_3 + m_2m_3 + n_2n_3 = 0\,;\end{aligned}\right\} \tag{1.33}$$

and

$$\begin{vmatrix} l_1 & m_1 & n_1 \\ l_2 & m_2 & n_2 \\ l_3 & m_3 & n_3 \end{vmatrix} \equiv J\,h_1h_2h_3 = 1\,. \tag{1.34}$$

by (1.29).

A: Integrability Conditions

The fundamental equations ensuring the orthogonality of the Roche coordinates ξ, η, ζ are those which require the vanishing of the mixed metric coefficients g_{ij}, and are expressed by the Equations (1.23)–(1.25). Moreover, the first one of them—namely, the Roche potential ξ—is already known to us in a closed form from Equation (1.6) of Chapter II; so the task confronting us is to solve Equations (1.23)–(1.25) for the corresponding coordinates η and ζ.

In order to do so, let us first ensure the compatibility of Equations (1.23)–(1.25) for the given form of ξ; and this can be done in the following manner. As is well known, the differential equation generating the equipotential surfaces $\xi(x, y, z)$ = constant is of the form

$$\xi_x\,dx + \xi_y\,dy + \xi_z\,dz = 0\,; \tag{1.35}$$

while the equations generating surfaces η = constant and ζ = constant which are orthogonal to those produced by setting ξ = constant are given by

$$\frac{dx}{\xi_x} = \frac{dy}{\xi_y} = \frac{dz}{\xi_z}\,. \tag{1.36}$$

Moreover, eliminating η between (1.23)–(1.25) we arrive (cf. Darboux, 1910) at the determinantal equation

$$\begin{vmatrix} \xi_x & \zeta_x & \zeta_x\xi_{xx} + \zeta_y\xi_{xy} + \zeta_z\xi_{xz} \\ \xi_y & \zeta_y & \zeta_x\xi_{xy} + \zeta_y\xi_{yy} + \zeta_z\xi_{yz} \\ \xi_z & \zeta_z & \zeta_x\xi_{xz} + \zeta_y\xi_{yz} + \zeta_z\xi_{zz} \end{vmatrix} = 0 \tag{1.37}$$

or, more explicitly,

$$A\zeta_x^2 + B\zeta_y^2 + C\zeta_z^2 + F\zeta_y\zeta_z + G\zeta_x\zeta_z + H\zeta_x\zeta_y = 0\,; \tag{1.38}$$

with the coefficients

$$\left.\begin{aligned} A &= \xi_y\xi_{xz} - \xi_z\xi_{xy}\,,\\ B &= \xi_z\xi_{xy} - \xi_x\xi_{yz}\,,\\ C &= \xi_x\xi_{yz} - \xi_y\xi_{xz}\,,\end{aligned}\right\} \tag{1.39}$$

and

$$\left.\begin{array}{rcl} F & = & \xi_x\xi_{yy} - \xi_y\xi_{xy} + \xi_z\zeta_{xz} - \xi_x\xi_{zz} , \\ G & = & \xi_y\xi_{yz} - \xi_z\xi_{yz} + \xi_x\zeta_{xy} - \xi_y\xi_{xx} , \\ H & = & \xi_z\xi_{xx} - \xi_x\xi_{zx} + \xi_y\zeta_{yz} - \xi_z\xi_{yy} , \end{array}\right\} \tag{1.40}$$

in terms of the respective partial derivatives of the Roche potential which is known to us in algebraic form.

The algebraic Equation (1.38) is homogeneous and quadratic in $\zeta_{x,y,z}$. Another such relation between $\zeta_{x,y,z}$ and $\xi_{x,y,z}$, of which we have made no use so far, is represented by the third integrability condition (1.24). Therefore, after division by (say) ζ_x it is possible to solve (1.24) and (1.38) simultaneously for the ratio $\zeta_y/\zeta_x = w_1$ from the quadratic equation

$$\mathcal{A}w_1^2 + \mathcal{B}w_1 + \mathcal{C} = 0 , \tag{1.41}$$

with the coefficients

$$\mathcal{A} = B + (\xi_y/\xi_z)^2 C - (\xi_y/\xi_z) F , \tag{1.42}$$

$$\mathcal{C} = A + (\xi_x/\xi_z)^2 C - (\xi_x/\xi_z) G , \tag{1.43}$$

and

$$\mathcal{B} = 2(\xi_x\xi_y/\xi_z^2) C + H - (\xi_x/\xi_z)F - (\xi_y/\xi_z) G + H . \tag{1.44}$$

Moreover, once the quadratic equation (1.42) has been solved for w_1, the ratio $\zeta_z/\zeta_x \equiv w_2$ follows directly from (1.24) in the form

$$w_2 = -(\xi_x/\xi_z) - (\xi_y/\xi_z) w_1 . \tag{1.45}$$

Therefore, it is possible to rewrite the ratios of the derivatives $\xi_{x,y,z}$ in the form

$$\zeta_x : \zeta_y : \zeta_z = U : V : W , \tag{1.46}$$

where

$$U = 2\mathcal{A} , \; V = -\mathcal{B} - \sqrt{\mathcal{B}^2 - 4\mathcal{A}\mathcal{C}} , \tag{1.47}$$

and

$$W = -(\xi_x/\xi_z) U - (\xi_y/\xi_z) V , \tag{1.48}$$

such that

$$\xi_x U + \xi_y V + \xi_z W = 0 . \tag{1.49}$$

In order that our function ζ may exist, it is both necessary and sufficient that the total differential equation

$$U\,dx + V\,dy + W\,dz = 0 \tag{1.50}$$

be uniquely integrable; and this will indeed be the case (cf. Darboux, 1910) provided that the functions U, V, and W satisfy the partial differential equation

$$U(V_z - W_y) + V(W_x - U_z) + W(U_y - V_x) = 0 , \tag{1.51}$$

of third order in ξ.

Is the foregoing equation actually satisfied by the Roche equipotentials $\xi(x, y, z)$ of partial derivates as given by Equations (2.20)–(2.22) and (2.56)–(2.61) of Chapter II? Let us prove first that these expressions satisfy indeed the integrability condition (1.51) when either ω or q is equal to zero. If $n > 0$, $q = 0$ (i.e., in the case of purely rotational distortion), it follows that

$$\mathcal{A}, \mathcal{C} = \pm\omega^2 Lxyz \quad \text{and} \quad \mathcal{B} = \omega^2 L(x^2 - y^2)z \,, \tag{1.52}$$

where we have abbreviated

$$L = r^{-9} - 3\omega^2 z^2 r^{-8} - 2\omega^2 r^{-6} + \omega^4 r^{-3} \,. \tag{1.53}$$

Accordingly, from Equations (1.47) and (1.48)

$$U = 2\omega^2 Lxyz \,, \quad V = -2\omega^2 Lx^2 z \,, \quad W = 0 \,, \tag{1.54}$$

which are indeed formed to satisfy the integrability condition (1.51) for any value of ω. Since, moreover, our entire argument which proves the existence of ζ can be reversed to do the same for η—by eliminating ζ instead of η from the fundamental system of Equations (1.23)–(1.25)—this latter system is clearly integrable to furnish a triply-orthogonal set of curvilinear Roche coordinates in the rotational case.

On the other hand, in the case of purely tidal distortion (i.e., $n = 0$, $q > 0$), Equations (2.20)–(2.22), together with (2.56)–(2.61), of Chapter II yield

$$\left.\begin{aligned} \mathcal{A} &= 0 \,, \\ \mathcal{B} &= -3q(y^2 + z^2)zM \,, \\ \mathcal{C} &= -3qxyzM + 3q^2y^2zN \,, \end{aligned}\right\} \tag{1.55}$$

where we have abbreviated

$$\begin{aligned} M \;=\; & r^{-9}r'^{-5} + q\{r^{-8} + 2r^{-6}r'^{-5} - 2r^{-8}r'^{-3} + r^{-6}r'^{-8} + r^{-8}r'^{-6}\} + \\ & + q^2\{r^{-3}r'^{-5} + r^{-5}r'^{-3} - 2r^{-5}r'^{-6} + 2r^{-3}r^{-8} + r^{-5}r'^{-9}\} + \\ & + q^3 r'^{-8} \,, \end{aligned} \tag{1.56}$$

and

$$\begin{aligned} N \;=\; & r^{-6}r'^{-8} - r^{-6}r'^{-5} - \\ & - q\{r^{-5} + 2r^{-3}r'^{-5} - 3r^{-5}r'^{-3} + 3r^{-5}r'^{-6} - 2r^{-3}r'^{-8} - r^{-5}r'^{-9}\} - \\ & - q^2\{r'^{-5} - r'^{-8}\} \,, \end{aligned} \tag{1.57}$$

which leads, by (1.47) and (1.48), to

$$U = 0 \,, \quad V = -2\mathcal{B} \,, \quad W = 2(y/z)\mathcal{B} \,,$$

likewise satisfying the integrability condition (1.51). Hence, in the case of a tidally-distorted Roche model, our fundamental Equations (1.23)–(1.25) can again be integrated to furnish a triply-orthogonal system of curvilinear coordinates.

However, when terms arising from both rotation and tides are present in the expressions for ξ and its derivatives, all three coefficients $\mathcal{A}, \mathcal{B}$ and $\mathcal{C}$ will *not* vanish. Inserting (2.56)–(2.61) of Chapter II in (1.42)–(1.44) we find (cf. Kopal and Ali, 1971) that the Roche potential arising from the combined effects of rotation and tides fails to satisfy the Darboux condition (1.51) of integrability; and, accordingly, numerical integrations (cf. Kitamura, 1970) offer the only avenue of approach to an investigation of their geometry.

B: Rotational Problem

In order to take up the first possibility in which the equations (1.23)–(1.25) governing the Roche orthogonal coordinates have been found to be analytically integrable, consider the case of purely rotational distortion in which the respective Roche potential, which for $q = 0$ and $n = \frac{1}{2}$ reduces to

$$\xi = \frac{1}{r} + \frac{r^2}{2}(1 - \nu^2), \tag{1.58}$$

where $\nu \equiv \cos\theta$; and in which the first partial derivatives of ξ with respect to x, y, z—already known from Equations (2.20)–(2.22) of Chapter II—now reduce to

$$\xi_x = -x(r^{-3} - 1), \tag{1.59}$$

$$\xi_y = -y(r^{-3} - 1), \tag{1.60}$$

$$\xi_z = -z(r^{-3}). \tag{1.61}$$

The lines that are orthogonal to the equipotentials ξ = constant, as defined by Equation (1.58), are known to be given by Equations (1.36). Since, however, by (1.61) and (1.60),

$$\frac{\xi_x}{\xi_y} = \frac{x}{y} \tag{1.62}$$

regardless of the extent of distortion, the first part of (1.38) readily yields

$$\frac{dx}{dy} = \frac{x}{y}, \tag{1.63}$$

integrating to

$$\frac{x}{y} = \text{constant} = \phi\,(\text{say}). \tag{1.64}$$

Accordingly, the surface η = constant turn out to be identical with the meridional cross-sections ϕ = constant of the rotationally-distorted Roche model (and the corresponding three-dimensional Roche coordinates prove to constitute what Darboux termed a "Lamé family" of the respective triply-orthogonal set.

Since, in such a case, $d\eta = 0$, the metric transformation (1.1) reduces then to

$$(ds)^2 = h_1^2(d\xi)^2 + h_3^2(d\zeta)^2 , \tag{1.65}$$

where, by (1.26) and (1.59)–(1.61)

$$h_1^{-2} = r^{-4}\{1 - 2r^3(1-\nu^2) + r^6(1-\nu^2)\} . \tag{1.66}$$

Moreover, the radius-vector r reduces to a particular case of the expansion on the r.h.s. of Equation (2.11) of Chapter II, in which we have set $q = 0$ and $n = 1/2$; of the form

$$\begin{aligned} r &= r_0\left\{1 + \frac{1}{2}r_0^3(1-\nu^2) + \frac{3}{4}r_0^6(1-\nu^2)^2 + \frac{3}{2}r_0^9(1-\nu^2)^3 + \ldots\right\} = \\ &= r_0\,{}_2F_1\left\{\frac{1}{3}, \frac{2}{3}; \frac{3}{2}; \frac{27}{8}r_0^3(1-\nu^2)\right\} , \end{aligned} \tag{1.67}$$

where ${}_2F_1$ stands for an ordinary hypergeometric series of the argument $(\frac{3}{2}r_0)^3\times \sin^2\theta$ (and, for $r_0 = \frac{2}{3}$, reduces for the radius-vector of the respective Roche limit—as given by Equation (2.44) of Chapter II).

If $\eta =$ constant, Equations (1.23) and (1.25) safeguarding the orthogonality of our Roche coordinates are identically satisfied; but the surface $\zeta =$ constant remain yet to be specified by means of (1.24); and its exact solution can likewise be constructed (cf. Kopal, 1971) in the following way. Let, by hypothesis,

$$\cos\zeta = \nu\sum_{j=0}^{\infty} r^{3j}X_j(\nu) , \tag{1.68}$$

where the X_j's are polynomials in $\nu \equiv \cos\theta$ whose form is to be determined.

In order to do so, differentiate the foregoing Equation (1.68) with respect to x, y, z; and remember (from Equations (1.5) of Chapter II) that $r_x \equiv \lambda$, $r_y \equiv \mu$, $r_z \equiv \nu$; while $r\nu_x = -\lambda\nu$, $r\nu_y = -\mu\nu$ and $r\nu_z = \;\; -\nu^2$): we obtain

$$\zeta_x \sin\zeta = -\frac{\lambda\nu}{r}\sum_{j=0}^{\infty} r^{3j}[3jX_j - (\nu X_j)'] , \tag{1.69}$$

$$\zeta_y \sin\zeta = -\frac{\mu\nu}{r}\sum_{j=o}^{\infty} r^{3j}[3jX_j - (\nu X_j)'] , \tag{1.70}$$

$$\zeta_z \sin\zeta = -\frac{1-\nu^2}{r}\sum_{j=0}^{\infty} r^{3j}\left[\frac{3j\nu^2X_j}{1-\nu^2} + (\nu X_j)'\right] , \tag{1.71}$$

where prime on X_j denotes differentiation with respect to ν. If, moreover, we cross-multiply the foregoing Equations (1.69)–(1.71) with the $\xi_{x,y,z}$'s as given by (1.61)–(1.61), we find that

$$\begin{aligned} &-(\xi_x\zeta_x + \xi_y\zeta_y + \xi_z\zeta_z)\sin\zeta = \\ &= \frac{(1-\nu^2)\nu}{r^3}\sum_{j=0}^{\infty}\left\{r^3\left[3jX_j - (\nu X_j)'\right] - \frac{3jX_j}{1-\nu^2}\right\} r^{3j} . \end{aligned} \tag{1.72}$$

If the coordinates ξ and ζ are to be orthogonal to each other, the left-hand side of the foregoing equation must necessarily vanish; and so will the right-hand side provided that the coefficients of successive powers of r^{3j} for $j = 0, 1, 2, \ldots$. This will indeed be the case provided that

$$X_0(\nu) = 1\,, \;\; X_1(\nu) = -\frac{1}{3}(1-\nu^2)\,; \tag{1.73}$$

while, for $j > 1$, all higher successive polynomials $X_j(\nu)$ are generated with the aid of the recursion formula

$$3jX_j + (1-\nu^2)\left\{(\nu X_{j-1})' - 3(j-1)X_{j-1}\right\} = 0\,. \tag{1.74}$$

If, moreover,

$$\lim_{j\to\infty} j\, r^{3j}\, X_j(\nu) = 0\,, \tag{1.75}$$

the series on the right-hand side of Equation (1.68) converges for $r < 1$ and represents the solution of our problem.

In particular,

$$\begin{aligned}\cos\zeta \;\; &= \;\; \nu\left\{1 - \frac{1}{3}r^3(1-\nu^2) - \frac{1}{9}r^6(1-\nu^2) - \right.\\ &\qquad \left. -\frac{1}{81}r^9(1-\nu^2)(5-3\nu^2) + \ldots\right\}\,;\end{aligned} \tag{1.76}$$

the inversion of which yields

$$\begin{aligned}&1-\nu^2 \equiv \sin^2\theta = \\ &= \left\{1 - \frac{2}{3}r_0^3\cos^2\zeta - \frac{2}{9}r_0^6(7 - 8\cos^2\zeta)\cos^2\zeta + \ldots\right\}\sin^2\zeta\,.\end{aligned} \tag{1.77}$$

Moreover, the metric coefficient h_3 will be given by (1.28) as

$$\begin{aligned}h_3^{-2}(r,\nu) \;\; &= \;\; r^{-2}\left\{1 - \frac{2}{3}r^3(1-2\nu^2) - \right.\\ &\qquad \left. -\frac{1}{9}r^6(1 - 12\nu^2 + 9\nu^4) + \ldots\right\}\end{aligned} \tag{1.78}$$

or, by (1.67) and (1.77),

$$\begin{aligned}-r_0^2h_1(r_0,\zeta) \;\; &= \;\; 1 + 2r_0^3\sin^2\zeta - \\ &\qquad - \frac{1}{12}r_0^6(22 - 85\sin^2\zeta)\sin^2\zeta + \ldots\end{aligned} \tag{1.79}$$

and

$$\begin{aligned}r_0^{-1}h_3(r_0,\zeta) \;\; &= \;\; 1 - \frac{1}{6}r_0^3(2 - 7\sin^2\zeta) + \\ &\qquad + \frac{1}{36}r_0^6(2 - 88\sin^2\zeta + 145\sin^4\zeta) + \ldots\,.\end{aligned} \tag{1.80}$$

Lastly, the direction cosines $l_{1,2,3}, m_{1,2,3}$ and $n_{1,2,3}$ as defined by Equations (1.30)–(1.32) turn out to be given by

$$l_1 = \lambda\left\{1 - r^3\nu^2 - \frac{3}{2}r^6\nu^2(1-\nu^2) + \ldots\right\}, \tag{1.81}$$

$$m_1 = \mu\left\{1 - r^3\nu^2 - \frac{3}{2}r^6\nu^2(1-\nu^2) + \ldots\right\}, \tag{1.82}$$

$$n_1 = \nu\left\{1 + r^3(1-\nu^2) + \frac{1}{2}r^6(1-\nu^2)(2-3\nu^2) + \ldots\right\}; \tag{1.83}$$

$$l_2 = -\sin\phi, \quad m_2 = +\cos\phi, \quad n_2 = 0; \tag{1.84}$$

and

$$l_3 = \nu\left\{1 + r^3(1-\nu^2) + \frac{1}{2}r^3(1-\nu^2)(2-3\nu^2) + \ldots\right\}\cos\phi, \tag{1.85}$$

$$m_3 = \nu\left\{1 + r^3(1-\nu^2) + \frac{1}{2}r^6(1-\nu^2)(2-3\nu^2) + \ldots\right\}\sin\phi, \tag{1.86}$$

$$n_2 = -\left\{1 - r^3\nu^2 - \frac{3}{2}r^6\nu^2(1-\nu^2) + \ldots\right\}\sin\theta; \tag{1.87}$$

the right-hand sides of which can be rewritten in terms of the Roche coordinates r_0 and ζ by means of the expansions (1.67) and (1.77) with equal ease.

C: Double-Star Problem

The general theory outlined in Section II.1A led us to expect that the Roche coordinates associated with the rotational problem admit of a solution in the closed form; and the contents of Section II.1B did not disappoint these expectations. Moreover, it follows from II.1A that the tidal problem should admit of a similar solution if we change over from the Z-axis for the X-axis (and the direction cosine ν for λ) as one of symmetry of the respective configuration. However, if both axial rotation and tidal action participate in causing our configuration to depart from the spherical form, the set of orthogonal coordinates associated with equipotentials so generalized no longer constitutes a Lamé family; and as we have seen in Section II.1A, the system of Equations (1.23)–(1.25) defining them ceases to be consistent, and no longer admits of a solution in a closed form. The arguments advanced in Section II.1A do not rule out that solutions of (1.23)–(1.25) cannot be constructed which are *asymptotic* in a *limited region* of space; and in what follows such a solution will be obtained possessing such asymptotic properties in the neighbourhood of the origin of coordinates $r = 0$.

In order to do so, let us return to Equations (1.1) and (1.36), which on division by the differential element dr can be rewritten as

$$\left(\frac{dx}{dr}\right)^2 + \left(\frac{dy}{dr}\right)^2 + \left(\frac{dz}{dr}\right)^2 = \left(\frac{ds}{dr}\right)^2 \tag{1.88}$$

and

$$\frac{dy}{dr} = \left(\frac{\xi_y}{\xi_x}\right) \frac{dx}{dr} \; , \; \frac{dz}{dr} = \left(\frac{\xi_z}{\xi_x}\right) \frac{dx}{dr} \; . \tag{1.89}$$

If, moreover, l_1, m_1, n_1 continue to stand for the direction cosines of a normal to the surface ξ = constant, it follows that

$$l_1 = \frac{dx}{ds} \, , \; m_1 = \frac{dy}{ds} \, , \; n_1 = \frac{dz}{ds} \, , \tag{1.90}$$

in which, to the first order in surficial distortion, we are entitled to set $d\sigma \equiv dr$. If so, and if in the equipotential ξ = constant terms of rotational as well as tidal origin are retained to the same degree of approximation, the direction cosines l_1, m_1, n_1 as given by Equations (1.30) are

$$l_1 \equiv \frac{dx}{dr} \;\; = \;\; \lambda - (1+q)\, r^3\, \lambda\nu^2 - q(1-\lambda^2) \sum_{j=2}^{4} r^{j+1}\, P_j'(\lambda) \, , \tag{1.91}$$

$$m_1 \equiv \frac{dy}{dr} \;\; = \;\; \mu - (1+q)\, r^3\, \mu\nu^2 + q\lambda\mu \sum_{j=2}^{4} r^{j+1}\, P_j'(\lambda) \, , \tag{1.92}$$

$$n_1 \equiv \frac{dz}{dr} \;\; = \;\; \nu + (1+q)\, r^3 \nu(1-\nu^2) + q\lambda\nu \sum_{j=2}^{4} r^{j+1}\, P_j'(\lambda) \, ; \tag{1.93}$$

where the primes on the Legendre polynomials $P_j(\lambda)$ signify derivatives with respect to λ.

On the other hand, by virtue of Equations (1.5) of Chapter II it follows also that

$$l_1 \equiv \frac{dx}{dr} = \frac{d}{dr}(r\lambda) = r\,\frac{d\lambda}{dr} + \lambda \, , \tag{1.94}$$

$$m_1 \equiv \frac{dy}{dr} = \frac{d}{dr}(r\mu) = r\,\frac{d\mu}{dr} + \nu \, , \tag{1.95}$$

$$n_1 \equiv \frac{dz}{dr} = \frac{d}{dr}(r\nu) = r\,\frac{d\nu}{dr} + \nu \, ; \tag{1.96}$$

which on insertion in (1.41)–(1.43) lead to

$$\frac{d\lambda}{dr} \;\; = \;\; -(1+q)\, r^2\, \lambda\nu^2 - q(1-\lambda^2) \sum_{j=2}^{4} r^j\, P_j'(\lambda) \, , \tag{1.97}$$

$$\frac{d\mu}{dr} \;\; = \;\; -(1+q)\, r^2\, \mu\nu^2 + q\lambda\nu \sum_{j=2}^{4} r^j\, P_j'(\lambda) \, . \tag{1.98}$$

$$\frac{d\nu}{dr} \;\; = \;\; +(1+q)\, r^2 \nu\,(1-\nu^2) + q\lambda\nu \sum_{j=2}^{4} r^j\, P_j'(\lambda) \, . \tag{1.99}$$

Integrate now both sides of the foregoing equations with respect to r: in doing so we find that

$$\lambda = -\frac{1}{3}(1+q)\,r^3\,\lambda\nu^2 - q(1-\lambda^2)\sum_{j=2}^{4}\frac{r^{j+1}}{j+1}P_j'(\lambda) + C_1\,, \tag{1.100}$$

$$\mu = -\frac{1}{3}(1+q)\,r^3\,\mu\nu^2 + q\,\lambda\mu\sum_{j=2}^{4}\frac{r^{j+1}}{j+1}P_j'(\lambda) + C_2\,, \tag{1.101}$$

$$\nu = \frac{1}{3}(1+q)\,r^3\,\nu\,(1-\nu^2) + q\,\lambda\nu\sum_{j=2}^{4}\frac{r^{j+1}}{j+1}P_j'(\lambda) + C_3\,, \tag{1.102}$$

where the integration constants $C_{1,2,3}$ which are independent of r can be set equal to

$$\left.\begin{aligned} C_1 &= \cos\eta\,\sin\zeta\,, \\ C_2 &= \sin\eta\,\sin\zeta\,, \\ C_3 &= \cos\zeta\,; \end{aligned}\right\} \tag{1.103}$$

the last of which is seen (for $q = 0$) to be identical with Equation (1.68) of the rotational case for $j = 1$. Needless to stress, the same process can be repeated for $j > 1$; but to do so must be left as an exercise for the interested reader.

III.2 Equations of Motion in Roche Coordinates

In the preceding section of this chapter we have introduced a three-dimensional system of special curvilinear coordinates, defined so that its angular components are orthogonal to the Roche equipotentials developed in Chapter II. The aim of the present section will be to apply these coordinates to the tasks to which they were intended to begin with: namely, to express in their terms the basic equations of motion of a mass-point in the gravitational field of a revolving Roche dipole. The main advantage of their use is the fact that the boundary conditions imposed on the surface of an equipotential can be made to depend on one coordinate only—namely, ξ—regardless of the extent to which these may deviate from a sphere.

The equations of motion to be so transformed are already known to us (at least, in their simplest form) from Section II.2B (in which the rectangular coordinates are referred to a doubly-primed set of axes revolving with the constant radius-vector R of our gravitational dipole). This form is physically the simplest, but mathematically unsatisfactory in one respect: namely, that the spatial coordinates occur in them as both the dependent as well as independent variables. This is, however, not necessary, but optional if we employ the coordinates in the role of independent variables, the result will generally be referred to as the *Eulerian equations* of our problem. whereas should we tolerate their presence only as dependent variables, the equations so obtained are referred to as *Lagrangian.*

Both such systems possess certain advantages as well as disadvantages. Thus the Eulerian equations are generally of lower order in independent variables than the Lagrangian; and the conservation of mass (i.e., the "equation of continuity") becomes one of lower degree in their terms. For these reasons, in what follows (and throughout this book) we shall use the equations of motion consistently in their Eulerian form; their principal characteristic being the fact that their time operator (denoted in Section II.2B by a dot), should be replaced by

$$\frac{D}{Dt} \equiv \frac{\partial}{\partial t} + u\frac{\partial}{\partial x} + v\frac{\partial}{\partial y} + w\frac{\partial}{\partial z}, \tag{2.1}$$

where

$$\dot{x} \equiv u\,,\; \dot{y} \equiv v\,,\; \dot{z} \equiv w\,. \tag{2.2}$$

In order to rewrite the Lagrangian time derivative (2.1) in terms of the Roche coordinates, differentiate the functions $\xi(x,y,z)$, $\eta(x,y,z)$ and $\zeta(x,y,z)$ to obtain

$$d\xi = \xi_x\,dx + \xi_y\,dy + \xi_z\,dz\,, \tag{2.3}$$

$$d\eta = \eta_x\,dx + \eta_y\,dy + \eta_z\,dz\,, \tag{2.4}$$

$$d\zeta = \zeta_x\,dx + \zeta_y\,dy + \zeta_z\,dz \tag{2.5}$$

and divide by dt: we find that

$$\dot{\xi} = \xi_x\dot{x} + \xi_y\dot{y} + \xi_z\dot{z} \equiv \xi_x u + \xi_y v + \xi_z w\,, \tag{2.6}$$
$$\dot{\eta} = \eta_x\dot{x} + \eta_y\dot{y} + \eta_z\dot{z} \equiv \eta_x u + \eta_y v + \eta_z w\,, \tag{2.7}$$
$$\dot{\zeta} = \zeta_x\dot{x} + \zeta_y\dot{y} + \zeta_z\dot{z} \equiv \zeta_x u + \zeta_y v + \zeta_z w \tag{2.8}$$

by (2.2); so that the Roche velocity components $u_{\xi,\eta,\zeta}$ in the direction of the Roche coordinates ξ, η, ζ can be expressed as

$$u_1 \equiv h_1\dot{\xi} = l_1 u + m_1 v + n_1 w\,, \tag{2.9}$$
$$u_2 \equiv h_2\dot{\eta} = l_2 u + m_2 v + n_2 w\,, \tag{2.10}$$
$$u_3 \equiv h_3\dot{\zeta} = l_3 u + m_3 v + n_3 w\,; \tag{2.11}$$

where the direction cosines l_j, m_j, n_j $(j = 1,2,3)$ continue to be given by Equations (1.30)–(1.32) subject to Equation (1.34); and the inversion of (2.9)–(2.11) yields

$$u = (m_2n_3 - m_3n_2)u_1 + (m_3n_1 - m_1n_2)u_2 + + (m_1n_2 - m_2n_1)u_3\,, \tag{2.12}$$
$$v = (l_3n_2 - l_2n_3)u_1 + (l_1n_3 - l_3n_1)u_2 + + (l_2n_1 - l_1n_2)u_3\,, \tag{2.13}$$
$$w = (l_2m_3 - l_3m_2)u_1 + (l_3m_1 - l_1m_3)u_2 + + (l_1m_2 - l_2m_1)u_3\,. \tag{2.14}$$

Moreover, the differential operators occurring on the r.h.s. of Equation (2.1) can be expressed as

$$\left\{ \begin{array}{c} \dfrac{\partial}{\partial x} \\ \dfrac{\partial}{\partial y} \\ \dfrac{\partial}{\partial z} \end{array} \right\} = \left\{ \begin{array}{ccc} \xi_x & \eta_x & \zeta_x \\ \xi_y & \eta_y & \zeta_y \\ \xi_z & \eta_z & \zeta_z \end{array} \right\} \left\{ \begin{array}{c} \dfrac{\partial}{\partial \xi} \\ \dfrac{\partial}{\partial \eta} \\ \dfrac{\partial}{\partial \zeta} \end{array} \right\} ; \tag{2.15}$$

which inserted together with (2.12)–(2.14) in (2.1) discloses that the Lagrangian time derivative can be expressed in terms of the Roche coordinates as

$$\frac{D}{Dt} \equiv \frac{\partial}{\partial t} + \frac{u_1}{h_1}\frac{\partial}{\partial \xi} + \frac{u_2}{h_2}\frac{\partial}{\partial \eta} + \frac{u_3}{h_3}\frac{\partial}{\partial \zeta} ; \tag{2.16}$$

and, accordingly (cf., e.g., Lamb, 1932; p.158)

$$\begin{aligned} \frac{Du_1}{Dt} &= \frac{\partial u_1}{\partial t} + \frac{u_1}{h_1}\frac{\partial u_1}{\partial \xi} + \frac{u_2}{h_2}\frac{\partial u_1}{\partial \eta} + \frac{u_3}{h_3}\frac{\partial u_1}{\partial \zeta} + \\ &+ \frac{u_1}{h_1}\left\{ \frac{u_1}{h_1}\frac{\partial h_1}{\partial \xi} + \frac{u_2}{h_2}\frac{\partial h_1}{\partial \eta} + \frac{u_3}{h_3}\frac{\partial h_1}{\partial \zeta} \right\} - \\ &- \frac{1}{h_1}\left\{ \frac{u_1^2}{h_1}\frac{\partial h_1}{\partial \xi} + \frac{u_2^2}{h_2}\frac{\partial h_2}{\partial \xi} + \frac{u_3^2}{h_3}\frac{\partial h_3}{\partial \xi} \right\} , \end{aligned} \tag{2.17}$$

$$\begin{aligned} \frac{Du_2}{Dt} &= \frac{\partial u_2}{\partial t} + \frac{u_1}{h_1}\frac{\partial u_2}{\partial \xi} + \frac{u_2}{h_2}\frac{\partial u_2}{\partial \eta} + \frac{u_3}{h_3}\frac{\partial u_2}{\partial \zeta} + \\ &+ \frac{u_2}{h_2}\left\{ \frac{u_1}{h_1}\frac{\partial h_2}{\partial \xi} + \frac{u_2}{h_2}\frac{\partial h_2}{\partial \eta} + \frac{u_3}{h_3}\frac{\partial h_2}{\partial \zeta} \right\} - \\ &- \frac{1}{h_2}\left\{ \frac{u_1^2}{h_1}\frac{\partial h_1}{\partial \eta} + \frac{u_2^2}{h_2}\frac{\partial h_2}{\partial \eta} + \frac{u_3^2}{h_3}\frac{\partial h_3}{\partial \eta} \right\} , \end{aligned} \tag{2.18}$$

$$\begin{aligned} \frac{Du_3}{Dt} &= \frac{\partial u_3}{\partial t} + \frac{u_1}{h_1}\frac{\partial u_3}{\partial \xi} + \frac{u_2}{h_2}\frac{\partial u_3}{\partial \eta} + \frac{u_3}{h_3}\frac{\partial u_3}{\partial \zeta} + \\ &+ \frac{u_3}{h_3}\left\{ \frac{u_1}{h_1}\frac{\partial h_3}{\partial \zeta} + \frac{u_2}{h_2}\frac{\partial h_3}{\partial \eta} + \frac{u_3}{h_3}\frac{\partial h_3}{\partial \zeta} \right\} - \\ &- \frac{1}{h_3}\left\{ \frac{u_1^2}{h_1}\frac{\partial h_1}{\partial \zeta} + \frac{u_2^2}{h_2}\frac{\partial h_2}{\partial \zeta} + \frac{u_3^2}{h_3}\frac{\partial h_3}{\partial \zeta} \right\} , \end{aligned} \tag{2.19}$$

Moreover, the components of the *gradient* of a scalar function will be given by

$$\nabla \equiv \frac{\mathbf{i}}{h_1}\frac{\partial}{\partial \xi} + \frac{\mathbf{j}}{h_2}\frac{\partial}{\partial \eta} + \frac{\mathbf{k}}{h_3}\frac{\partial}{\partial \zeta} , \tag{2.20}$$

where $\underline{i}, \underline{j}, \underline{k}$ are unit vectors in the direction of increasing coordinates ξ, η, ζ. The divergence Δ of the velocity vector assumes the form

$$\Delta \equiv \frac{1}{h_1 h_2 h_3} \left\{ \frac{\partial}{\partial \xi}(h_2 h_3 u_1) + \frac{\partial}{\partial \eta}(h_3 h_1 u_2) + \frac{\partial}{\partial \zeta}(h_1 h_2 u_3) \right\} , \tag{2.21}$$

while the Laplacian operator

$$\begin{aligned} \nabla &\equiv \frac{1}{h_1 h_2 h_3} \left\{ \frac{\partial}{\partial \xi}\left(\frac{h_2 h_3}{h_1}\frac{\partial}{\partial \xi}\right) + \frac{\partial}{\partial \eta}\left(\frac{h_3 h_1}{h_2}\frac{\partial}{\partial \eta}\right) + \frac{\partial}{\partial \zeta}\left(\frac{h_1 h_2}{h_3}\frac{\partial}{\partial \zeta}\right) \right\} \equiv \\ &\equiv J \left\{ \frac{\partial}{\partial \xi}\left(\frac{1}{h_1^2 J}\frac{\partial}{\partial \xi}\right) + \frac{\partial}{\partial \eta}\left(\frac{1}{h_2^2 J}\frac{\partial}{\partial \eta}\right) + \frac{\partial}{\partial \zeta}\left(\frac{1}{h_3^2 J}\frac{\partial}{\partial \zeta}\right) \right\} \end{aligned} \tag{2.22}$$

by (1.29); with the Jacobian determinant j as given by (1.11).

The (conservative) vector form of the Eulerian equations of motion for the vector velocity

$$\mathbf{V} = \mathbf{i}\, u_1 + \mathbf{j}\, u_2 + \mathbf{k}\, u_3 \tag{2.23}$$

can be expressed as

$$\frac{D\mathbf{V}}{Dt} + 2(\vec{\omega} \times \mathbf{V}) = \mathrm{grad}\Psi \tag{2.24}$$

where the explicit form of the total potential Ψ (gravitational plus centrifugal) has already been investigated in Chapter II, and the vector product $\vec{\omega} \times \mathbf{V}$ stands for the effects of the Coriolis force arising from the rotation of (primed) body-axes with respect to the inertial (unprimed) space-axes, such that the vector angular velocity

$$\vec{\omega} \equiv \mathbf{i}\omega_x + \mathbf{j}\omega_y + \mathbf{k}\omega_z , \tag{2.25}$$

with respect to the space (inertial axes). If, moreover, $\omega_x = \omega_y = 0$ while $\omega_z \equiv \omega =$ constant, the components of the Coriolis force will be

$$\vec{\omega} \times \mathbf{V} = (-\mathbf{i}v + \mathbf{j}u + \mathbf{k}0)\omega \tag{2.26}$$

in the rectangular coordinates. However, in the Roche coordinates,

$$\begin{aligned} \vec{\omega} \times \mathbf{V} &= \mathbf{i}(-l_1 v + m_1 u)\,\omega + \\ &\quad + \mathbf{j}(-l_2 v + m_2 u)\,\omega + \\ &\quad + \mathbf{k}(-l_3 v + m_3 u)\,\omega , \end{aligned} \tag{2.27}$$

in which we can insert for u and v from (2.12) and (2.13). Doing so we find that

$$\left. \begin{aligned} -l_1 v + m_1 u &= \qquad\quad -n_3 u_2 \; + n_2 u_3 , \\ -l_2 v + m_2 u &= n_3 u_1 \qquad\qquad - n_1 u_3 , \\ -l_3 v + m_3 u &= -n_2 u_1 \; + n_1 u_2 \end{aligned} \right\} \tag{2.28}$$

Moreover, by Equation (1.7) of Chapter II,

$$\mathrm{grad}\Psi = \frac{\mathbf{i}}{h_1}\left(\frac{Gm_1}{R}\right) + \frac{\mathbf{j}}{h_2}(0) + \frac{\mathbf{k}}{h_3}(0) ; \tag{2.29}$$

so that the vector equation (2.24) of this chapter can be rewritten as a simultaneous system of three scalar equations of the form

$$\frac{Du_1}{Dt} + 2(-n_3 u_2 + n_2 u_3)\omega = \frac{Gm_1}{h_1 R}, \tag{2.30}$$

$$\frac{Du_2}{Dt} + 2(n_3 u_1 - n_1 u_3)\omega = 0, \tag{2.31}$$

$$\frac{Du_3}{Dt} + 2(-n_2 u_1 + n_1 u_2)\omega = 0, \tag{2.32}$$

for the velocity components $u_{1,2,3}$, in which the Lagrangian time-derivatives are given, more explicitly, by Equations (2.17)–(2.19); the metric coefficients $h_{1,2,3}$ by Equations (1.26)–(1.28); and the direction cosines $n_{1,2,3}$ by the last of Equations (1.30)–(1.32).

A: Geometry of the Roche Coordinates

In the concluding part of the preceding section concerned with the double-star problem, the only solution of Equations (1.23)–(1.25) defining the Roche coordinates, which we have been able to construct by analytic means, holds good only in the proximity of the mass centres of our binary configuration, where it tends asymptotically towards the spherical polars. Outside the limits of this restriction, however, the general geometrical properties of Roche coordinates can be investigated—numerically or otherwise—only in such cases for which the system (1.23)–(1.25) can be reduced to ordinary differential equations tractable by more elementary means; and a possibility to do so arises at the intersection of the planes $X = 0$, $Y = 0$ or $Z = 0$ with the surfaces $\xi =$ constant, $\eta =$ constant and $\zeta =$ constant, where the respective curves of intersection can be constructed by numerical integration.

Exception arises, of course, with the radial coordinate ξ which is known to us as a function of x, y, z in a closed form. Schematic sketches of the equipotential profiles in the xy-plane have already been presented in the preceding chapters; and the accompanying Figures III.1 to III.4 exhibit their more detailed presentations, drawn to scale for 4 different values of the mass-ratios $q = 1, 0.8, 0.6$ and 0.4; while Figure III.5 shows in perspective a three-dimensional view of the surfaces passing through different Lagrangian points (see Section II.2D) of our configuration.

Moreover, in the same plane (i.e., $Z = 0$) the orthogonality conditions (1.23)–(1.25) reduce to

$$\xi_x \eta_x + \xi_y \eta_y = 0, \tag{2.33}$$

disclosing that the curves $\eta(x, y, 0) =$ constant in the XY-plane can be generated by one-parameter solutions of the ordinary differential equation

$$\frac{dx}{dy} = \frac{\xi_x}{\xi_y} = \frac{x(1 - r_1^3)r_2^3 - q(1 - x)r_1^3(1 - r_2^3)}{y[(1 - r_1^3)r_2^3 + qr_1^3(1 - r_2^3)]} \tag{2.34}$$

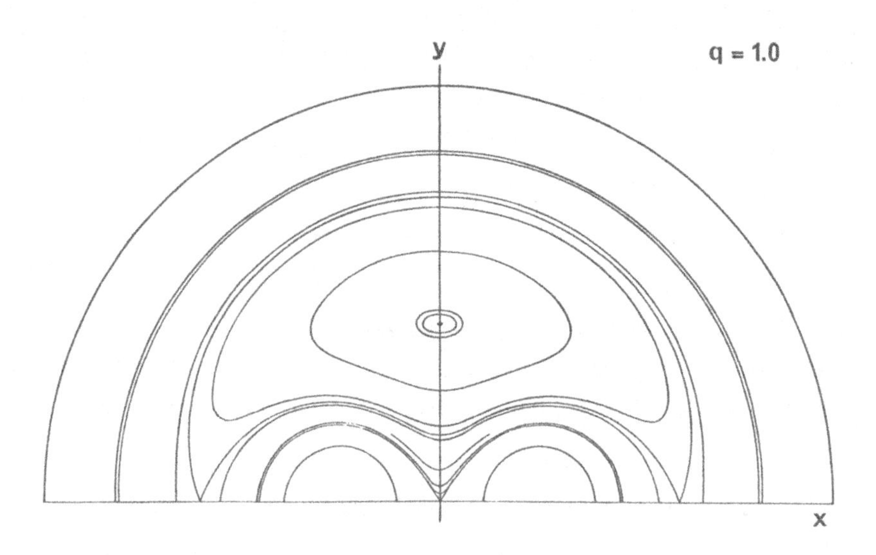

Figure III.1: Cross-sections of a family of Roche equipotentials ξ = constant with the xy-plane for $y \geq 0$, corresponding to 12 different values of the potential and the mass-ratio $q = 1$ (after Kopal, 1959).

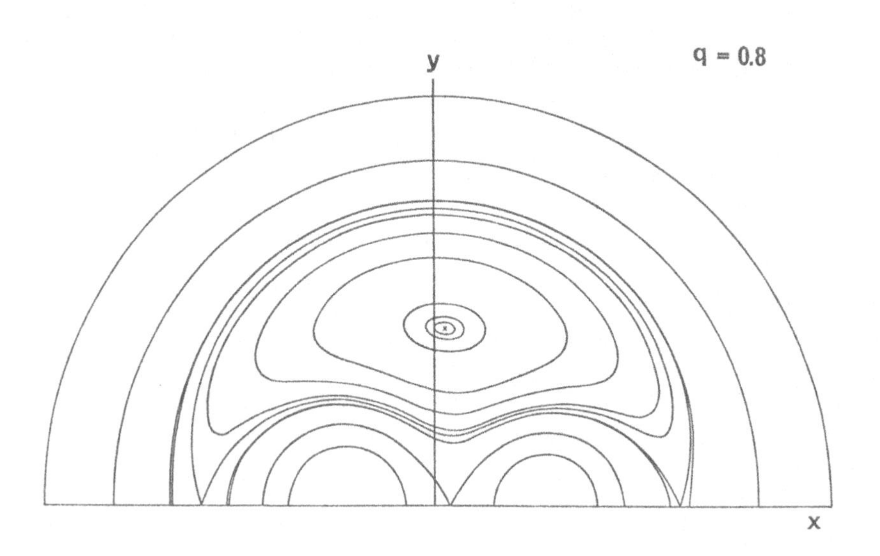

Figure III.2: Cross-sections of a family of Roche equipotentials ξ = constant with the xy-plane for $y \geq 0$, corresponding to 11 different values of the potential, and the mass-ratio $q = 0.8$ (after Kopal, 1959).

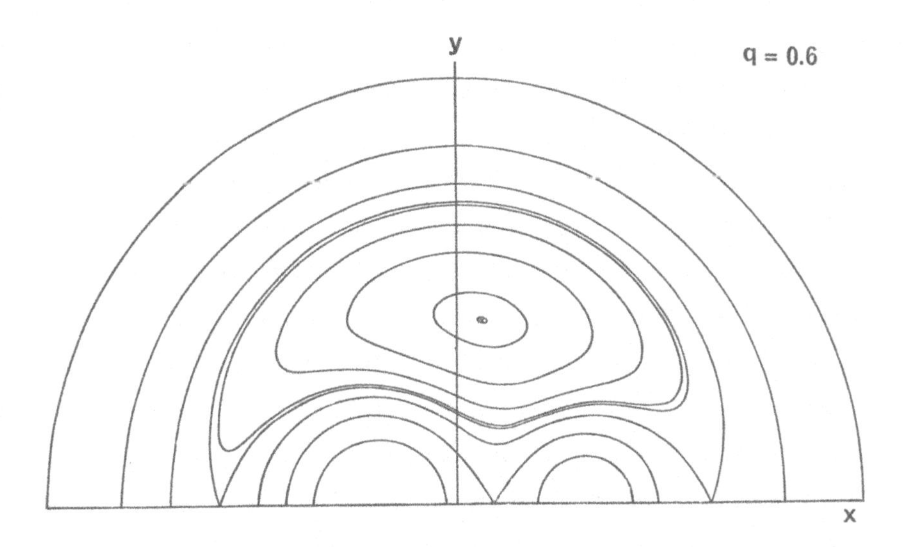

Figure III.3: Cross-section of a family of the Roche equipotentials ξ = constant with the xy-plane for $y \geq 0$, corresponding to 10 different values of the potential, and the mass-ratio $q = 0.6$ (after Kopal, 1959).

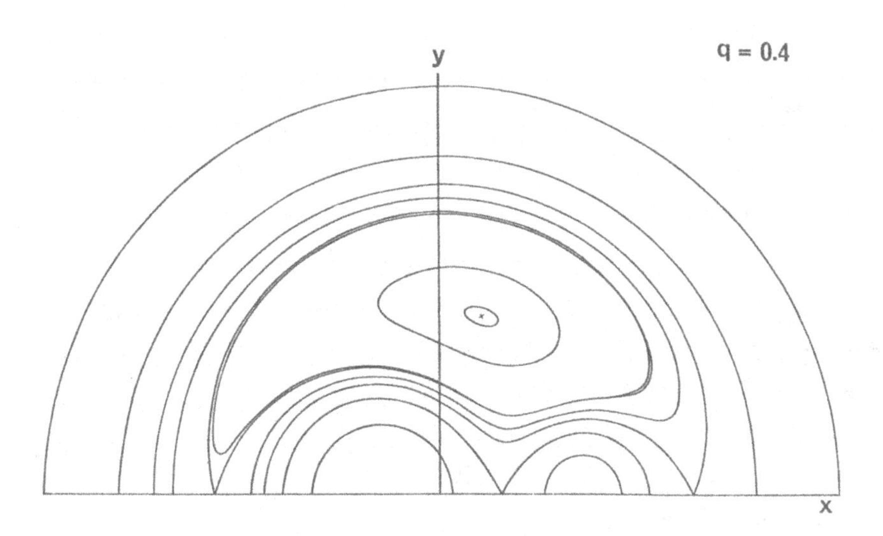

Figure III.4: Cross-sections of a family of the Roche equipotentials ξ = constant with the xy-plane for $y \geq 0$, corresponding to 8 different values of the potential, and the mass-ratio $q = 0.4$ (after Kopal, 1959).

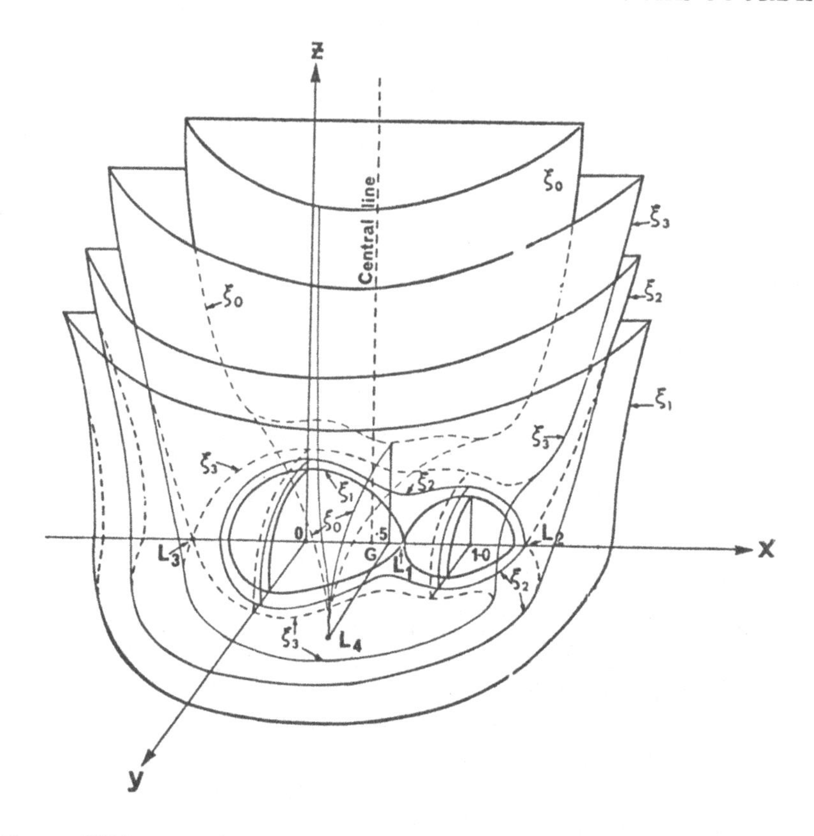

Figure III.5: A schematic view of the Roche equipotentials ξ = constant in three-dimensional perspective, corresponding to the values which the potential assumes at different Lagrangian points of the configuration (after Kitamura, 1970).

by Equations (2.20)–(2.21) of Chapter II, where

$$r_1^2 = x^2 + y^2 \text{ and } r_2^2 = (1-x)^2 + y^2 ; \tag{2.35}$$

the integration constant of which becomes identical with the Roche coordinate η. Moreover, replacing in (2.34) the variables x and y by r_1 and r_2, we can rewrite the latter in the alternative form

$$\frac{dr_2^2}{dr_1^2} = \frac{2fr_2^2 - (1 - r_1^2 - r_2^2)}{2r_1^2 - f(1 - r_1^2 - r_2^2)} , \tag{2.36}$$

where we have abbreviated

$$f \equiv q \frac{1 - r_2^{-3}}{1 - r_1^{-3}} > 0 ; \tag{2.37}$$

and which (unlike (2.34)) involves only integral powers of each variable.

The differential equations (2.34) or (2.36) are of first order, but do not admit of the separation of variables to make them solvable by quadratures; and numerical integrations remain the only avenue of approach. These have been extensively carried out by Kitamura (1970); and disclosed that, for large values of ξ (i.e., in the proximity of the two finite mass centres, or at large distances from our gravitational dipole), the curves η = constant depart but slightly from straight lines. However, in the intermediate range, these lines curve along the envelope defined by the equation

$$2r_1^2 = f(1 - r_1^2 - r_2^2) \tag{2.38}$$

which represents a pole of the differential equation (2.36). All pass through the Lagrangian triangular points $L_{4,5}$; while the collinear points $L_{1,2,3}$ can be reached by them along the limiting lines coinciding with the X- or Y-axes regardless of the mass ratio.

These results are graphically shown on the accompanying Figures III.6 to III.13, drawn to scale for eight values of the mass-ratios q ranging from 1 to 0.001. The mapping of the positions of the Lagrangian points in the $\xi - \eta$ plane is shown on Figure III.14. The mass-points $m_{1,2}$ as well as the potential "curtain" surrounding the entire system is represented by the abscissae $\xi > \xi_1$, and coalesce at the Lagrangian point L_1, characterized in the $\xi - \eta$ plane by the coordinates $\xi = \xi_1$ and $\eta = 0$. The remaining collinear points $L_{2,3}$ transfer to the $\xi - \eta$ plane as points with the coordinates $(\xi_2, 0)$ and (ξ_3, π) while the equilateral-triangle points $L_{4,5}$ map on the same plane as a straight line $\xi = \xi_0$.

If we turn now to the plane $Y = 0$, the curves $\zeta(x, 0, z)$ in the XZ must satisfy the orthogonality condition

$$\xi_x \zeta_x + \xi_z \zeta_z = 0\ ; \tag{2.39}$$

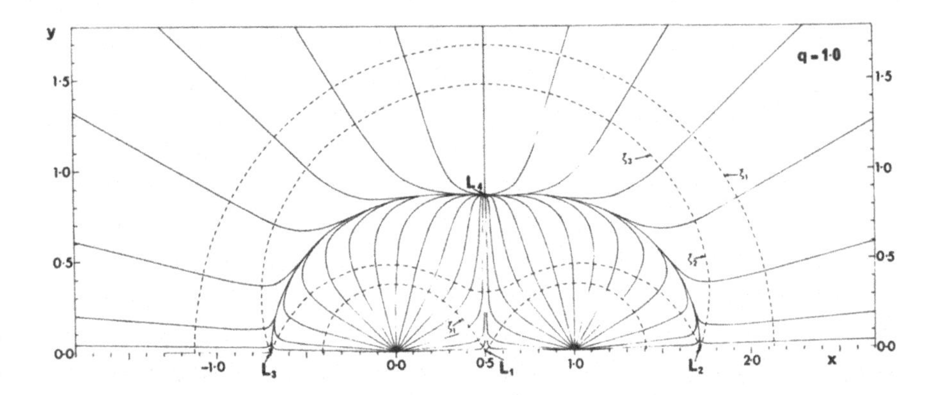

Figure III.6: The geometry of the curves η = constant in the xy-plane for $y \geq 0$, corresponding to the mass-ratio $q = 1$. The trajectories of $\eta(x, y, 0)$ are marked with full curves; while the dotted curves represent the equipotentials ξ = constant for the values which the potential assumes at the Lagrangian collinear points $L_{1,2,3}$ (after Kitamura, 1970)

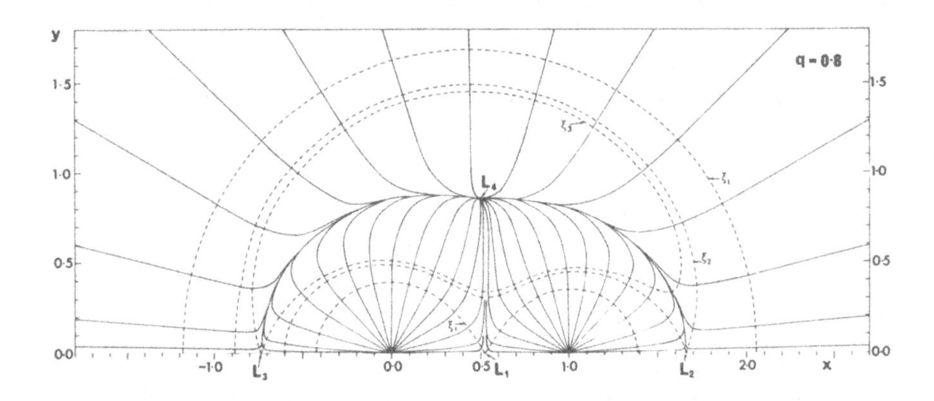

Figure III.7: The geometry of the curves η = constant in the xy-plane for $y \geq 0$, corresponding to the mass-ratio $q = 0.8$. The trajectories of $\eta(x, y, 0)$ are marked with full curves; while the dotted curves represent the equipotentials ξ = constant for the values which the potential assumes at the Lagrangian collinear points $L_{1,2,3}$ (after Kitamura, 1970).

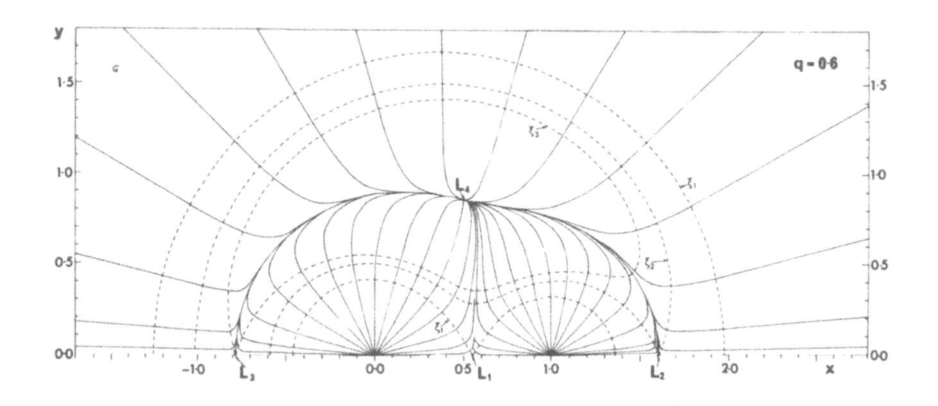

Figure III.8: The geometry of the curves η = constant in the xy-plane for $y \geq 0$, corresponding to the mass-ratio $q = 0.6$. The trajectories of $\eta(x, y, 0)$ are marked with full curves; while the dotted curves represent the equipotentials ξ = constant for the values which the potential assumes at the Lagrangian collinear points $L_{1,2,3}$ (after Kitamura, 1970).

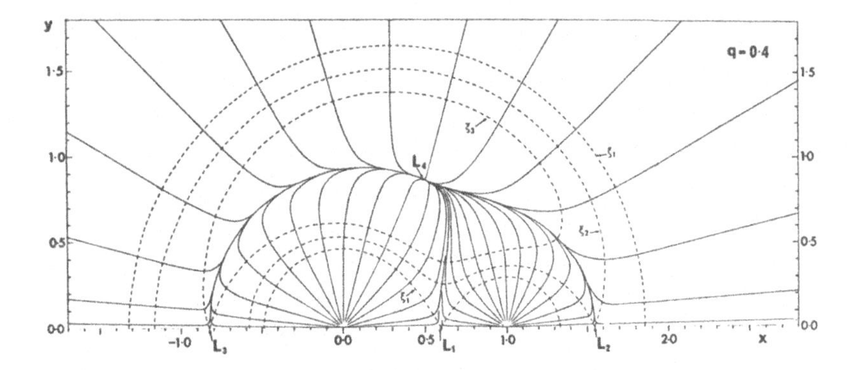

Figure III.9: The geometry of the curves η = constant in the xy-plane for $y \geq 0$, corresponding to the mass-ratio $q = 0.4$. The trajectories of $\eta(x, y, o)$ are marked with full curves; while the dotted curves represent the equipotentials ξ = constant for the values which the potential assumes at the Lagrangian collinear points $L_{1,2,3}$ (after Kitamura, 1970).

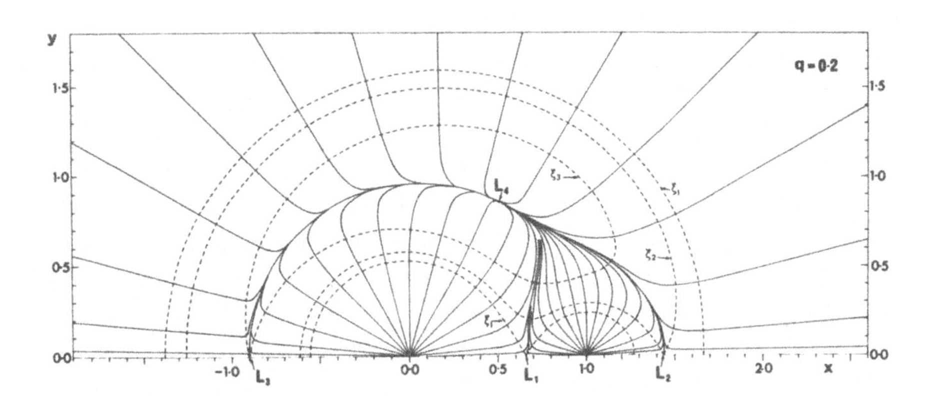

Figure III.10: The geometry of the curves η = constant in the xy-plane for $y \geq 0$, corresponding to the mass-ratio $q = 0.2$. The trajectories of $\eta(x, y, 0)$ are marked with full curves; while the dotted curves represent the equipotentials ξ = constant for the values which the potential assumes at the Lagrangian collinear points $L_{1,2,3}$ (after Kitamura, 1970).

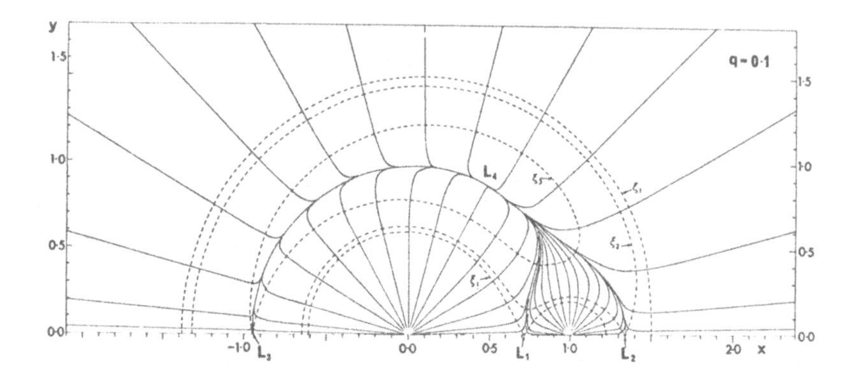

Figure III.11: The geometry of the curves η = constant in the xy-plane for $y \geq 0$, corresponding to the mass-ratio $q = 0.1$. The trajectories of $\eta(x, y, 0)$ are marked with full curves; while the dotted curves represent the equipotentials ξ = constant for the values which the potential assumes at the Lagrangian collinear points $L_{1,2,3}$ (after Kitamura, 1970).

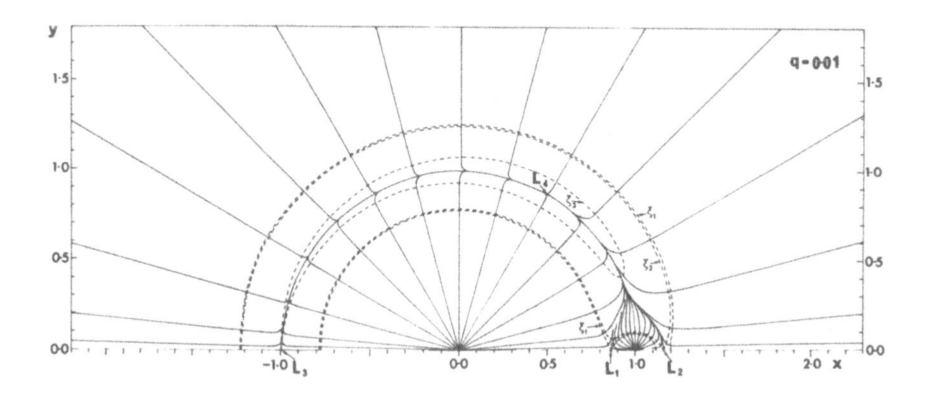

Figure III.12: The geometry of the curves η = constant in the xy-plane for $y \geq 0$, corresponding to the mass-ratio $q = 0.01$. The trajectories of $\eta(x, y, 0)$ are marked with full curves; while the dotted curves represent the equipotentials ξ = constant for the values which the potential assumes at the Lagrangian collinear points $L_{1,2,3}$ (after Kitamura, 1970).

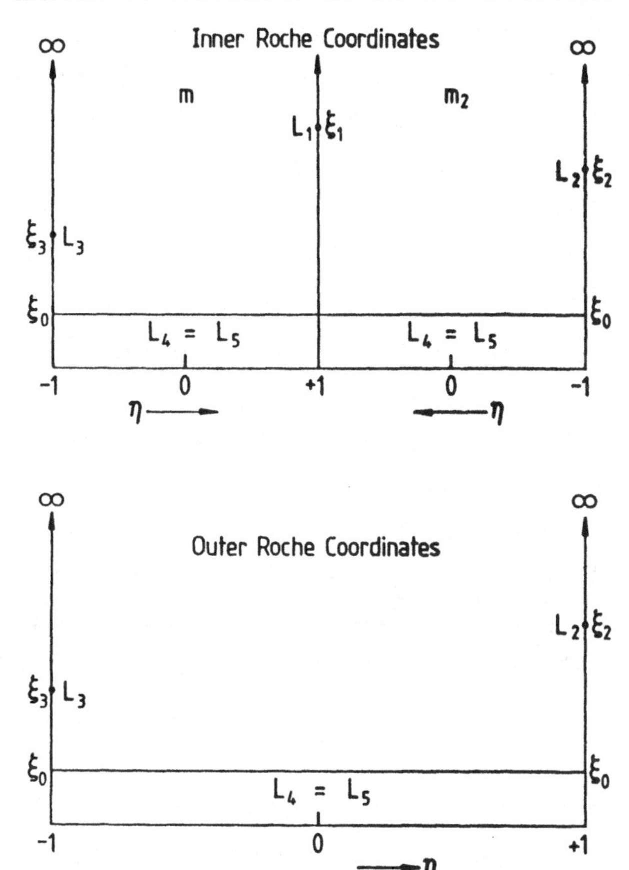

Figure III.13: Mapping of the positions of the Lagrangian points $L_{1,\ldots,5}$ of the XY-plane on to the $\xi\eta$-plane in the Roche curvilinear coordinates (a) outside, and (b) within, the envelope of the η = constant curves (after Kitamura, 1970).

and, as such, will correspond to the integration constants of a first-order differential equation

$$\frac{dx}{dz} = \frac{\xi_x}{\xi_z} = \frac{x(1-r_1^3)r_2^3 - q(1-x)r_1^3(1-r_2^3)}{z(1+qr_1^3)} \tag{2.40}$$

where

$$r_1^2 = x^2 + z^2 \quad \text{and} \quad r_2^2 = (1-x)^2 + z^2 \, . \tag{2.41}$$

This equation too has been integrated numerically by Kitamura (1970); and the profiles of the corresponding curves $\zeta(x, 0, z)$ in the XZ-plane are shown on the accompanying Figures III.15 to III.20; while Figure III.21 represents the surfaces $\zeta(x, y, z)$ = constant in three-dimensional perspective.

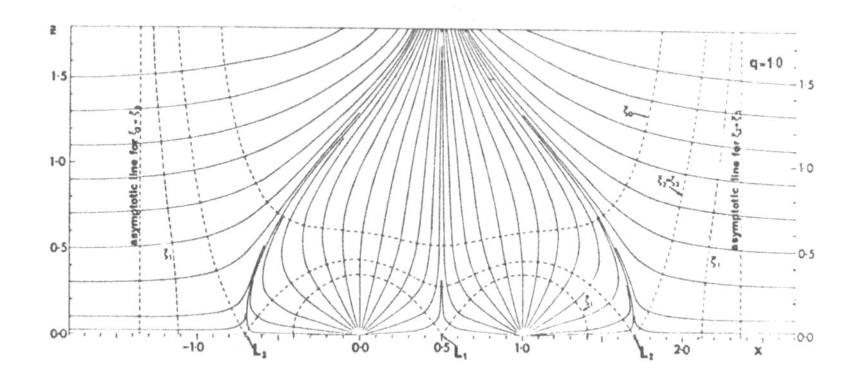

Figure III.14: The geometry of the curves ζ = constant in the xz-plane for $z \geq 0$, corresponding to the mass-ratio $q = 1$. The trajectories of $\zeta(x, 0, z)$ are marked with full curves; while dotted curves represent the equipotentials ξ = constant for the values which the potential assumes at the Lagrangian collinear points $L_{1,2,3}$ (after Kitamura, 1970).

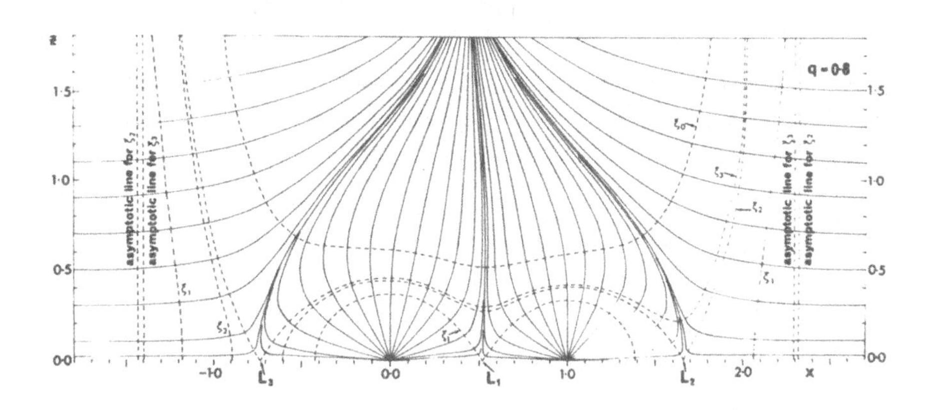

Figure III.15: The geometry of the curves ζ= constant in the xz-plane for $z \geq 0$, corresponding to the mass-ratio $q = 0.8$. The trajectories of $\zeta(x, 0, z)$ are marked with full curves; while dotted curves represent the equipotentials ξ = constant for the values which the potential assumes at the Lagrangian collinear points $L_{1,2,3}$ (after Kitamura, 1970).

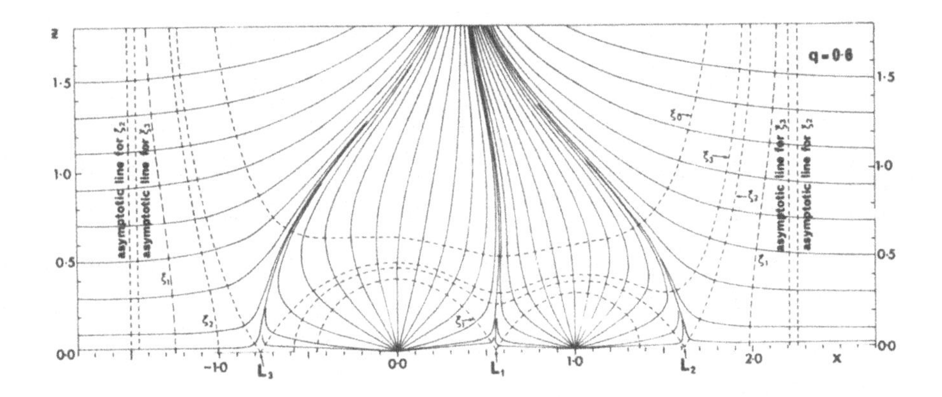

Figure III.16: The geometry of the curves ζ = constant in the xz-plane for $z \geq 0$, corresponding to the mass-ratio $q = 0.6$. The trajectories of $\zeta(x,0,z)$ are marked with full curves; while dotted curves represent the equipotentials ξ = constant for the values which the potential assumes at the Lagrangian collinear points $L_{1,2,3}$ (after Kitamura, 1970).

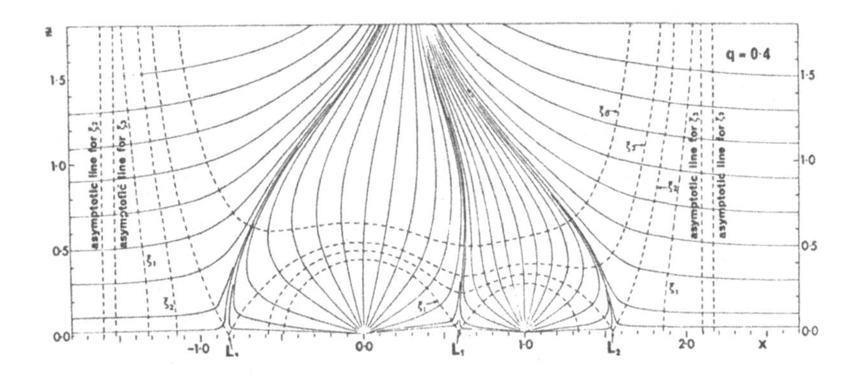

Figure III.17: The geometry of the curves ζ = constant in the xz-plane for $z \geq 0$, corresponding to the mass-ratio $q = 0.4$. The trajectories of $\zeta(x,0,z)$ are marked with full curves; while dotted curves represent the equipotentials ξ = constant for the values which the potential assumes at the Lagrangian collinear points $L_{1,2,3}$ (after Kitamura, 1970).

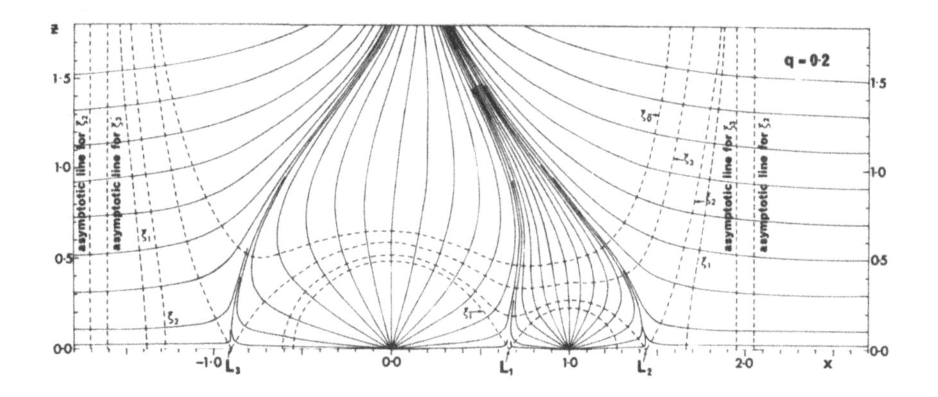

Figure III.18: The geometry of the curves ζ = constant in the xz-plane for $z \geq 0$, corresponding to the mass-ratio $q = 0.2$. The trajectories of $\zeta(x, 0, z)$ are marked with full curves; while dotted curves represent the equipotentials ξ = constant for the values which the potential assumes at the Lagrangian collinear points $L_{1,2,3}$ (after Kitamura, 1970).

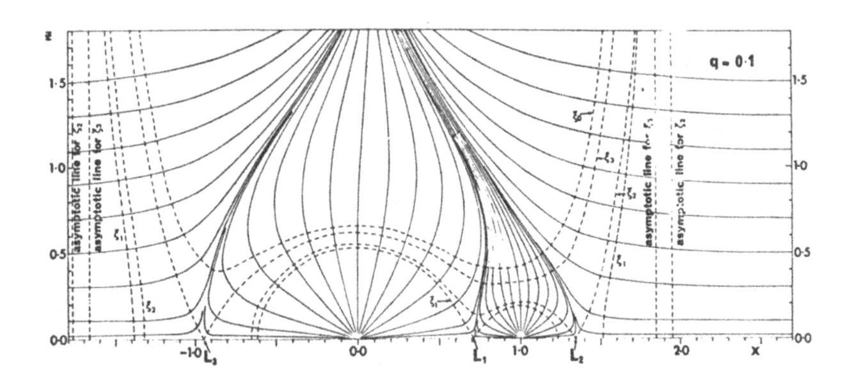

Figure III.19: The geometry of the curves ζ = constant in the xz-plane for $z \geq 0$, corresponding to the mass-ratio $q = 0.1$. The trajectories of $\zeta(x, 0, z)$ are marked with full curves; while dotted curves represent the equipotentials ξ = constant for the values which the potential assumes at the Lagrangian collinear points $L_{1,2,3}$ (after Kitamura, 1970).

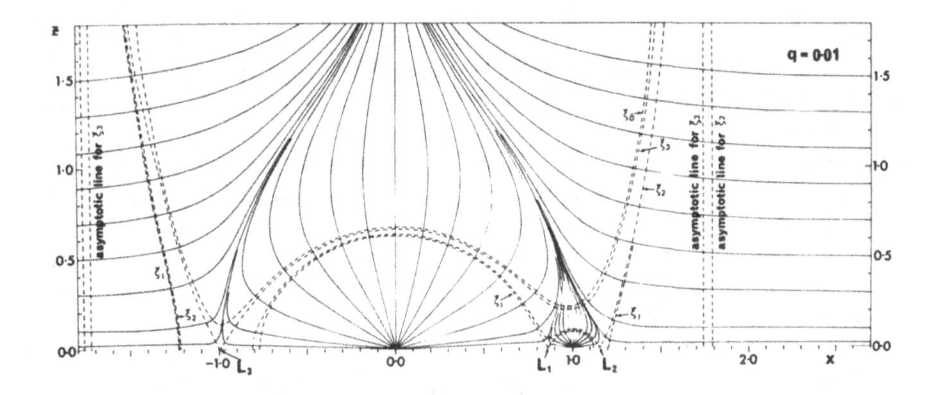

Figure III.20: The geometry of the curves ζ = constant in the xz-plane for $z \geq 0$, corresponding to the mass-ratio $q = 0.01$. The trajectories of $\zeta(x,0,z)$ are marked with full curves; while dotted curves represent the equipotentials ξ = constant for the values which the potential assumes at the Lagrangian collinear points $L_{1,2,3}$ (after Kitamura, 1970).

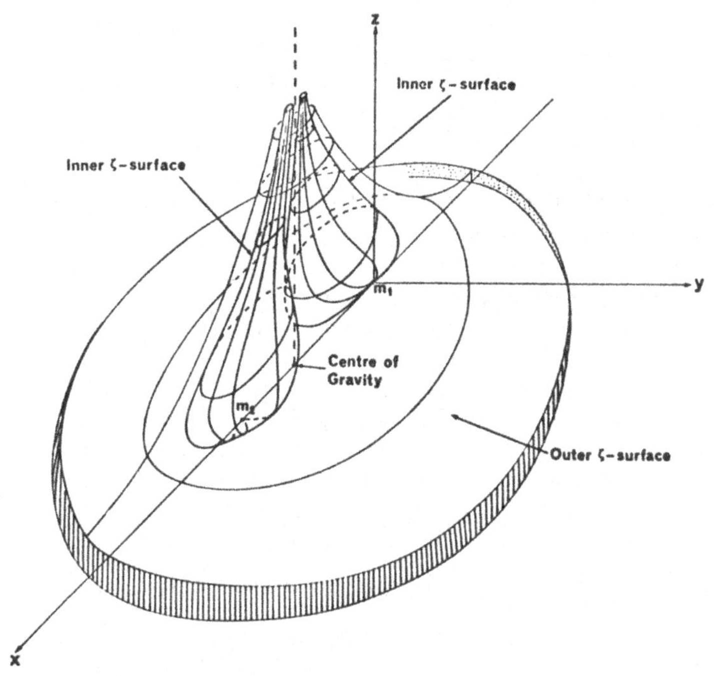

Figure III.21: Three-dimensional representation of the surfaces $\zeta(x,y,z)$ = constant (after Kitamura, 1970).

B: Motion in the Equatorial Plane

In conclusion of the present chapter we wish to employ the Roche coordinates introduced in the preceding section to investigate the motion of a mass point in the field of the gravitational dipole m_1, m_2 in the equatorial plane characterized by the coordinates ξ and η.

In order to do so, let us return to the metric transformation (1.1) and note that, for $d\zeta = 0$, this transformation contains only two metric coefficients (1.26) and (1.27), reducing in the equatorial plane to

$$h_1^2 = \frac{1}{\xi_1^2 + \xi_y^2} \quad \text{and} \quad h_2^2 = \frac{1}{\eta_x^2 + \eta_y^2} \,. \tag{2.42}$$

Moreover, for $Z = 0$ the velocity component $u_3 = 0$; while the direction cosines $n_j\,(j = 1, 2, 3)$ as given by the last ones of Equations (1.30)–(1.32) are

$$n_1 = n_2 = 0, \;\; n_1 = 1 \,. \tag{2.43}$$

Therefore, the equations (2.30)–(2.32) of motion will reduce to the simultaneous system

$$\frac{Du_1}{Dt} - 2u_2\omega \;=\; \frac{Gm_1}{h_1 R} \,, \tag{2.44}$$

$$\frac{Du_2}{Dt} + 2u_1\omega \;=\; 0 \,; \tag{2.45}$$

where the Lagrangian time-derivatives of $u_{1,2}$ on the left-hand sides by virtue of Equations (2.17)–(2.19) become

$$\frac{Du_1}{Dt} = \frac{\partial u_1}{\partial t} + \frac{u_1}{h_1}\frac{\partial u_1}{\partial \xi} + \frac{u_2}{h_2}\frac{\partial u_1}{\partial \eta} + \frac{u_1 u_2}{h_2} H_1 - \frac{u_2^2}{h_1} H_2 \,, \tag{2.46}$$

$$\frac{Du_2}{Dt} = \frac{\partial u_2}{\partial t} + \frac{u_1}{h_1}\frac{\partial u_2}{\partial \xi} + \frac{u_2}{h_2}\frac{\partial u_2}{\partial \eta} + \frac{u_1 u_2}{h_1} H_2 - \frac{u_1^2}{h_2} H_1 \,, \tag{2.47}$$

where

$$H_{1,2} \equiv \frac{1}{h_{1,2}} \frac{\partial h_{1,2}}{\partial \eta, \xi} \tag{2.48}$$

are the respective Christoffel symbols representing the curvature of the coordinate axes.

In order to evaluate these quantities in terms of the local coordinate, let us differentiate the preceding Equations (2.42) with respect to η and ξ. In particular, it follows readily that

$$\frac{\partial h_2^2}{\partial \xi} = 2h_2 \frac{\partial h_2}{\partial \xi} = -\frac{2}{(\eta_x^2 + \eta_y^2)^2} \left\{ \eta_x \frac{\partial \eta_x}{\partial \xi} + \eta_y \frac{\partial \eta_y}{\partial \xi} \right\} , \tag{2.49}$$

where the operator

$$\frac{\partial}{\partial \xi} = x_\xi \frac{\partial}{\partial x} + y_\xi \frac{\partial}{\partial y} = h_1^2 \left\{ \xi_x \frac{\partial}{\partial x} + \xi_y \frac{\partial}{\partial y} \right\} ; \tag{2.50}$$

and if so,

$$H_2 = -h_1^2 h_2^2 \{ \xi_x [\eta_x \eta_{xx} + \eta_y \eta_{yy}] + \xi_y [\eta_x \eta_{xy} + \eta_y \eta_{yy}] \} . \tag{2.51}$$

Next, let us set out to eliminate η from the preceding expression by means of the orthogonality condition (2.33). Differentiating the latter with respect to x and y we find that

$$\eta_x \xi_{xx} + \eta_{xx} \xi_x + \eta_{xy} \xi_y + \eta_y \xi_{xy} = 0 \tag{2.52}$$

and

$$\eta_x \xi_{xy} + \eta_{xy} \xi_x + \eta_{yy} \xi_y + \eta_y \xi_{yy} = 0 , \tag{2.53}$$

from which it follows that

$$-\xi_x(\eta_x \eta_{xx} + \eta_y \eta_{xy}) - \xi_y(\eta_x \eta_{xy} + \eta_y \eta_{yy}) = \eta_x^2 \xi_{xx} + 2\eta_x \eta_y \xi_{xy} + \eta_y^2 \xi_{yy} ; \tag{2.54}$$

so that

$$H_2 = \frac{\xi_x + 2(\eta_y/\eta_x)\xi_{xy} + (\eta_y/\eta_x)^2 \xi_{yy}}{(\xi_x^2 + \xi_y^2)[1 + (\eta_y/\eta_x)^2]} . \tag{2.55}$$

An insertion of (2.33) in (2.55) readily discloses that the first one of our Christoffel symbols (2.48) can be expressed symmetrically in terms of the partial derivatives of ξ with respect to x and y by the equation

$$\begin{aligned} H_2 &= \frac{\xi_y^2 \xi_{xx} - 2\xi_x \xi_y \xi_{xy} + \xi_x^2 \xi_{yy}}{(\xi_x^2 + \xi_y^2)^2} = \\ &= h_1^4 (\xi_y^2 \xi_{xx} - 2\xi_x \xi_y \xi_{xy} + \xi_x^2 \xi_{yy}) , \end{aligned} \tag{2.56}$$

where ξ_x and ξ_y have already been given by Equations (2.20) and (2.21) of Chapter II; while, in accordance with (2.56)–(2.57) and (2.59) of that chapter,

$$\xi_{xx} = (3x^2 - r^2) r^{-5} + q\{3(1-x)^2 - r'^2\} r'^{-5} + q + 1 , \tag{2.57}$$

$$\xi_{yy} = (3y^2 - r^2) r^{-5} + q\{3y^2 - r'^2\} r'^{-5} + q + 1 , \tag{2.58}$$

and

$$\xi_{xy} = 3y\{r^{-5} - q(1-x) r'^{-5}\} \tag{2.59}$$

for $z = 0$.

In order to evaluate the second Christoffel symbol defined by (2.48), let us differentiate the first of Equations (2.42) for h_1 with respect to η to obtain

$$\begin{aligned} H_1 &= -h_1^2 \left\{ \xi_x \frac{\partial \xi_x}{\partial \eta} + \xi_y \frac{\partial \xi_y}{\partial \eta} \right\} = \\ &= -h_1^2 \{ \xi_x [\xi_{xx} x_\eta + \xi_{xy} y_\eta] + \xi_y [\xi_{xy} x_\eta + \xi_{yy} y_\eta] \} , \end{aligned} \tag{2.60}$$

which on insertion for x_η and y_η from (1.8) and (1.9) transforms into

$$H_1 = -h_1^2 h_2^2 \{\eta_x(\xi_x\xi_{xx} + \xi_y\xi_{xy}) + \eta_y(\xi_x\xi_{xy} + \xi_y\xi_{yy})\} . \tag{2.61}$$

However, if we multiply now (2.52) by ξ_x, (2.53) by ξ_y, and add, the result discloses that

$$-\eta_x(\xi_x\xi_{xx} + \xi_y\xi_{xy}) - \eta_y(\xi_x\xi_{xy} + \xi_y\xi_{yy}) = \xi_x^2\eta_{xx} + 2\xi_x\xi_y\eta_{xy} + \xi_y^2\eta_{yy} , \tag{2.62}$$

which by use of the orthogonality condition (2.34) eventually leads to the expression

$$\begin{aligned} H_1 &= \frac{\eta_y^2\eta_{xx} - 2\eta_x\eta_y\eta_{xy} + \eta_x^2\eta_{yy}}{(\eta_x^2 + \eta_y^2)^2} = \\ &= h_2^4(\eta_y^2\eta_{xx} - 2\eta_x\eta_y\eta_{xy} + \eta_x^2\eta_{yy}) . \end{aligned} \tag{2.63}$$

The reader may notice that this latter expression for the second one of our Christoffel symbols involved in Equations (2.46)–(2.47) is of exactly the same form in terms of η as the first one was in terms of ξ. Since, however, it is the function $\xi(x, y)$ which is known to us in a closed form—while $\eta(x, y)$ must be deduced from it by integration (numerically or otherwise) of the orthogonality condition (2.33)—it is of interest to rewrite Equation (2.60) as far as possible in terms of the derivatives of ξ rather than η. In order to do so, let us return to Equation (2.61) and combine it with (2.33); in doing so we find that

$$H_1 = h_1^2\xi_x^{-1}\{\xi_x\xi_y(\xi_{xx} - \xi_{yy}) - (\xi_x^2 - \xi_y^2)\xi_{xy}\}h_2^2\eta_y . \tag{2.64}$$

Since, moreover, the function $\xi(x, y)$ as given by (1.6) of Chapter II satisfies the differential equation[1]

$$y(\xi_{xx} + \xi_{yy}) + \xi_y = 3(1 + q)y , \tag{2.65}$$

the first term on the right-hand side of Equation (2.64) can be simplified further.

Having determined the Christoffel symbols $H_{1,2}$ as defined by Equation (2.48) in terms of the Roche coordinates ξ, η, let us return now to Equations (2.44)–(2.45) of motion in the plane to ask ourselves the following question: under which conditions can these equations be compatible with an assumption that either u_1 or u_2 vanish?

If

$$u_1 = 0 , \tag{2.66}$$

Equation (2.44) combined with (2.46) reduces to the algebraic equation

$$H_2u_2^2 + 2\omega h_1 u_2 + Gm_1/R = 0 \tag{2.67}$$

[1]This is indeed Laplace's equation in the plane, valid by virtue of the fact that ξ remains constant over equipotentials.

while (2.45) with (2.47) simplifies to

$$\frac{\partial u_2}{\partial t} + \frac{u_2}{h_2}\frac{\partial u_2}{\partial \eta} = 0 . \tag{2.68}$$

This latter equation readily integrates to

$$u_2 = \text{constant} \equiv k(\xi) ; \tag{2.69}$$

but can satisfy (2.67) only if the metric coefficient h_1 and the Christoffel symbol H_2 depend on ξ but not on η as well — which can be true only in the immediate vicinity of the origin. Therefore, the motions along the equipotential curves in the plane are possible only if the orbits are circles of a radius which is either infinitesimally small (so that the perturbations from the companion stars can be ignored), or again so large that the gravitational attraction of the rotating dipole simulates that of a single mass.

If, on the other hand,

$$u_2 = 0 , \tag{2.70}$$

Equations (2.45) and (2.47) will reduce to

$$\frac{\partial u_1}{\partial t} + \frac{u_1}{h_1}\frac{\partial u_1}{\partial \xi} = \frac{Gm_1}{h_1 R} \tag{2.71}$$

and

$$\left(2\omega - \frac{u_1}{h_2} H_1\right) u_1 = 0 . \tag{2.72}$$

This latter equation (being algebraic) can be fulfilled only if

$$u_1 = 0 , \tag{2.73}$$

which together with (2.70) would correspond to a state of no motion; or if

$$u_1 = 2\omega h_2 / H - 1 , \tag{2.74}$$

to a steady-state motion.

If so, however, Equation (2.71) would reduce to

$$\frac{\partial u_1^2}{\partial \xi} = \frac{2Gm_1}{R} , \tag{2.75}$$

integrating to

$$u_1^2 = 2\xi + \text{constant} \tag{2.76}$$

which is, however, inconsistent with (2.74). Therefore, the curves representing gradients to the equipotentials can represent the actual trajectories of mass-particles of the restricted plane problem of three bodies no more than do the equipotentials themselves.

Moreover, if both velocity components $u_{1,2}$ are different from zero, but remain small enough for their squares and cross-products to be ignorable, Equations (2.44)–(2.47) reduce to

$$\frac{\partial u_1}{\partial t} - 2\omega u_2 = \frac{Gm_1}{h_1 R}, \tag{2.77}$$

$$\frac{\partial u_2}{\partial t} + 2\omega u_1 = 0; \tag{2.78}$$

which (for a "circular" problem of three bodies) can be satisfied by a complete primitive of the form

$$u_1(\xi, \eta; t) = A(\xi, \eta) \cos\{2\omega t + B(\xi, \eta)\} \tag{2.79}$$

and

$$u_2(\xi, \eta; t) = (Gm_1/2\omega h_1 R) - A(\xi, \eta) \sin\{2\omega t + B(\xi, \eta)\}, \tag{2.80}$$

in which the amplitude $A(\xi, \eta)$ as well as phase $B(\xi, \eta)$ are functions of the Roche coordinates.

In conclusion, certain additional regions exist in which the Roche coordinates introduced in this section can prove useful for a construction of the solutions of the restricted problem of three bodies: namely, in the proximity of the Lagrangian points $L_{1,\ldots 5}$ in the XY-plane. In order to demonstrate this, let us recall (cf. Section II.1D) that their positions are defined by the equations

$$\xi_x = \xi_y = 0; \tag{2.81}$$

and if so, by (2.42), the metric coefficient

$$h_1 = \infty. \tag{2.82}$$

In the proximity of this pole, the coefficient h_1 (though no longer infinite) continues to act as a very large divisor—a fact which permits us to approximate the exact equations (2.44) and (2.45) by a homogeneous simultaneous system

$$\frac{\partial u_1}{\partial t} - 2\omega u_2 + \frac{u_2}{h_2}\frac{\partial u_1}{\partial \eta} + \frac{u_1 u_2}{h_2} H_1 = 0 \tag{2.83}$$

and

$$\frac{\partial u_2}{\partial t} + 2\omega u_1 + \frac{u_2}{h_2}\frac{\partial u_2}{\partial \eta} - \frac{u_1^2}{h_2} H_1 = 0, \tag{2.84}$$

an analysis of which (whether analytical or numerical) should enable us to study the motions of a mass point in the proximity of the initial conditions $u_1 = u_2 = 0$.

III.3 Bibliographical Notes

The curvilinear Roche coordinates, which are orthogonal to the Roche equipotential surfaces discussed in the preceding chapter, have their origin in three papers by the present author (Kopal, 1969, 1970, 1971) followed by Kopal and Ali (1971). In 1969, such a system of coordinates was developed in the plane of the symmetry of the respective system; and in 1970, generalized to three dimensions.

The material in the first part of this chapter is mostly new. The conditions under which Equations (1.23)-(1.25) defining the Roche coordinates are integrable were, however, investigated by Kopal and Ali in 1971 (cf. also Hadrava, 1987); and the reader will find the essential parts of the argument reproduced in Section III.2A of this chapter. Section III.1B contains their exact analytical solution for the rotational problem; while an alternative approximate solution of a more general double-star problem (in which the effects of rotation as well as tides have been taken into account) is given in Section III.1C to the first order in small quantities.

For a discussion of the "Roche harmonics", representing a separable solution of Laplace's equation $\nabla^2\Psi = 0$ in the Roche coordinates see Kopal (1972) in the $\xi - \zeta$ plane; and by Roach (1975, 1981) or Langebartel (1979) in three-dimensional space.

A long and valuable memoir by Kitamura (1970) contains the most extensive set of numerical integrations of Equations (2.33) and (2.39) in the planes $X = 0$ and $Z = 0$ carried out so far; the reader will find the graphical representations of some on Figures III.5 to III.12 and Figures III.14 to III.21, reproduced in Section III.2A.

The equations of motion of the restricted problem of three bodies in plane Roche coordinates of Section III.2B go back to Kopal (1970). Of previous work on this problem cf. that of Prendergast (1960) who attempted to set up a cylindrical system of coordinates ξ, η, Z, which for $z = 0$ become identical with that developed in Section III.2B of the present volume. Should $z \neq 0$, Prendergast's equations of motion are incomplete because of his neglect of the non-orthogonality of his cylindrical coordinates in space. A plane Roche problem was also considered a few years later by Szebehely (1963) in Cartesian xy-coordinates; though it is of interest to note a close relation between his work in xy-coordinates, and that in the $\xi\eta$-coordinates introduced by the present writer in 1970. In fact, Szebehely's Equation (8) is identical with our Equation (2.67) of the preceding chapter in a different coordinate system; and the reader may notice other similarities as well.

Chapter IV

Continuous Mass Distribution: Clairaut's Theory

In Chapter II of this book we formulated heuristically the closed form of the Roche potential, the gravitational field of which is governed by that of a dipole of two point-masses m_1 and m_2, revolving around the common centre of gravity with an angular velocity ω. The fact that the entire mass of each component was assumed to be confined to a point made it unnecessary for us to concern ourselves about the internal structure of each component—beyond a tacit assumption (underlying the Roche model) that the "mean-free-path" of the infinitesimal mass particle moving in the gravitational field of our rotating dipole is infinite. The aim of the present chapter will, however, be to remove this latter assumption, and to generalize our problem by allowing the mass to be distributed *continuously* throughout the interior of the respective configuration: i.e., that its internal structure is characterized by a continuous distribution of density ρ, not exhibiting a pole at the centre. Moreover, a restriction to the mean-free-path of infinitesimal mass particles to become finite (even though the latter may be long in comparison with the distance separating the finite masses m_1 and m_2) will give rise to a finite pressure P. In such a case the equations of *particle mechanics* at the basis of our discussion of Section II-3B should be replaced by Eulerian equations of *hydrodynamics*, which will constitute the physical generalization of the processes underlying Chapters II and III. In particular, the aim of the present chapter should be to outline a theory of the *equilibrium* form of the components of close binary systems of *arbitrary* structure, as a consequence of the fact that if no element of fluid constituting them moves relative to the elements which are adjacent to it, the surface of equal density ρ—of which the external form corresponding to $\rho = 0$ constitutes a limiting case—must be *equipotentials*.

IV.1 Equipotential Surfaces

In order to prove that this is the case, let us depart from the fundamental equations of hydrostatics, of the form

$$\operatorname{grad} P = \rho \operatorname{grad} \Psi \,, \tag{1.1}$$

where the total potential

$$\Psi = \Omega + V' \tag{1.2}$$

consists of a sum of its components arising from self-attraction (Ω), and of the disturbing potential (V') due to rotation or tides.

In scalar form, Equation (1.1) can be rewritten as

$$\frac{\partial P}{\partial x} = \rho \frac{\partial \Psi}{\partial x}, \quad \frac{\partial P}{\partial y} = \rho \frac{\partial \Psi}{\partial y}, \quad \frac{\partial P}{\partial z} = \rho \frac{\partial \Psi}{\partial z}. \tag{1.3}$$

Equating the mixed second derivatives $\partial^2 P/\partial x \partial y$, $\partial^2 P/\partial x \partial z$ and $\partial^2 P/\partial y \partial z$ obtained by appropriate differentiation of Equations (1.3) we find (cf. p.140 of Jeans, 1919) that

$$\frac{\partial \rho}{\partial x}\frac{\partial \Psi}{\partial y} = \frac{\partial \rho}{\partial y}\frac{\partial \Psi}{\partial x}, \quad \frac{\partial \rho}{\partial x}\frac{\partial \Psi}{\partial z} = \frac{\partial \rho}{\partial z}\frac{\partial \Psi}{\partial x}, \quad \frac{\partial \rho}{\partial y}\frac{\partial \Psi}{\partial z} = \frac{\partial \rho}{\partial z}\frac{\partial \Psi}{\partial y}. \tag{1.4}$$

from which it follows that

$$\frac{\frac{\partial \rho}{\partial x}}{\frac{\partial \Psi}{\partial x}} = \frac{\frac{\partial \rho}{\partial y}}{\frac{\partial \Psi}{\partial y}} = \frac{\frac{\partial \rho}{\partial z}}{\frac{\partial \Psi}{\partial z}}. \tag{1.5}$$

The structure of these equations discloses at once that the surfaces ρ = constant coincide necessarily with those of Ψ = constant. Under these circumstances, Equations (1.5) can be rewritten in the form of a single total differential equation

$$dP = \rho d\Psi, \tag{1.6}$$

implying that if Ψ is a function of ρ only, so must be P — a fact which requires the existence of an "equation of state" of the form

$$P \equiv f(\rho) \tag{1.7}$$

where $f(\rho)$ stands for an arbitrary function of the density. Any surface over which ρ and P are constant must, therefore, be an equipotential

$$\Psi = \text{constant}. \tag{1.8}$$

This statement contains, in fact, a complete specification of our problem (for its more rigorous discussion, cf. p.9 of Poincaré, 1902); but in an implicit form requiring explicit development. While this task will be completed, in some detail, in subsequent Sections IV.2 to IV.4 for deformations due to axial rotation and tides (or their combination), the aim of the present section will be to prepare the ground for such a solution by a specification of the *gravitational potential* of distorted self-gravitating bodies of arbitrary structure.

In order to do so, let us fix our attention to an arbitrary point M in the interior of our configuration, in a position specified by the coordinates

$$\left.\begin{aligned} x &= r \cos\phi \sin\theta, \\ y &= r \sin\phi \sin\theta, \\ z &= r \cos\theta, \end{aligned}\right\} \tag{1.9}$$

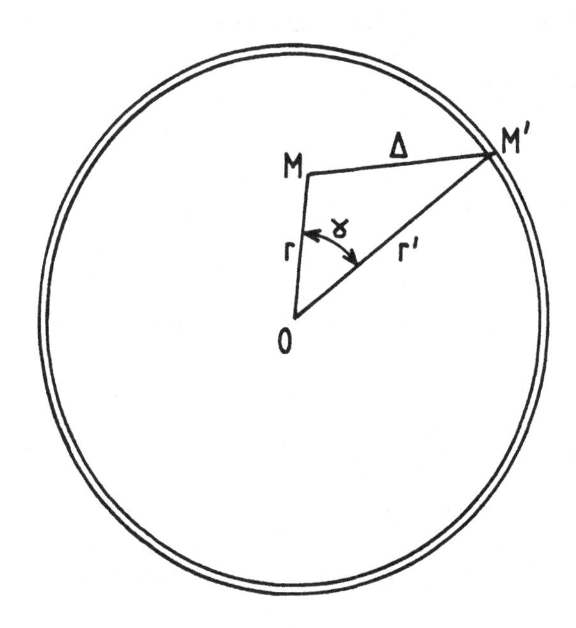

Figure IV.1: Geometry for the determination of the potential

acted upon by the attraction of a stratum comprised between the radii $r = r_0$ and r_1. Let, moreover, $M'(r', \theta', \phi')$ be an arbitrary point of this stratum (see Figure IV.1. If so, the *interior potential* U at M will evidently be given by the integral

$$U = G \int_{r_0}^{r_1} \frac{dm'}{\Delta}, \tag{1.10}$$

where G denotes the gravitation constant; dm', the mass element

$$\rho\, dx'\, dy'\, dz' = dm' = \rho r'^2\, dr' \sin\theta\, d\theta'\, d\phi' ; \tag{1.11}$$

and from the triangle OMM' (cf. again Figure IV.1)

$$\Delta^2 = r^2 + r'^2 - 2rr' \cos\gamma , \tag{1.12}$$

where

$$\cos\gamma = \cos\theta \cos\theta' + \sin\theta \sin\theta' \cos(\phi - \phi') . \tag{1.13}$$

Moreover, the *exterior potential* V will be given by an analogous expression of the form

$$V = G \int_{0}^{r_0} \frac{dm'}{\Delta}, \tag{1.14}$$

similar to Equation (1.10), in which the limits of integration extend from the origin to r_0; and the sum

$$U + V = \Omega \tag{1.15}$$

constitutes the total potential of our configuration arising from its mass.

To evaluate the two constituents of Ω, let us expand Δ^{-1} in terms of the Legendre polynomials $P_n(\cos\gamma)$ of ascending integral order n in a well-known series of the form

$$\begin{aligned} \frac{1}{\Delta} &= \frac{1}{r}\sum_{n=0}^{\infty}\left(\frac{r'}{r}\right)^n P_n(\cos\gamma)\,,\ r' < r\,, \\ &= \frac{1}{r}\sum_{n=0}^{\infty}\left(\frac{r}{r'}\right)^n P_n(\cos\gamma)\,,\ r' > r\,, \end{aligned} \tag{1.16}$$

which on insertion in Equations (1.10) and (1.14) permits us to express U and V as

$$U = \sum_{n=0}^{\infty} r^n U_n \tag{1.17}$$

and

$$V = \sum_{n=0}^{\infty} r^{-n-1} V_n\,, \tag{1.18}$$

where

$$U_n = G\int_{r_0}^{r_1}\int_0^{\pi}\int_0^{2\pi} \rho(r')^{1-n}\, P_n(\cos\gamma)dr' \sin\theta'\, d\theta'\, d\phi' \tag{1.19}$$

and

$$V_n = G\int_0^{r_0}\int_0^{\pi}\int_0^{2\pi} \rho(r')^{2+n}\, P_n(\cos\gamma)dr' \sin\theta'\, d\theta'\, d\phi'\,. \tag{1.20}$$

Next, let in the foregoing expression

$$r' \equiv r(a,\, \theta',\, \phi') \tag{1.21}$$

represent the parametric equation of an equipotential surface of constant density and pressure. By definition, the parameter a = constant over the surface Ψ = constant; and provided that the surfaces so defined do not intersect, the parameter a can be treated as a radial coordinate, related with the total potential Ψ. However, can such a potential be regarded as a single-valued function of the spatial coordinates x, y and z? An affirmative answer to this question was given already by Dirichlet in the 19th century (cf. Dirichlet, 1850), who proved that (1) at all points at which the potential Ω arising from the mass (and its first two derivatives) are (at least piecewise) continuous, and (2) Ω satisfies the Poisson equation $\nabla^2\Omega = -4\pi G\rho$ for $\rho > 0$ (zero for $\rho = 0$), the function Ω does admit of a unique solution (for a rigorous proof of this theorem see Poincaré, 1899). In consequence, for any particular value of a only one equipotential surface can pass through any point; and since the density must remain constant, it follows that

$\rho \equiv \rho(a)$, such that $\rho > 0$ for $0 \leq a < a_1$ and if $\rho(a_1) = 0$, $a = a_1$ stands for the surface of the respective configuration.

The fact that the density ρ can thus be regarded as the function of a single variable a suggests that it should be of advantage to change over from r' to a as the new variable of integration on the r.h.s. of Equations (1.19) and (1.20). Inasmuch as the Jacobian J of the transformation from (r', θ', ϕ') to (a, θ', ϕ') is of the form

$$J = \begin{vmatrix} \frac{\partial r'}{\partial a} & \frac{\partial r'}{\partial \theta'} & \frac{\partial r'}{\partial \phi'} \\ 0 & 1 & 0 \\ 0 & 0 & 1 \end{vmatrix} = \frac{\partial r'}{\partial a} , \tag{1.22}$$

a transition from r' to a can be effected simply by setting

$$dr' = \frac{\partial r'}{\partial a} da . \tag{1.23}$$

If so, however, then evidently

$$U_n = \frac{G}{2-n} \int_{a_0}^{a_1} \rho \frac{\partial}{\partial a} \left\{ \int_0^{\pi} \int_0^{2\pi} (r')^{2-n} P_n(\cos \gamma) \sin \theta' \, d\theta' \, d\phi' \right\} da \tag{1.24}$$

for $n \gtrless 2$, and

$$U_2 = G \int_{a_0}^{a_1} \rho \frac{\partial}{\partial a} \left\{ \int_0^{\pi} \int_0^{2\pi} (\log r') \, P_2(\cos \gamma) \sin \theta' \, d\theta' \, d\phi' \right\} da ; \tag{1.25}$$

while, similarly,

$$V_n = \frac{G}{n+3} \int_0^{a_0} \rho \frac{\partial}{\partial a} \left\{ \int_0^{\pi} \int_0^{2\pi} (r')^{n+3} P_n(\cos \gamma) \sin \theta' \, d\theta' \, d\phi' \right\} da \tag{1.26}$$

for any value of n. If $n = 0$,

$$U_0 = 4\pi G \int_{a_0}^{a_1} \rho a \, da \tag{1.27}$$

and

$$V_0 = G \int_0^{a_0} dm' = Gm(a_0) , \tag{1.28}$$

where $m(a_0)$ denotes the mass of our configuration interior to a_0.

In order to proceed further, let us assume that the radius vector r' of an equipotential surface can be expanded in a series of the form

$$r' = a \left\{ 1 + \sum_{j=0}^{\infty} Y_j(a, \theta', \phi') \right\} , \tag{1.29}$$

where the Y_j's stand for solid harmonic functions of the respective coordinates, satisfying the partial differential equation

$$\frac{1}{\sin\theta'}\frac{\partial}{\partial\theta'}\left(\sin\theta'\frac{\partial Y_j}{\partial\theta'}\right)+\frac{1}{\sin^2\theta'}\frac{\partial^2 Y_j}{\partial\phi'^2}+j(j+1)Y_j=0\,, \tag{1.30}$$

where j is zero or a positive integer. If, moreover, the solid harmonics $Y_j(a,\theta',\phi')$ can be factorized in the form

$$Y_j(a,\theta',\phi')=f_j(a)\,P_j(\theta',\phi')\,, \tag{1.31}$$

where the P_j's are the same zonal harmonics which occur on the r.h.s. of Equation (1.16), it is possible to assert, by a well-known orthogonality theorem, that

$$\begin{aligned}\int_0^\pi\int_0^{2\pi} P_n(\cos\gamma)\,Y_j(a,\theta',\phi') \quad & \sin\theta'\,d\theta'\,d\phi' = \\ = 0 \quad & \text{if } j\neq n\,, \\ =\frac{4\pi}{2j+1}Y_j(a,\theta,\phi) \quad & \text{if } j=n\,. \end{aligned} \tag{1.32}$$

With the aid of this theorem it should be possible—on insertion for r' from Equation (1.29)—to rewrite the expansion of Equations (1.17) and (1.18) for U and V in the form

$$U=4\pi G\sum_{j=0}^{\infty}\frac{E_j(a)}{2j+1}\,r^j\,P_j(\theta,\phi) \tag{1.33}$$

and

$$V=4\pi G\sum_{j=0}^{\infty}\frac{F_j(a)}{2j+1}\,r^{-j-1}\,P_j(\theta,\phi)\,, \tag{1.34}$$

where $E_j(a)$ and $f_j(a)$ are appropriate functions of a only.

An explicit evaluation of these functions will be undertaken in the next section. In this place we wish to note that although the sum $\Omega=U+V$ of the foregoing expansions of Equations (1.33) and (1.34) constitute the complete potential arising from the mass of our configuration, it does not yet represent the total potential Ψ; for, in order to specify the latter we must adjoin to Ω the *disturbing potential* V', whose action will cause this configuration from spherical form. In point of fact, the total potential $\Psi=U+V+V'$ must satisfy the Poisson equation

$$\nabla^2\Psi+4\pi G\rho=2\omega^2\,, \tag{1.35}$$

where ω denotes the angular velocity of axial rotation that may, but need not, be constant [1]. The partial sum $\Omega=U+V$ represents the particular integral and

[1] An assumption of constant angular velocity is not essential for the validity of our procedure. The latter would continue to hold good—though in more complicated form—as long as $\omega(a,\theta,\phi)$ remains expansible in terms of spherical harmonics $P_j(\theta,\phi)$.

V', the complementary function of this equation; the form of which remains yet to be specified.

If our configuration were nonrotating, the disturbing (centrifugal) potential V' would represent a solution of the homogeneous (Laplace) equation

$$\nabla^2 V' = 0 , \tag{1.36}$$

which is known to be of the form

$$V' = \sum_j \{c_j r^j + d_j r^{-j-1}\} P_j(\theta, \phi) , \tag{1.37}$$

where c_j and d_j are arbitrary constants. The values of these constants depend, in turn, on the nature of the disturbing forces; and the regularity of V' at the origin—if required—necessitates that $d_j = 0$.

Let us first confine our attention to the case of *rotational distortion* of our configuration, arising from uniform rotation with constant angular velocity ω about one (say, Z-) axis whose direction is fixed in space. In such a case, the differential equation for V' is of the form

$$\nabla^2 V' = 2\omega^2 . \tag{1.38}$$

and its regular solution reduces to

$$V' = \frac{1}{2}\omega^2 r^2 \sin^2\theta = c_2 r^2 \{1 - P_2(\cos\theta)\} , \tag{1.39}$$

where

$$c_2 = \frac{1}{3}\omega^2 . \tag{1.40}$$

If Equations (1.33), (1.34) and (1.39) are combined, the total potential of a self-gravitating configuration whose distortion derives from the centrifugal potential Equation (1.39) will assume the form

$$\begin{aligned} \Psi(r,\theta,\phi) &= 4\pi G \sum_{j=0}^{\infty} \frac{r^j E_j(a) + r^{-j-1} F_j(a)}{2j+1} P_j(\cos\theta) + \\ &+ \frac{1}{3}\omega^2 r^2 \{1 - P_2(\cos\theta)\} , \end{aligned} \tag{1.41}$$

valid to any arbitrary degree of accuracy. Therefore, over an equipotential surface specified by the radius vector r',

$$\begin{aligned} \Psi(r/,\theta,\phi) &= 4\pi G \sum_{j=0}^{\infty} \frac{(r')^j E_j(a) + (r')^{-j-1} F_j(a)}{2j+1} P_j(\cos\theta) + \\ &+ \frac{1}{3}\omega^2 (r')^2 \{1 - P_2(\cos\theta)\} , \end{aligned} \tag{1.42}$$

where r' continues to be given by Equations (1.29)–(1.31) and where, accordingly, the last term can be expanded as

$$\begin{aligned}\frac{1}{3}\omega^2(r')^2 &= \frac{1}{3}\omega^2 a^2(1-f_2^2) - \\ &+\frac{2}{3}\omega^2\left\{f_2 - \frac{3}{7}f_2^2\right\}(r')^2 P_2(\cos\theta) + \\ &+\frac{\omega^2}{a^2}\left\{\frac{2}{3}f_4 - \frac{18}{35}f_2^2\right\}(r')^4 P_4(\cos\theta) + \ldots \end{aligned} \tag{1.43}$$

which, if we regard ω^2 itself as a small quantity of first order, represents an expansion of the centrifugal potential $V'(r')$ correctly to terms of third order.

Suppose next that we expand the r.h.s. of Equation (1.42) in a Neumann series of the form

$$\Psi(r',\theta,\phi) = \sum_{j=0}^{\infty} \alpha_j(a) P_j(\theta,\phi)\,, \tag{1.44}$$

with coefficients defined by

$$a_j(a) = \frac{2j+1}{4\pi}\int_0^{\pi}\int_0^{2\pi} \Psi(r',\theta,\phi)P_j(\theta,\phi)\sin\theta\, d\theta\, d\phi\,. \tag{1.45}$$

If now—consistent with Equation (1.2)—the total potential Ψ is to remain constant over a surface of the form Equation (1.29), it follows that *all terms on the r.h.s. of Equation* (1.44) *factored by* $P_j(\theta,\phi)$ *for* $j > 0$ *must necessarily vanish*; and this can be true only if we set

$$a_j(a) = 0, \quad j > 0\,, \tag{1.46}$$

leaving us with α_0 as the constant value characterizing the respective equipotential of mean radius a.

A determination of the gravitational potential of *tidally-distorted* configurations can be made to follow a closely parallel course. Their only difference rests on the fact that the disturbing potential $V'(t)$ must then be identified with that arising from the presence of an external mass. As an exterior potential V' will continue to satisfy Laplace's equation (1.36); but the explicit form of spherical harmonics $P_j(\theta,\phi)$ as well as of the coefficients c_j factoring them will depend on the magnitude and position of the disturbing body in space. Therefore, the second term on the r.h.s. of Equation (1.37) will, for tidal distortion, be of a form different from (1.39), but still expansible in a Neumann series (1.44) with coefficients given by (1.45). In the sections which follow we shall, accordingly, proceed to develop the explicit form of Equations (1.46) in the case of the distortion arising from pure rotation and tides, as well as from their mutual interction.

In other words—whatever the cause of the distortion—if the distorted surface is to represent an equipotential, n equations of the form (1.46) can be set up to specify n amplitudes $f_j(a)$, $j = 1, 2, \ldots n$ of the expansion

$$r' = a\left\{1 + \sum_{j=0}^{\infty} f_j(a) P_j(\theta', \phi')\right\}, \tag{1.47}$$

representing Equation (1.21); and once these amplitudes have been determined, we are in a position (cf. Section V.2) to specify also the explicit forms of $\alpha_0(a)$ from the requirement that the mass of the configuration should be uninfluenced by distortion.

IV.2 Rotational Distortion

The aim of the present section will be to establish the explicit form of the amplitudes $f_j(a)$ in the expansion (1.47) for the shape of an equipotential surface of a rotating configuration distorted by centrifugal force. If this rotation takes place about the Z-axis fixed in space, it is evident (on account of symmetry) that this expansion can contain only harmonics $P_j \cos\theta)$ of even orders. Moreover, if the amplitude $f_2(a)$ on the r.h.s. of Equation (1.47) represents a quantity of first order in surficial distortion, $f_4(a)$ will be of the order of f_2^2 or of second order; $f_6(a)$ of third order, etc.

Now let us, in what follows, set out to describe the shape of rotating configurations to quantities of *third* order in surficial distortion—a scheme within which Equation (1.47) will be restricted to the terms

$$\begin{aligned} r' &= a\{1 + f_0 + f_2 P_2(\cos\theta) + f_4 P_4(\cos\theta) + f_6 P_6(\cos\theta) + \ldots\} \\ &\equiv a\{1 + \Sigma\}\,. \end{aligned} \tag{2.1}$$

Within the scheme of our approximation,

$$\begin{aligned} (r')^{2-n} &= a^{2-n}\left\{1 - (n-2)\Sigma + \frac{1}{2}(n-1)(n-2)\Sigma^2 - \right. \\ &\quad \left. - \frac{1}{6}n(n-1)(n-2)\Sigma^3 + \ldots\right\} \end{aligned} \tag{2.2}$$

for $n \gtrless 2$ and, for $n = 2$,

$$\log r' = \log a + \Sigma - \frac{1}{2}\Sigma^2 + \frac{1}{3}\Sigma^3 - \ldots, \tag{2.3}$$

while

$$\begin{aligned} (r')^{n+3} &= a^{n+3}\left\{1 + (n+3)\Sigma + \frac{1}{2}(n+3)(n+2)\Sigma^2 + \right. \\ &\quad \left. + \frac{1}{6}(n+3)(n+2)(n+1)\Sigma^3 + \ldots\right\} \end{aligned} \tag{2.4}$$

for any value of n.

Let us decompose next the powers and cross-products of Legendre polynomials occurring in different powers of Σ on the r.h.s. of Equations (2.2)–(2.4) into their linear combinations by use of the well-known formula which asserts that, for $m \leq n$,

$$P_m P_n = \sum_{j=0}^{m} \frac{A_{m-j} A_j A_{n-j}}{A_{m+n-1}} \left\{ \frac{2m+2n+1-4j}{2m+2n+1-2j} \right\} P_{m+n-2j} \,, \tag{2.5}$$

where $A_0 = 1$ and, for $j > 0$,

$$A_j = \frac{1,3,5,\ldots(2j-1)}{j!} \,. \tag{2.6}$$

The orthogonality properties of the P_n's are such that

$$\begin{aligned} \int_{-1}^{1} P_m P_n \, d \cos \theta \quad &= 0 \qquad \text{if } m \neq n \,, \\ &= \frac{2}{2n+1} \quad \text{if } m = n \,. \end{aligned} \tag{2.7}$$

If so, then the insertion of (2.1)–(2.7) together with the use of the orthogonality theorem (1.32) in Equations (1.24)–(1.26) for U_n and V_n should enable us to express the latter in the forms

$$(2n+1)U_n = 4\pi G E_n(a) P_n(\cos\theta) \,, \tag{2.8}$$

$$(2n+1)V_n = 4\pi G F_n(a) P_n(\cos\theta) \,, \tag{2.9}$$

the amplitude of which we shall now proceed to evaluate.

To begin, we note that V_0 as defined by (1.26) can be written as

$$\begin{aligned} F_0 &= \frac{1}{12\pi} \int_0^a \rho \frac{\partial}{\partial a} \left\{ \int_0^\pi \int_0^{2\pi} (r')^3 \sin\theta \, d\theta \, d\phi \right\} da \\ &= \int_0^a \rho \frac{\partial}{\partial a} \left\{ a^3 \left[\frac{1}{3} + f_0 + \frac{1}{5} f_2^2 + \frac{2}{105} f_2^3 + \ldots \right] \right\} da \end{aligned} \tag{2.10}$$

(in which the zero subscript of the upper limit will hereafter be dropped) and represents the mass of our configuration interior to a. Since this mass must be independent of distortion, it follows that the zero-harmonic amplitude f_0 is constrained to satisfy the equation

$$f_0 + \frac{1}{5} f_2^2 + \frac{2}{105} f_2^3 + \ldots = 0 \,, \tag{2.11}$$

which yields

$$f_0 = -\frac{1}{5} f_2^2 - \frac{2}{105} f_2^3 - \ldots \tag{2.12}$$

correctly to quantities of third order.

Taking advantage of this fact, we establish—after some algebra—that, to the same order of accuracy, the coefficients

$$E_0 = \int_a^{a_1} \rho \frac{\partial}{\partial a}\left\{a^2\left[\frac{1}{2} - \frac{1}{10}f_2^2 - \frac{2}{105}f_2^3\right]\right\} da\,, \tag{2.13}$$

$$E_2 = \int_a^{a_1} \rho \frac{\partial}{\partial a}\left\{f_2 - \frac{1}{7}f_2^2 + \frac{12}{35}f_2^3 - \frac{2}{7}f_2 f_4\right\} da\,, \tag{2.14}$$

$$E_4 = \int_a^{a_1} \rho \frac{\partial}{\partial a}\left\{\frac{1}{a^2}\left[f_4 - \frac{27}{35}f_2^2 + \frac{216}{385}f_2^3 - \frac{60}{77}f_2 f_4\right]\right\} da\,, \tag{2.15}$$

$$E_6 = \int_a^{a_1} \rho \frac{\partial}{\partial a}\left\{\frac{1}{a^4}\left[f_6 + \frac{90}{77}f_2^3 - \frac{25}{11}f_2 f_4\right]\right\} da\,; \tag{2.16}$$

and, similarly,

$$F_0 = \int_0^a \rho a^2\, da\,, \tag{2.17}$$

$$F_2 = \int_0^a \rho \frac{\partial}{\partial a}\left\{a^5\left[f_2 + \frac{4}{7}f_2^2 + \frac{2}{35}f_2^3 + \frac{8}{7}f_2 f_4\right]\right\} da\,, \tag{2.18}$$

$$F_4 = \int_0^a \rho \frac{\partial}{\partial a}\left\{a^7\left[f_4 + \frac{54}{35}f_2^2 + \frac{108}{77}f_2^3 + \frac{120}{77}f_2 f_4\right]\right\} da\,, \tag{2.19}$$

$$F_6 = \int_0^a \rho \frac{\partial}{\partial a}\left\{a^9\left[f_6 + \frac{24}{11}f_2^3 + \frac{40}{11}f_2 f_4\right]\right\} da\,. \tag{2.20}$$

As the next step of our procedure, let us establish the explicit form of Equations (1.47) for $j = 2, 4$ and 6. On evaluating the integrals on the r.h.s. of Equation (1.45) by the same method, we find that, for $j = 2, 4, 6$ Equations (1.47) assume the more explicit forms

$$\begin{aligned} &\frac{a^2 E_2}{5}\left\{1 + \frac{4}{7}f_2 + \frac{4}{7}f_4 + \frac{1}{35}f_2^2\right\} + \frac{8}{63}a^4 f_2 E_4 - \\ &- \frac{F_0}{a}\left\{f_2 - \frac{2}{7}f_2^2 + \frac{29}{35}f_2^3 - \frac{4}{7}f_2 f_4\right\} + \\ &+ \frac{F_2}{5a^3}\left\{1 - \frac{6}{7}f_2 - \frac{6}{7}f_4 + \frac{111}{35}f_2^2\right\} - \frac{10}{63}\frac{f_2}{a^5}F_4 = \\ &= \frac{\omega^2 a^2}{12\pi G}\left\{1 - \frac{10}{7}f_2 - \frac{9}{35}f_2^2 + \frac{4}{7}f_4\right\}\,; \end{aligned} \tag{2.21}$$

$$\begin{aligned} &\frac{a^2 E_2}{5}\left\{\frac{36}{35}f_2 + \frac{40}{77}f_4 + \frac{108}{385}f_2^2\right\} + \frac{a^4 E_4}{9}\left\{1 + \frac{80}{77}f_2\right\} - \\ &- \frac{F_0}{a}\left\{f_4 - \frac{18}{35}f_2^2 + \frac{108}{385}f_2^3 - \frac{40}{77}f_2 f_4\right\} - \\ &- \frac{F_2}{5a^3}\left\{\frac{54}{35}f_2 + \frac{60}{77}f_4 - \frac{648}{385}f_2^2\right\} + \frac{F_4}{9a^5}\left\{1 - \frac{100}{77}f_2\right\} = \\ &= \frac{\omega^2 a^2}{6\pi G}\left\{\frac{18}{35}f_2 - \frac{57}{77}f_4 - \frac{9}{77}f_2^2\right\}\,; \end{aligned} \tag{2.22}$$

$$\frac{a^2E_2}{5}\left\{\frac{10}{11}f_4+\frac{18}{77}f_2^2\right\}+\frac{20}{99}a^4f_2E_4+\frac{1}{13}a^6E_6-$$
$$-\frac{F_0}{a}\left\{f_6+\frac{18}{77}f_2^3-\frac{10}{11}f_2f_4\right\}-\frac{3}{11}\frac{F_2}{a^3}\left\{f_4-\frac{36}{35}f_2^2\right\}-$$
$$-\frac{25}{99}\frac{f_2f_4}{a^5}+\frac{F_6}{13a^7}=\frac{\omega^2a^2}{6\pi G}\left\{\frac{5}{11}f_4+\frac{9}{77}f_2^2\right\}. \tag{2.23}$$

To reduce these equations to more symmetrical forms, let us note that, to the *first* order in small quantities, Equation (2.22) reduces to

$$\frac{a^2E_2}{5}+\frac{F_2}{5a^3}-\frac{f_2F_0}{a}=\frac{\omega^2a^2}{12\pi G} \tag{2.24}$$

and, with its aid, Equations (2.22) and (2.23) yield

$$\frac{a^2E_2}{5}+\frac{F_2}{5a^3}\left\{1-\frac{10}{7}f_2\right\}-\frac{f_2F_0}{a}\left\{1-\frac{6}{7}f_2\right\}=\frac{\omega^2a^2}{12\pi G}\{1-2f_2\} \tag{2.25}$$

and

$$\frac{a^4E_4}{9}+\frac{F_4}{9a^5}-\frac{18}{35}\frac{f_2F_2}{a^3}-\frac{F_0}{a}\left\{f_4-\frac{54}{35}f_2^2\right\}=0\,, \tag{2.26}$$

correctly to quantities of *second* order. By insertion from the foregoing Equations (2.24)–(2.26) in (2.22)–(2.23) for terms which are multiplied by small quantities, it is possible to rewrite (2.22)–(2.23) correctly to terms of third order in the following alternative forms

$$\begin{aligned}\frac{a^2E_2}{5}+\frac{F_2}{5a^3}-\frac{f_2F_0}{a} &= -\frac{F_0}{a}\left\{\frac{6}{7}f_2^2-\frac{748}{245}f_2^3+\frac{16}{7}f_2f_4\right\}+\\ &\quad+\frac{F_2}{5a^3}\left\{\frac{10}{7}f_2-\frac{338}{49}f_2^2+\frac{10}{7}f_4\right\}+\\ &\quad+\frac{2f_2}{7a^5}F_4+\frac{a^2}{3}\left(\frac{\omega^2}{4\pi G}\right)\left\{1-2f_2+\frac{6}{7}f_2^2\right\},\end{aligned} \tag{2.27}$$

$$\begin{aligned}\frac{a^4E_4}{9}+\frac{F_4}{9a^5}-\frac{f_4F_0}{a} &= -\frac{F_0}{a}\left\{\frac{54}{35}f_2^2-\frac{1890}{2695}f_2^3+\frac{160}{77}f_2f_4\right\}+\\ &\quad+\frac{F_2}{5a^3}\left\{\frac{18}{7}f_2-\frac{2988}{539}f_2^2+\frac{100}{77}f_4\right\}+\\ &\quad+\frac{20f_2}{77a^5}F_4-\frac{\omega^2a^2}{4\pi G}\left\{\frac{2}{3}f_4-\frac{18}{35}f_2^2\right\},\end{aligned} \tag{2.28}$$

and

$$\begin{aligned}\frac{a^6E_6}{13}+\frac{F_6}{13a^7}-\frac{f_6F_0}{a} &= \frac{F_0}{a}\left\{\frac{216}{77}f_2^3-\frac{40}{11}f_2f_4\right\}+\\ &\quad+\frac{F_2}{a^3}\left\{\frac{5}{11}f_4-\frac{90}{77}f_2^2\right\}+\frac{5f_2}{11a^5}F_4\,.\end{aligned} \tag{2.29}$$

The same equations would have been obtained had we expanded the r.h.s. of Equation (1.44) for $\Psi(r'\theta, \phi)$ in terms of the products $r'^j P_j(\cos\theta)$ and equated their coefficients for $j = 2, 4, 6\ldots$ to zero.

The foregoing Equations (2.28)–(2.29) contain the unknown amplitudes f_6 both in front of, and behind, the integral signs in the expressions for E_j and F_j as given by Equations (2.13)–(2.20). To lure them out from behind the integral signs, multiply Equations (2.28)–(2.29) by a^j $(j = 2, 4, 6)$, respectively (so as to render the coefficients of E_j on the left-hand sides constant), and differentiate with respect to a. The derivatives of E_j and F_j are merely equal to the integrands on the r.h.s. of Equations (2.13)–(2.16) taken with the negative sign. If, subsequently, we eliminate the terms F_n for $n \neq j$ factored by small quantities with the aid of Equations (2.24)–(2.26) valid to the accuracy of lower orders, it is possible to express the F_j's for $j = 2, 4, 6$ in the form

$$\begin{aligned} F_2(a) &= a^2 F_0 \Big\{ 3f_2 - af_2' + \frac{12}{7}f_2^2 - \frac{4}{7}f_2(af_2') + \frac{2}{7}(af_2')^2 - \\ &- \frac{102}{35}f_2^3 - \frac{2}{7}f_2^2(af_2') - \frac{4}{7}f_2(af_2')^2 - \frac{8}{35}(af_2')^3 + \\ &+ \frac{52}{7}f_2f_4 - \frac{4}{7}a(f_2f_4' + f_4f_2') + \frac{4}{7}a^2 f_2'f_4' \Big\} + \\ &+ \frac{2}{3}\left(\frac{\omega^2 a^5}{4\pi G}\right)\Big\{ af_2' + \frac{4}{7}f_2(af_2') - \frac{2}{7}(af_2')^2 \Big\}, \end{aligned} \tag{2.30}$$

$$\begin{aligned} F_4(a) &= a^4 F_0 \Big\{ f_4 - af_4' + \frac{108}{35}f_2^2 - \frac{72}{35}f_2(af_2') + \frac{18}{35}(af_2')^2 + \\ &+ \frac{972}{385}f_2^3 - \frac{612}{385}f_2^2(af_2') + \frac{324}{385}f_2(af_2')^2 - \frac{108}{385}(af_2')^3 + \\ &+ \frac{520}{77}f_2f_4 - \frac{80}{77}a(f_2f_4' + f_4f_2') + \frac{40}{77}a^2 f_2'f_4' \Big\} \\ &+ 2\left(\frac{\omega^2 a^7}{4\pi G}\right)\Big\{ -\frac{2}{3}f_4 + \frac{1}{3}af_4' + \frac{18}{35}f_2^2 + \frac{24}{35}f_2(af_2') - \frac{6}{35}(af_2')^2 \Big\}; \end{aligned} \tag{2.31}$$

and

$$\begin{aligned} F_6(a) &= a^6 F_0 \Big\{ 7f_6 - af_6' + \frac{270}{77}f_2^3 - \frac{18}{7}f_2^2(af_2') + \\ &+ \frac{90}{77}f_2(af_2')^2 - \frac{18}{77}(af_2')^3 + \frac{130}{11}f_2f_4 - \\ &- \frac{30}{11}a(f_2f_4' + f_4f_2') + \frac{10}{11}a^2 f_2'f_4' \Big\}, \end{aligned} \tag{2.32}$$

where primes denote differentiation with respect to a, correctly to terms of third order.

As the last step of our analysis, let us differentiate the foregoing Equations (2.31)–(2.32) once more with respect to a, and insert for $F_j'(a)$ from (2.17)–(2.20);

the outcome discloses that the amplitudes $f_j(a)$ for $j = 2, 4, 6$ should satisfy the following second-order differential equations:

$$\begin{aligned}
a^2 f_2'' + 6D(af_2' + f_2) - 6f_2 &= \frac{2}{7}\{2\eta_2(\eta_2+9) - 9D\eta_2(\eta_2+2)\} f_2^2 - \\
&- \frac{4}{35}\left\{(7\eta_2^3 + 33\eta_2^2 + 180\eta_2 + 66) + 3D(2\eta_2^3 - 15\eta_2^2 - 27\eta_2 + 5)\right\} f_2^3 + \\
&+ \frac{4}{7}\{2(\eta_2\eta_4 + 15\eta_2 + 8\eta_4) - 3D(3\eta_2\eta_4 + 3\eta_2 + 3\eta_4 - 7)\} f_2 f_4 + \\
&+ \frac{3\omega^2}{\pi G\overline{\rho}}(1-D)\left\{f_2 + af_2' + \frac{6}{7}f_2(af_2') + \frac{3}{7}(af_2')^2\right\} + \\
&+ \frac{1}{6}\left(\frac{3\omega^2}{\pi G\overline{\rho}}\right)^2 (1-D)(\eta_2+1)f_2\,,
\end{aligned} \tag{2.33}$$

$$\begin{aligned}
a^2 f_4'' + 6D(af_4' + f_4) - 20f_4 &= \frac{18}{35}\left\{2\eta_2(\eta_2+2) - 3D(3\eta_2^2 + 6\eta_2 + 7)\right\} f_2^2 + \\
&+ \frac{36}{385}\left\{2(3-\eta_2)(1-5\eta_2) - 3D(3\eta_2^3 + 9\eta_2^2 + 12\eta_2 + 4)\right\} f_2^3 + \\
&+ \frac{44}{77}\{(2\eta_2\eta_4 + 23\eta_2 + 9\eta_4) - 9D(\eta_2\eta_4 + \eta_2 + \eta_4)\} f_2 f_4 + \\
&+ \frac{3\omega^2}{\pi G\overline{\rho}}(1-D)\left\{f_4 + af_4' + \frac{9}{35}[7f_2^2 + 6f_2(af_2') + 3(af_2')^2]\right\}
\end{aligned} \tag{2.34}$$

and

$$\begin{aligned}
a^2 f_6'' &+ 6D(af_6' + f_6) - 42f_6 = \\
&= \frac{18}{77}\left\{4(3-\eta_2)(\eta_2+2) - 3D(\eta_2^3 + 3\eta_2^2 + 15\eta_2 + 5)\right\} f_2^3 + \\
&+ \frac{10}{11}\{2(\eta_2\eta_4 + 6\eta_2 - \eta_4) - 3D(3\eta_2\eta_4 + 3\eta_2 + 3\eta_4 + 11)\} f_2 f_4,
\end{aligned} \tag{2.35}$$

equivalent to (2.22)–(2.23), where

$$\eta_j \equiv \frac{a}{f_j}\frac{\partial f_j}{\partial a}\,, \tag{2.36}$$

and where we have abbreviated

$$\overline{\rho} = \frac{3}{a^3}\int_0^a \rho a^2\,da\,, \quad D = \frac{\rho}{\overline{\rho}}\,. \tag{2.37}$$

The *boundary conditions* necessary for complete specification of the particular solutions of the foregoing equations, which are to represent the amplitudes $f_j(a)$ of the individual harmonic terms on the r.h.s. of the expansion (2.1) for r', are imposed partly at the centre and partly at the boundary of our configuration. As, at the centre, all the $f_j(a)$'s are to be a minimum, the necessary condition for this to be so is that, for $a = 0$,

$$f_j'(0) = 0 \text{ for } j = 2, 4, 6, \ldots\,. \tag{2.38}$$

On the other hand, at the boundary $a = a_1$, all $E_j(a_1)$'s as defined by Equations (2.16)–(2.16) are equal to zero and for $j > 0$, the F_j's continue to be given by (2.31)–(2.32); whereas, for $j = 0$, $4\pi F_0(a) = m_1$ by (2.20), where m_1 denotes the total mass of our configuration. Inserting them in Equations (2.28)–(2.29), we find that, for $a = a_1$,

$$\begin{aligned} 2f_2 + af_2' + \frac{5}{3}\left(\frac{\omega^2 a_1^3}{Gm_1}\right) &= \frac{2}{3}\left(\frac{\omega^2 a_1^2}{Gm_1}\right)\left\{(\eta_2+5)f_2 - \frac{1}{7}(2\eta_2^2+6\eta_2+15)f_2^2\right\} + \\ &+ \frac{2}{35}\left\{5(\eta_2^2+3\eta_2+6)f_2^2 - \right\} \\ &- (4\eta_2^3+30\eta_2^2+60\eta_2+76)f_2^3 + \\ &+ 5(2\eta_2\eta_4+3\eta_2+3\eta_4+266)f_2f_4\} \;, \end{aligned} \tag{2.39}$$

$$\begin{aligned} 4f_4 + af_4' &= \frac{2}{3}\left(\frac{\omega^2 a_1^3}{Gm_1}\right)\left\{(\eta_4+7)f_4 - \frac{9}{35}(2\eta_2^2+10\eta_2+21)f_2^2\right\} + \\ &+ \frac{18}{35}(\eta_2^2+5\eta_2+6)f_2^2 - \frac{36}{385}(3\eta_2^3+18\eta_2^2+44\eta_2+54)f_2^3 + \\ &+ \frac{20}{77}(2\eta_2\eta_4+5\eta_2+5\eta_4+26)f_2f_4 \;, \end{aligned} \tag{2.40}$$

and

$$\begin{aligned} 6f_6 + af_6' &= \frac{18}{77}(\eta_2+2)(\eta_2^2+6\eta_2+12)f_2^3 + \\ &+ \frac{5}{11}(2\eta_2\eta_4+7\eta_2+26)f_2f_4 \;. \end{aligned} \tag{2.41}$$

A construction (numerically or otherwise) of the desired particular solutions of the simultaneous system of differential equations (2.33)–(2.35) specified by the boundary conditions (2.38) and (2.39)–(2.41) can be accomplished by successive approximations in the following manner. Within the scheme of a first-order approximation (2.33) reduces to the well-known equation

$$a^2 f_2'' + 6D(af_2' + f_2) = 6f_2 \;, \tag{2.42}$$

which for

$$D = 1 - \lambda a^2 + \ldots \tag{2.43}$$

admits, in the proximity of the origin, of a solution varying as

$$f_2 = k_2\left(1 + \frac{3}{7}\lambda a^2 + \ldots\right) \;, \tag{2.44}$$

where k_2 stands for an arbitrary constant. Integrating Equation (2.42) can proceed hereafter until $a = a_1$, at which point the l.h.s. of Equation (2.39) discloses that

$$2f_2 + af_2' + \frac{5}{3}\left(\frac{\omega^2 a_1^3}{GM_1}\right) = 0 \;. \tag{2.45}$$

and this (algebraic) equation can be used to specify the value of $\omega^2 a_1^3/GM_1$ corresponding to the initially adopted value of k_2.

With a first order approximation to $f_2(a)$ thus in our hands, we can now proceed to the *second* approximation, which consists of finding a solution of the equations

$$\begin{aligned} a^2 f_2'' + 6D(af_2' + f_2) - 6f_2 &= \frac{2}{7}\{2\eta_2(\eta_2+9) - 9\eta_2(\eta_2+2)\}f_2^2 + \\ &+ \frac{3\omega^2}{\pi G\bar{\rho}}(1-D))(\eta_2+1)f_2 \end{aligned} \tag{2.46}$$

and

$$\begin{aligned} a^2 f_4'' + 6D(af_4' + f_4) - 20f_4 &= \frac{18}{35}\{2\eta_2(\eta_2+2) - \\ &- 3D(3\eta_2^2 + 6\eta_2 + 7)\}f_2^2 \,, \end{aligned} \tag{2.47}$$

subject to the boundary conditions requiring that, at the centre,

$$f_2'(0) = f_4'(0) = 0 \,, \tag{2.48}$$

while on the surface $a = a_1$,

$$\begin{aligned} 2f_2 + a_1 f_1' + \frac{5}{3}\left(\frac{\omega^2 a_1^3}{Gm_1}\right) &= \frac{1}{7}(2\eta_2^2 + 6\eta_2 + 12)f_2^2 + \\ &+ \frac{2}{3}\left(\frac{\omega^2 a_1^2}{Gm_1}\right)(\eta_2+5)f_2^2 \,, \end{aligned} \tag{2.49}$$

and

$$4f_4 + a_1 f_4' = \frac{118}{35}(\eta_2^2 + 5\eta_2 + 6)f_2^2 \,. \tag{2.50}$$

As Equation (2.46) is independent of f_4, its solution satisfying (2.48) can, near the origin, be expanded in a series in ascending even powers of a. If, moreover, we note that

$$\frac{3\omega^2}{\pi G\rho} = \frac{D}{\rho/\rho_c}\left(\frac{3\omega^2}{\pi G\rho_c}\right) \tag{2.51}$$

where $\rho_c \equiv \rho(0)$ and, consistent with (2.43),

$$\frac{\rho}{\rho_c} = 1 - \frac{5}{2}\lambda a^2 + \ldots \,, \tag{2.52}$$

we find that, correctly to terms of the order of the squares of surficial distortion,

$$\begin{aligned} f_2 &= k_2\left\{1 + \frac{3}{7}(1+v)\lambda a^2 + \right. \\ &\left. + \frac{3}{14}\left(1 - \frac{11}{5}\right)\lambda^2 a^3 - \frac{4k_2}{147}\lambda^2 a^4 + \ldots\right\} \,, \end{aligned} \tag{2.53}$$

where

$$v = \frac{\omega^2}{2\pi G \rho_c} \tag{2.54}$$

denotes a constant [2]

Moreover, the structure of Equation (2.47) for f_4, solved in a similar way discloses that, near the origin,

$$\begin{aligned} f_4 &= k_4 a^2 \left\{ 1 + \frac{18}{55}\lambda a^2 + \ldots \right\} + \\ &+ \frac{27}{35} k_2^2 \left\{ 1 - \frac{82}{77}\lambda^2 a^4 + \frac{8}{99} k_2 \lambda^2 a^4 + \right\} + \\ &+ \frac{156}{385} k_2^{-1} (1 + v) \lambda^2 a^4 + \ldots \right\} , \end{aligned} \tag{2.55}$$

consisting of two parts: the first (factored by k_4) represents the "complementary function" of the homogeneous version of (2.47) with its left-hand side equated to zero, while the second (factored by k_2^2) stands for a "particular integral" arising from the non-vanishing right-hand side. The constant k_4 introduced through the complementary function is new, and its value must be specified (after integration has been completed) from the boundary condition (2.50)—just as k_2 needs to be re-computed from (2.49).

With a second-order approximation to $f_2(a)$ near the origin as represented by Equation (2.53), we can proceed to evaluate the r.h.s. of Equations (2.46) amd (2.47) correctly to quantities of the third (or any higher) order. In general, the structure of the differential equations (2.46)–(2.47) governing the $f_j(a)$'s makes it evident that, near the origin, the complementary function of each f_j will vary as $k_j a^{j-2}$, while its particular integral will be factored by $k_2^{j/2}$. If, therefore, we set

$$f_2(0) = k_2 , \tag{2.56}$$

it follows from (2.33)–(2.35) that (for $\eta_j(0) = 0$),

$$\begin{aligned} f_4(0) &= \frac{27}{35} f_2^2(0) + \frac{108}{2695} f_2^3(0) + \ldots \\ &= \frac{27}{35} \left\{ k_2^2 + \frac{4}{77} k_2^3 + \ldots \right\} \end{aligned} \tag{2.57}$$

and

$$\begin{aligned} f_6(0) &= \frac{5}{6} f_2(0) f_4(0) - \frac{9}{154} f_2^3(0) + \ldots \\ &= \frac{45}{77} k_2^3 + \ldots \end{aligned} \tag{2.58}$$

[2] Strictly speaking, one should augment the r.h.s. of (2.53) still by quantities arising from possible bi-quadratic terms on the r.h.s. of Equations (2.43) or (2.52) which are not spelled out explicitly in the latter; but their inclusion may be left as an exercise for the interested reader.

etc., while the lowest derivatives of the f_j's which do not vanish at the origin are given by

$$f_2^2(0) = \frac{6}{7}k_2(1+v)\lambda \tag{2.59}$$

and, for $j > 2$,

$$f_j^{(j-2)}(0) = (j-2)!k_j\,, \quad j = 4, 6, \ldots\,. \tag{2.60}$$

The constants k_2, k_4, k_6, ... (not to be confused with those defined later by Equation (6.4)) constitute a set of the "eigen-parameters" of our rotational problem; and their values for any given value of $v \equiv \omega^2/2\pi G\rho_0$ must be determined algebraically from the boundary conditions (2.39)–(2.41) at $a = a_1$. Inasmuch as the right-hand sides of Equations (2.33)–(2.35) for f_j are known algebraic functions of f_2, $f_4 \ldots f_{j-2}$, their system readily lends itself for a solution by successive approximations: first we solve (2.43) for f_2 to accuracy of first order; next (2.46) and (2.47) for f_2 and f_4 to quantities of second order; and, eventually, all three for f_2, f_4, f_6 to third order. Moreover, in case of need the same process can evidently be extended to attain the accuracy of any order. Such an extension consistent to quantities of *fourth* order has indeed been already worked out by Kopal and Kamala Mahanta (1974). Its details are beyond the scope of this section; but can be found in the paper just quoted.

The results established in this section describe the explicit form of the amplitudes $f_j(a)$ in the expansion (2.1) for the equipotential surface of a rotating configuration in terms of the spherical harmonics of the form $P_2(\cos\,\theta)$. A choice of the latter for the angle-dependent part of the solid harmonics $Y_j(a,\theta,\phi)$ on the r.h.s. of Equation (2.1) has behind it a tradition of more than two hundred years established by Legendre (1793) and Laplace (1825); and all subsequent investigators of the subject were content to follow in this respect in their footsteps. Yet this strategy is neither unique, nor necessary, for the treatment of the problem. A choice of the form of the harmonics on the r.h.s. of the expansion (2.1) is, to be sure, intimately connected with those involved in the disturbing potential V' causing distortion. But an equally valid representation of the centrifugal potential (1.39) can be written up as expansions in terms of harmonics other than zonal—sectorial harmonics, for instance; or others—and the radius-vector r' of the respective equipotential surfaces expanded accordingly.

IV.3 Tidal Distortion

A determination of the gravitational potential and form of a fluid configuration of mass m_1 distorted by the attraction of a body of mass m_2 in its neighbourhood raising tides on m_1 can follow a course so parallel with the one developed already in the preceding section that only its gist needs to be given in this place.

The principal distinguishing feature of our present problem will be a different form of the disturbing potential V'. In order to specify the potential arising from tidal interaction, let the positions of the masses $m_{1,2}$ be described by the

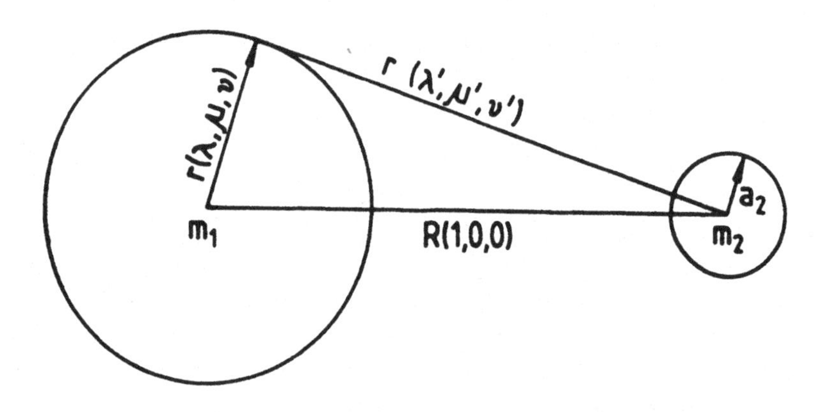

Figure IV.2:

rectangular coordinates $x_{1,2}$, $y_{1,2}$, $z_{1,2}$ and let, in what follows, $x_1 = y_1 = z_1 = 0$ (making the centre of mass m_1 the origin of coordinates) and $x_2 = R$, $y_2 = 0$, $z_2 = 0$ (assigning m_2 a position on the X-axis at a distance R from the origin; cf. Figure IV.2). Let, furthermore, r denote an arbitrary radius vector, of orientation specified by the direction cosines

$$\left.\begin{aligned} \lambda &= \cos\phi \sin\theta , \\ \mu &= \sin\phi \sin\theta , \\ \nu &= \cos\theta . \end{aligned}\right\} \tag{3.1}$$

If, moreover, $a_{1,2}$ denote the mean radii of the two components and $(k_j)_{1,2}$, the coefficients specifying the internal structure of the two stars[3], it is well known that the exterior potential $V(r)$ of mass m_1 can be expressed as

$$V(r) = \frac{Gm_1}{r}\left\{1 + \frac{m_2}{m_1}\sum_{j=2}^{\infty}(k_j)_1\left(\frac{a_1}{R}\right)^{j+1}\left(\frac{a_1}{r}\right)^{j} P_j(\lambda)\right\} , \tag{3.2}$$

in conformity with the particular solution (1.37) of Laplace's equation (1.36) for $d_j \neq 0$, where $r > a_1$ and λ represents the cosine of an angle between the radii-vectors $r(\lambda, \mu, \nu)$ and $R(1, 0, 0)$.

The exterior potential $V'(r)$ arising from the mass m_2 of the secondary component can be likewise obtained by an appropriate permutation of indices in (3.2) as

$$V'(r) = \frac{Gm_2}{r_2}\left\{1 + \frac{m_1}{m_2}\sum_{j=2}^{\infty}(k_j)_2\left(\frac{a_2}{R}\right)^{j+1}\left(\frac{a_2}{r_2}\right)^{j} P_j(\lambda_2)\right\} , \tag{3.3}$$

[3] For their specification in terms of the density distribution inside the mass m_j cf. Section IV.5.

where r_2 denotes the distance of an arbitrary element of mass m_1 from the centre of m_2, and λ_2 is the cosine of an angle between r_2 and R.

From the triangles marked on Figure IV.2 it is evident that the radii-vectors r_1, r_2 and R are connected with the direction cosines λ and λ_2 by the equations

$$r^2 = R^2 + r_2^2 - 2Rr_2\lambda_2 , \tag{3.4}$$

$$r_2^2 = R^2 + r^2 - 2Rr\lambda . \tag{3.5}$$

Eliminating r_2 between them we find that

$$\begin{aligned} \lambda_2 &= \frac{R - \lambda r}{\sqrt{R^2 + r^2 - 2Rr\lambda}} = \left\{1 - \frac{r\lambda}{R}\right\} \sum_{j=0}^{\infty} \left(\frac{r}{R}\right)^j P_j(\lambda) = \\ &= 1 + \sum_{j=1}^{\infty} \frac{j}{2j+1} \{P_{j+1} - P_{j-1}\} \left(\frac{r}{R}\right)^{j+1} , \end{aligned} \tag{3.6}$$

where $P_j \equiv P_j(\lambda)$. On the other hand, by eliminating λ_2 between Equations (3.5) and (3.5) we find that

$$\frac{1}{r_2} = \frac{1}{\sqrt{R^2 + r^2 - 2Rr\lambda}} = \frac{1}{R} \sum_{j=0}^{\infty} \left(\frac{r}{R}\right)^j P_j(\lambda) . \tag{3.7}$$

An inspection of the structure of Equations (3.2) or (3.3) discloses that the leading terms of the summations on the right-hand sides are of the order of $a_{1,2}^5$; their squares being of the order of $a_{1,2}^{10}$, etc. Consequently, terms which are of the first order in surficial distortion correspond to $j = 2, 3, 4$; those of second order, to $j = 5, 6, 7$; etc. *let us, in what follows, limit ourselves to consider terms in V' of no higher than the second order.* If so, then by noting that, within this scheme of approximation,

$$\left(\frac{R}{r_2}\right)^3 = 1 + 3\left(\frac{r}{R}\right) P_1(\lambda) + \left(\frac{r}{R}\right)^2 \{5P_2(\lambda) + 1\} + \dots \tag{3.8}$$

and

$$P_2(\lambda_2) = 1 + \left(\frac{r}{R}\right)^2 \{P_2(\lambda) - 1\} + \dots , \tag{3.9}$$

while, for $j > 2$,

$$P_j(\lambda_2) = 1 , \tag{3.10}$$

we find from (3.5) that — correctly to quantities of second order — the disturbing potential $V'(r)$ which raises tides on the configuration of mass m_1 can be expressed as

$$V'(r) = G\frac{m_2}{R}\left\{1 + \sum_{j=2}^{7} \left(\frac{r}{R}\right)^j P_j(\lambda)\right\} +$$

$$+ G\frac{m_1}{R}(k_2)_2\left(\frac{a_2}{R}\right)^5\left\{1+3\left(\frac{r}{r}\right)P_1(\lambda)+6\left(\frac{r}{R}\right)^2 P_2(\lambda)+\ldots\right\}+$$

$$+ G\frac{m_1}{R}(k_3)_2\left(\frac{a_2}{R}\right)^7\{1+\ldots\} = \sum_{j=0}^{7} c_j r^j P_j(\lambda)\,, \tag{3.11}$$

in accordance with the general form of Equation (1.37); where the constants

$$c_0 = G\frac{m_2}{R} + G\frac{m_1}{R}(k_2)_2\left(\frac{a_2}{R}\right)^5 + G\frac{m_1}{R}(k_3)_2\left(\frac{a_2}{R}\right)^7 + \ldots\,, \tag{3.12}$$

$$c_1 = + 3G\frac{m_1}{R^2}(k_2)_2\left(\frac{a_2}{R}\right)^5 + \ldots\,, \tag{3.13}$$

$$c_2 = G\frac{m_2}{R^3} + 6G\frac{m_1}{R^3}(k_2)_2\left(\frac{a_2}{R}\right)^5 + \ldots\,; \tag{3.14}$$

and, for $j > 2$,

$$c_j = G\frac{m_2}{r^{j+1}}\,. \tag{3.15}$$

If the secondary (disturbing) component of mass m_2 could be regarded as a mass-point—either because $a_2 = 0$ (zero size) or $(k_j)_2 = 0$ (infinite density concentration)—the entire expression for $V'(r)$ on the r.h.s. of Equation (3.11) would reduce to its first term; all others represent the effects of the departure of the secondary's behaviour from that of a mass-point.

The occurrence in the expression (3.11) for $V'(r)$ of harmonics $P_j(\lambda)$ for $j = 1(1)7$ to describe all effects of first and second order necessitates the radius-vector $r'(a,\,\theta'\,\phi')$ of the corresponding equipotential surface to be expressible as

$$r' = a\left\{1+\sum_{j=0}^{7} f_j(a)\,P_j(\lambda')\right\}\,, \tag{3.16}$$

where

$$\lambda' = \cos\phi'\,\sin\theta'\,, \tag{3.17}$$

correctly to quantities of second order—in contrast with the third-order Equation (2.1) of the rotational problem. The amplitudes $f_j(a)$ on the r.h.s. of Equation (3.16) represent, however, now the amplitudes of the respective partial tides invoked by the disturbing potential V'; and the principal aim of the present section will be to specify them in terms of the star's structure and the properties of the external field of force.

The products $f_j(a)P_j(\lambda')$ continue to be solid harmonics of the form $Y_j(a,\theta',\phi')$, satisfying the differential equation (1.30) and the orthogonality conditions (1.34) of Section IV.1. Therefore, raising Equation (3.16) to $(2-n)$th and $(n+3)$th power by use of the binomial theorem, and making use of the decomposition theorem (2.5) for Legendre polynomials we find that, correctly to quantities of second order,

$$(r')^{2-n} = a^{2-n}\left\{1-(n-2)\sum_{j=0}^{7}[f_j-(n-1)X_j]P_j\right\} \tag{3.18}$$

and

$$(r')^{n+3} = a^{n+3}\left\{1+(n+3)\sum_{j=0}^{7}[f_j+(n+2)X_j]P_j\right\}, \tag{3.19}$$

where

$$X_0 = \frac{1}{10}(f_2^2+\frac{5}{7}f_3^2+\ldots), \tag{3.20}$$

$$X_1 = \frac{9}{35}f_2f_3+\ldots, \tag{3.21}$$

$$X_2 = \frac{1}{7}(f_2^2+\frac{2}{3}f_3^2+2f_2f_4+\ldots), \tag{3.22}$$

$$X_3 = \frac{4}{15}f_2f_3+\ldots, \tag{3.23}$$

$$X_4 = \frac{9.}{35}(f_2^2+\frac{5}{11}f_3^2+\frac{100}{99}f_2f_4+\ldots), \tag{3.24}$$

$$X_5 = \frac{10}{21}f_2f_3+\ldots, \tag{3.25}$$

$$X_6 = \frac{50}{231}(f_3^2+\frac{21}{10}f_2f_4+\ldots); \tag{3.26}$$

and all X_j's for $j > 6$ are identically equal to zero. Like in Section 1, an expansion of the form (3.18) continues to be valid for $n \lessgtr 2$; while, for $n = 2$, the coefficient of P_2 in the expansion of $\log r'$ becomes equal to $f_2 - X_2$.

If we insert now the foregoing expansions (3.18) and (3.19) with the coefficients X_j as given by (3.26)–(3.26) in Equations (1.36) or (1.27) and (1.28) and take advantage of the orthogonality property of our harmonics as given by Equation (1.34), the interior and exterior potentials U and V of tidally-distorted configurations can again be expressed in the forms

$$U(r) = 4\pi G\sum_{j=0}^{\infty}\frac{r^jE_j(a)}{2j+1}P_j(\lambda), \tag{3.27}$$

$$V(r) = 4\pi G\sum_{j=0}^{\infty}\frac{f_j(a)P_j(\lambda)}{(2j+1)r^{j+1}}, \tag{3.28}$$

analogous to Equations (1.35) and (1.36) where (again correctly to quantities of second order)

$$E_0(a) = \int_a^{a_1}\rho\frac{\partial}{\partial a}\left\{a^2\left(\frac{1}{2}+f_0+X_0\right)\right\}da, \tag{3.29}$$

and

$$F_0(a) = \int_a^{a_1} \rho \frac{\partial}{\partial a} \left\{ a^3 \left(\frac{1}{3} + f_0 + 2X_0 \right) \right\} da \, ; \tag{3.30}$$

while, for $j > 0$,

$$E_j(a) = \int_a^{a_1} \rho \frac{\partial}{\partial a} \{ a^{2-j} [f_j + (1-j) X_j] \} \, da \tag{3.31}$$

and

$$F_j(a) = \int_0^a \rho \frac{\partial}{\partial a} \{ a^{j+3} [f_j + (j+2) X_j] \} \, da \, , \tag{3.32}$$

respectively.

The amplitude f_0 of the zero-order harmonic on the r.h.s. of Equation (3.16) can—as in the rotational case—be readily specified from the requirement that the total mass of our configuration must be independent of distortion—a requirement leading by Equation (2.10) to the condition

$$f_0 = -2X_0 = -\frac{1}{5} f_2^2 - \frac{1}{7} f_3^2 - \dots \, , \tag{3.33}$$

analogous to Equation (2.12) of the rotational problem. Moreover, in order to determine the f_j's for $j > 0$, recourse must again be made to the fact that the level surfaces represented by Equation (3.16) are equipotentials, over which the total potential

$$\Psi(r', \lambda) = U(r', \lambda) + V(r', \lambda) + V'(r', \lambda)) \tag{3.34}$$

must remain constant for the tide-generating potential V' as given by Equation (3.11).

Suppose that, as before, we expand the total potential $\Psi(r', \lambda)$ in a Neumann series of the form

$$\Psi(r', \lambda) = \sum_{j=0}^{\infty} \alpha_j(a) \, P_j(\lambda) \, ; , \tag{3.35}$$

with coefficients defined by

$$\alpha_j(a) = \frac{2j+1}{2} \int_{-1}^{1} \Psi(r', \lambda) \, P_j(\lambda) \, d\lambda \, . \tag{3.36}$$

If so, the constancy of $\Psi(r', \lambda)$ over a level surface will evidently be ensured by setting

$$\alpha_j(a) = 0 \text{ for } j > 0 \, ; \tag{3.37}$$

and j equations of this form constitute the necessary as well as sufficient conditions for a specification of the respective amplitudes f_j.

A determination of the explicit forms of Equation (3.37) is rather involved; but the results sum out to be relatively simple; for it can be shown (cf. pp. 46–48 of Kopal, 1960) that, for $j = 1(1)7$,

$$\frac{aE_1}{3} + \frac{F_1}{3a^2} - \frac{f_1 F_0}{a} + \frac{c_1 a}{4\pi G} = -\frac{9}{5} f_2 f_3 \frac{F_0}{a^2} + \frac{9}{35} f_3 \frac{F_2}{a^3} + \frac{9}{35} f_2 \frac{F_3}{a^4}, \quad (3.38)$$

$$\frac{a^2 E_2}{5} + \frac{F_2}{5a^3} - \frac{f_2 F_0}{a} + \frac{c_2 a^2}{4\pi G} = -\left\{\frac{6}{7} f_2^2 + \frac{16}{21} f_3^2 + \frac{16}{7} f_2 f_4\right\} \frac{F_0}{a} +$$

$$+ \left\{\frac{2}{7} f_2 + \frac{2}{7} f_4\right\} \frac{F_2}{a^3} + \frac{4}{21} f_3 \frac{F_3}{a^4} + \frac{2}{7} f_2 \frac{F_4}{a^5}, \quad (3.39)$$

$$\frac{a^3 E_3}{7} + \frac{F_3}{7a^4} - \frac{f_3 F_0}{a} + \frac{c_3 a^3}{4\pi G} = -\frac{9}{5} f_2 f_3 \frac{F_0}{a} + \frac{4}{15} f_3 \frac{F_2}{a^3} + \frac{4}{15} \frac{F_3}{a^4}, \quad (3.40)$$

$$\frac{a^4 E_4}{9} + \frac{F_4}{9a^5} - \frac{f_4 F_0}{a} + \frac{c_4 a^4}{4\pi G} = -\left\{\frac{54}{35} f_2^2 + \frac{72}{77} f_3^2 + \frac{80}{77} f_2 f_3\right\} \frac{F_0}{a} +$$

$$+ \left\{\frac{18}{35} f_2 + \frac{20}{77} f_4\right\} \frac{F_2}{a^3} + \frac{18}{77} f_3 \frac{F_3}{a^4} + \frac{20}{77} f_2 \frac{F_4}{a^5}, \quad (3.41)$$

$$\frac{a^5 E_5}{11} + \frac{F_5}{11a^6} - \frac{f_5 F_0}{a} + \frac{c_5 a^5}{4\pi G} = -\frac{10}{3} f_2 f_3 \frac{F_0}{a} + \frac{10}{21} f_3 \frac{F_2}{a^3} + \frac{10}{21} f_2 \frac{F_3}{a^4}, \quad (3.42)$$

$$\frac{a^6 E_6}{13} + \frac{F_6}{13a^7} - \frac{f_0 F_0}{a} + \frac{c_0 a^6}{4\pi G} = -\left\{\frac{400}{231} f_3^2 + \frac{40}{11} f_2 f_4\right\} \frac{F_0}{a} +$$

$$+ \frac{5}{11} f_4 \frac{F_2}{a^3} + \frac{100}{231} f_3 \frac{F_3}{a^4} + \frac{5}{11} f_2 \frac{F_4}{a^5}, \quad (3.43)$$

$$\frac{a^7 E_7}{15} + \frac{F_7}{15a^8} - \frac{f_7 F_0}{a} + \frac{c_7 a^7}{4\pi G} = 0; \quad (3.44)$$

where the constants c_j continue to be given by Equations (3.13)–(3.15), and

$$F_0 = \int_0^a \rho a^2 \, da \, . \quad (3.45)$$

The foregoing Equations (3.38)–(3.44) constitute the tidal equivalent of Equations (2.28)–(2.29) relevant to the case of the rotational problem. Like in this former case, the left-hand sides consist of terms which are of both first and second orders. Those on the right are, however, all of second order—consisting as they do of the products of various integrals $F_j (j = 0, 2, 3, 4)$ with different combinations of the amplitudes f_2, f_3, and f_4 of the most important partial tides. The latter are, moreover, known to possess leading terms of first order, satisfying the equation

$$\frac{a^j}{2j+1} \int_a^{a_1} \rho \frac{\partial}{\partial a}(a^{2j-j}) da + \frac{1}{(2j+1)a^{j+1}} \int_0^a \rho \frac{\partial}{\partial a}(a^{j+3} f_j) da +$$

$$+ \frac{f_j}{a} \int_0^a \rho a^2 \, da = \frac{c_j a^j}{4\pi G} \quad (3.46)$$

for j = 2, 3, and 4. Divide now this equation by a^j, differentiate with respect to a, and multiply subsequently by a^{2j+1}; in doing so we find that

$$\int_0^a \rho \frac{\partial}{\partial a}(a^{j+3} f_j) da = a^{j+1} \left\{ \frac{j+1}{a} f_j - \frac{\partial f_j}{\partial a} \right\} \int_0^a \rho a^2 \, da \, , \quad (3.47)$$

or, correctly to quantities of first order,

$$F_j(a) = a^{j+1}\left\{\frac{j+1}{a}f_1 - \frac{\partial f_j}{\partial a}\right\} F_0 , \tag{3.48}$$

corresponding to Equations (2.31) and (2.32) of Section 2.

If we insert now this latter equation on the right-hand sides of Equations (3.38)–(3.44) we can express the latter, more concisely, for $j = 1(1)7$ as integral equations for f_j, of the form

$$\begin{aligned} f_j \int_0^a \rho a^2\,da \; - \; \frac{1}{(2j+1)a^j}\int_0^a \rho\frac{\partial}{\partial a}(a^{j+3}f_j)da \; - \\ - \frac{a^{j+1}}{2j+1}\int_a^{a_1} \rho\frac{\partial}{\partial a}(a^{2-j}f_j)da = \mathcal{R}_j(a) + \frac{c_j a^{j+1}}{4\pi G} , \end{aligned} \tag{3.49}$$

where

$$\begin{aligned} \mathcal{R}_j(a) \;&=\; a\frac{\partial X_j}{\partial a}\int_o^a \rho a^2\,da + \frac{j+2}{(2j+1)a^j}\int_0^a \rho\frac{\partial}{\partial a}(a^{j+3}X_j)da \; - \\ &- \frac{(j-1)a^{j+1}}{2j+1}\int_a^{a_1} \rho\frac{\partial}{\partial a}(a^{2-j}X_j)da \end{aligned} \tag{3.50}$$

in terms of the X_j's as given by Equations (3.20)–(3.26).

Equations (3.49) define the amplitudes f_j in integral form. In order to reduce them to differential equations (in which they are more amenable to numerical solution), differentiate (3.49) with respect to a and divide by a^{2j}; then differentiate the quotient once more with respect to a and multiply by a^{j+2}. The result of these operations discloses that

$$\begin{aligned} a^2 f_j'' + 6D(af_j' + f_j) - j(j+1)f_j = \\ = F_0^{-1}\{a^2\mathcal{R}_j'' - j(j+1)\mathcal{R}_j\} \equiv T_j(a) , \end{aligned} \tag{3.51}$$

where primes denote differentiation with respect to a;

$$D \equiv \frac{\rho}{\overline{\rho}} = \frac{\rho a^3}{3\int_0^a \rho a^2\,da} = \frac{\rho a^3}{3F_0} , \tag{3.52}$$

in which $\overline{\rho}$ denotes the mean density of our configuration interior to a; and

$$T_1(a) = \frac{18}{35}\{2(\eta_2\eta_3 + 10\eta_2 + 7\eta_3) - 3D(3\eta_2\eta_3 + 3\eta_2 + 3\eta_3 - 5)\}f_2 f_3 , \tag{3.53}$$

$$\begin{aligned} T_2(a) \;&=\; \frac{2}{7}\{2\eta_2(\eta_2+9) - 9D\eta_2(\eta_2+4)\}f_2^2 + \\ &+ \frac{4}{21}\{2\eta_3(\eta_3+21) - 9D(\eta_3^2 + 2\eta_3 - 2)\}f_2^3 + \\ &+ \frac{4}{7}\{2(\eta_2\eta_4 + 15\eta_2 + 8\eta_4) - \\ &- 3D(3\eta_2\eta_4 + 3\eta_2 + 3\eta_4 - 7)\}f_2 f_4 \end{aligned} \tag{3.54}$$

$$T_3(a) = \frac{8}{15}\{2\eta_2\eta_3 + 15\eta_2 - 9\eta_3 - 9D(\eta_2\eta_3 + \eta_2 + \eta_3)\} f_2 f_3 , \tag{3.55}$$

$$\begin{aligned} T_4(a) &= \frac{18}{35}\{2\eta_2(\eta_2 + 2) - 3D(3\eta_2^2 + 6\eta_2 + 7)\} f_2^2 + \\ &+ \frac{18}{77}\{2\eta_3(\eta_3 + 14) - 3D(3\eta_3^2 + 6\eta_3 + 1)\} f_2^2 + \\ &+ \frac{40}{77}\{2\eta_2\eta_4 + 23\eta_2 + 9\eta_4 - 9D(\eta_2\eta_4 + \eta_2 + \eta_4)\} f_2 f_4 , \end{aligned} \tag{3.56}$$

$$T_5(a) = \frac{20}{21}\{2\eta_2(\eta_3 + 3) - 3D(3\eta_2\eta_3 + 3\eta_2 + 3\eta_3 + 5)\} f_2 f_3 , \tag{3.57}$$

$$\begin{aligned} T_6(a) &= \frac{100}{231}\{2\eta_3^2(\eta_3 + 3) - 9D(\eta_3^2 + 2\eta_3 + 4)\} f_3^2 + \\ &+ \frac{10}{11}\{2\eta_2\eta_4 + 6\eta_2 - \eta_4) - \\ &- 3D(3\eta_2\eta_4 + 3\eta_2 + 3\eta_4 + 11)\} f_2 f_4 , \end{aligned} \tag{3.58}$$

$$T_7(a) = 0 , \tag{3.59}$$

where we have abbreviated

$$\eta_j = \frac{a}{f_j}\frac{\partial f_j}{\partial a} \tag{3.60}$$

and taken advantage of the fact that—in accordance with (3.51)—the function $\eta_j (j = 2, 3, 4)$ satisfies the differential equation

$$a\eta_j' + 6D(\eta_j + 1) + \eta_j(\eta_j - 1) = j(j+1) \tag{3.61}$$

correctly to quantities of first order.

Equations (3.51) with their right-hand sides as defined by (3.53)–(3.59) constitute second-order analogies to Equations (2.33)–(2.35) of the rotational problem. In the present case, Equations (3.51) constitute a simultaneous system of seven ($j = 1(1)7$) second-order differential equations for the f_j's, subject to the boundary conditions requiring that, at the centre ($a = 0$),

$$f_j(0) = 0 \text{ for } j = 1(1)7 ; \tag{3.62}$$

while at the boundary $a = a_1$ (cf. Kopal, 1960; pp. 49-50)

$$f_j(a_1) = \frac{(2j+1)c_j a_1^{j+1}}{G_(j + \eta_j)_1\, m_1} + \left\{\frac{S_j(a)}{j + \eta_1}\right\}_{a_1} , \tag{3.63}$$

where

$$S_1 = \frac{9}{35} b_1^{(2,3)} f_2 f_3 , \tag{3.64}$$

$$S_2 = \frac{1}{7}b_2^{(2,2)}f_2^2 + \frac{2}{21}b_2^{(2,3)}f_3^2 + \frac{2}{7}b_2^{(2,4)}f_2f_4 , \tag{3.65}$$

$$S_3 = \frac{4}{15}b_3^{(2,3)}f_2f_3 , \tag{3.66}$$

$$S_4 = \frac{9}{35}b_4^{(2,2)}f_2^2 + \frac{9}{77}b_4^{(3,3)}f_3^2 + \frac{20}{77}b_4^{(2,4)}f_2f_4 , \tag{3.67}$$

$$S_5 = \frac{10}{21}b_5^{(2,3)}f_2f_3 , \tag{3.68}$$

$$S_6 = \frac{50}{231}b_6^{(3,3)}f_3^2 + \frac{5}{11}b_6^{(2,4)}f_2f_4 , \tag{3.69}$$

$$S_7 = 0 , \tag{3.70}$$

where we have abbreviated

$$b_j^{(i,k)} = 2\eta_i\eta_k + (j+1)(\eta_i+\eta_k) + i(i+1) + k(k+1) , \tag{3.71}$$

to be evaluated at the boundary $a = a_1$.

For arbitrary models of the stars—i.e., for arbitrary distribution of density ρ throughout the interior—the desired particular solutions of Equations (3.51)–(3.61) subject to the boundary conditions (3.62) and (3.63)–(3.71) can be constructed only by numerical methods; and their construction can be left as an exercise for the interested reader. While a determination of the amplitudes $f_j(a)$'s of the respective partial tides may, therefore, become a matter of some complexity, the symmetry of a configuration deformed by them is simple: namely, the crest of each tide is always in the direction of the radius-vector R connecting the mass centre of the two stars (i.e., for an observer on the crest of each tide the centre of mass m_2 is always in his zenith). This is, however, true only of the equilibrium tides investigated in this chapter. When we turn our attention to the dynamical tides in close binary systems in Chapter VI, we shall encounter a very different situation.

In conclusion of the present section, one additional consideration of some interest should be pointed out. In the case of rotational distortion treated previously in Section 2, the structure of the centrifugal potential (1.39) implied that the expansion (2.1) for the radius-vector r' contained only even harmonics of the respective angular variable cos θ; and, in such a case, the centre of mass of the rotating configuration is bound to coincide with the origin of our coordinate system. For if x, y, z denote the coordinates of the centre of gravity of mass m_1 arbitrarily placed, it follows by definition of the centre of mass that (cf. Equations

(1.20) and (1.26)

$$\left\{\begin{array}{c} x \\ y \\ z \end{array}\right\} m_1 = \int_0^{r_1}\int_0^{\pi}\int_0^{2\pi} \rho r'^2 \left\{\begin{array}{l} r' \sin\theta' \cos\phi' \\ r' \sin'\theta' \sin\phi' \\ r' \cos\theta' \end{array}\right\} dr' \sin\theta' \, d\theta' \, d\phi'$$

$$= \int_0^{a_1} \rho \frac{\partial}{\partial a} \left\{ \int_0^{\pi}\int_0^{2\pi} \frac{r'^4}{4} \left[\begin{array}{l} \sin\theta' \cos\phi' \\ \sin\theta' \sin\phi' \\ \cos\theta' \end{array}\right] \sin\theta' \, d\theta' \, d\phi' \right\} da \, . \tag{3.72}$$

The column matrix on the right-hand side of the foregoing equations consists of the components of the solid harmonic $Y_j(a, \theta, \phi)$ of order one, known to be of the form

$$Y_1(a, \theta', \phi') = A \sin\theta' \cos\phi' + B \sin\theta' \sin\phi' + C \cos\theta' \, , \tag{3.73}$$

where A, B, C are constants or functions of a. Therefore, unless the expansion of r'^4 itself contains a solid harmonic of the type Y_1, the orthogonality properties of spherical harmonics will compel all terms on the right-hand side of (3.72) to vanish identically; and if so, the left-hand side must similarly vanish—which it can do only if $x = y = z = 0$.

If the expansion on the right-hand side of (3.16) were to consist only of the harmonics of even orders—as was the case for purely rotational distortion—the term r'^4 would not contain any harmonic of order one; and, therefore, the centre of gravity would coincide with the origin of our coordinate system regardless of the extent of the distortion. However, if—as in the case of mutual tidal distortion—the expansion on the right-hand side contains harmonics of even as well as odd orders, the expression for r'^4 will contain products of the form $Y_j Y_{j-1}$; and each of these on decomposition will furnish a non-zero term varying as Y_1. In such a case, the right-hand side of Equation (3.72) will no longer vanish; and neither can then the left one—a phenomenon indicating that not all three coordinates x, y, z of the centre of mass of m_1 can be set equal to zero in our coordinate system. In point of fact, the positions of the centres of gravity and symmetry begin gradually to drift apart with increasing degree of tidal distortion until the initial single configuration may actually break up in two. The emergence of this tendency can be traced to terms factored by the powers and cross-products of the spherical harmonics contained in the tide-generating potential V', and introduced through it into the expansion (3.16) for r'. Whether or not a trend so initiated may, under sufficiently extreme conditions, result in a fission into two bodies of comparable mass constitutes a problem to which no satisfactory answer can so far be given. It may, however, be of interest to note that the first mathematical symptoms of a situation which may lead to this end appear with the introduction of terms which are of second order in surficial distortion.

IV.4 Interaction Between Rotation and Tides

In the preceding sections of this chapter we have investigated the rotational as well as tidal distortion of self-gravitating configurations of arbitrary structure, and established the explicit form of differential equations governing the first seven spherical-harmonic deformations ($j = 1$ to 7). Within the framework of a consistent theory including all tides with amplitudes of the order of the squares of the surficial distortion, seven harmonics are present with non-vanishing coefficients; while in the case of rotational distortion (containing no odd harmonics as long as the configuration remains spheroidal) the use of the first three even harmonics has permitted us to attain the accuracy of third order. The leading terms of the rotational as well as tidal distortion are, however, both of first order. Therefore, no second-order theory of either distortion can be regarded as complete until the effects of the cross-terms between first-order rotational and tidal distortion have been investigated. This constitutes a task to which we propose to address ourselves in this section.

Before we proceed to tackle this delicate task, we should clearly point out the limitations imposed by the need to maintain the equilibrium form of our configuration. In the preceding section we developed an equilibrium theory of tides as a "two-centre" problem—with the tide-generating mass m_2 kept stationary on the X-axis. In actual binary systems both components of masses $m_{1,2}$ must, to be sure, revolve in a plane orbit around the common centre of gravity - so that the X-axis actually rotates in space, with the motion of the radius-vector R of the relative orbit of the two stars. As long as this motion is uniform, and the radius-vector constant, tidal distortion will produce no motion in the fluid mass m_1 rotating with the Keplerian angular velocity $\omega_K^2 = G(m_1+m_2)/R^3$; and the results based on the equilibrium theory continue to be directly applicable to reality.

Suppose, however, that the radius-vector R ceases to be constant (on account of a finite eccentricity of the relative orbit); if so, the height of a j-th partial tide should vary as R^{-j-1} in the course of each orbital cycle, thus setting the fluid in motion. Moreover, if the distorted component also rotates about an axis which is not perpendicular to the orbital plane, and its angular velocity ω of rotation is different from ω_K, the fact that the crest of each tide follows (in the absence of dissipative forces) the radius-vector R will give rise to finite velocity components in the rotating system of coordinates—constituting dynamical tides.

Such tides will become the subject of appropriate discussion in the subsequent Chapter VI of this book. If we wish to remain in the framework of an equilibrium theory, and to avoid the emergence of velocity components arising from differential motions within the fluid, the following conditions must be satisfied:

1) the equator of the rotating configuration must be coplanar with the relative orbit of the disturbing body; and the orbit itself circular (i.e., $R =$ constant); and

2) the (constant) angular velocity ω of the distorted configuration must be

identical with the (constant) Keplerian angular velocity ω_K orbital revolution—in other words, the rotation and revolution must be synchronous.

A breakdown of either one of these conditions is bound to give rise to *dynamical tides*, a study of which is being postponed for the next chapter. In the present section we shall assume that the foregoing conditions (1) and (2) are indeed fulfilled; so that the cross-terms between rotation and tides which we wish to consider will not give rise to any motion in the rotating frame of reference.

In order to investigate the nature of such terms, let the direction cosines of the vectors r and r' diagramatically shown in Figure IV.1 be given by

$$\left.\begin{aligned} \lambda &= \cos\phi \sin\theta \\ \mu &= \sin\phi \sin\theta \\ \nu &= \cos\theta \end{aligned}\right\} \tag{4.1}$$

and

$$\left.\begin{aligned} \lambda' &= \cos\phi' \sin\theta' \\ \mu' &= \sin\phi' \sin\theta' \\ \nu' &= \cos\theta' \end{aligned}\right\} \tag{4.2}$$

respectively; so that the angle γ between them (cf. Equation 1.13) is given by the equation

$$\cos\gamma = \lambda\lambda' + \mu\mu' + \nu\nu' \, . \tag{4.3}$$

Let, moreover, the direction of the tide-generating body in the same coordinates be specified by the direction cosines λ'', μ'', ν''; so that the angles Θ, Θ' between the radii-vectors r, r' and the line (of length R) joining the centres of the two mutually attracting masses $m_{1,2}$ will be given by

$$\cos\Theta = \lambda\lambda'' + \mu\mu'' + \nu\nu'' \tag{4.4}$$

and

$$\cos\Theta' = \lambda'\lambda'' + \mu'\mu'' + \nu'\nu'' \, , \tag{4.5}$$

respectively. Within the framework of the equilibrium theory of tides, the X-axis of our rectangular system becomes identical with the orbital radius-vector, and the Z-axis with that of rotation of the respective configuration. If so, however, then

$$\lambda' = 1 \text{ and } \mu'' = \nu'' = 0 \, ; \tag{4.6}$$

and, accordingly,

$$\cos\Theta = \lambda \text{ and } \cos\Theta' = \lambda' \, . \tag{4.7}$$

A development of the radius-vector r' of the distorted surface consistent to quantities of second order is bound to contain also cross-terms varying as $P_i(\nu')P_j(\lambda')$ for $j = 2$, 3 and 4 within the scheme of second-order approximation, which must be added to the summation on the r.h.s. of Equation (3.16), together with purely rotational terms investigated already in Section IV.2.

Suppose that we do so, and include the terms of this form for the expression for r' used to evaluate the U_j's and V_j's constituting our potential expansion in Section IV.1. Doing so we find it, however, necessary to consider—in addition to the classical orthogonality theorem (2.7)—also more general orthogonality conditions for the *triple* products $P_i(\nu')P_j(\lambda')P_n(\cos\gamma)$ the integrals of which must be evaluated over the whole sphere in terms of the unprimed angular coordinates θ and ϕ.

In order to do so, let us observe that, in the case of synchronism between rotation and revolution with co-planar equator and orbit, the addition theorem for spherical harmonics discloses that

$$\begin{aligned} P_n(\cos\phi'\sin\theta') &= P_n(0)P_n(\cos\theta') + \\ &+ 2\sum_{i=1}^{n}\frac{(n-i)!}{(n+i)!}P_n^i(0)P_n^i(\cos\theta')\cos i\phi' , \end{aligned} \tag{4.8}$$

where

$$P_{2j}(0) = (-1)^j\,\frac{1.3.5\ldots(2j-1)}{2.4.6\ldots 2j} \tag{4.9}$$

if n happens to be an even integer, and

$$P_{2j+1}(0) = 0 \tag{4.10}$$

if it is odd.

With the aid of this addition theorem and the decomposition formula (2.5) it is possible to establish that

$$\int_0^{\pi}\int_0^{2\pi} P_j(\lambda')P_2(\nu')P_n(\cos\gamma)\sin\theta'\,d\theta'\,d\phi' = \frac{4\pi}{2n+1}\mathcal{B}_n^{(j)}(\theta,\phi) , \tag{4.11}$$

where the only terms non-vanishing for $j = 2(1)4$ assume (in terms of the Clebsch-Gordan coefficients) the explicit forms

$$\mathcal{B}_0^{(2)} = -\frac{1}{10} , \tag{4.12}$$

$$\mathcal{B}_2^{(2)} = -\frac{2}{7}\{P_2(\lambda)+P_2(\nu)\} , \tag{4.13}$$

$$\mathcal{B}_4^{(2)} = \frac{1}{10}+\frac{2}{7}\{P_2(\lambda)+P_2(\nu)+P_2(\lambda)P_2(\nu) ; \tag{4.14}$$

$$\mathcal{B}_1^{(3)} = -\frac{9}{70}P_1(\lambda) , \tag{4.15}$$

$$\mathcal{B}_3^{(3)} = -\frac{2}{15}P_1(\lambda)-\frac{1}{3}P_3(\lambda)-\frac{2}{3}P_1(\lambda)P_2(\nu) , \tag{4.16}$$

$$\mathcal{B}_5^{(3)} = \frac{11}{42}P_1(\lambda)+\frac{1}{3}P_2(\lambda)+\frac{2}{3}P_1(\lambda)P_3(\nu)+P_3(\lambda)P_2(\nu) ; \tag{4.17}$$

and

$$\mathcal{B}_2^{(4)} = \frac{1}{21}P_2(\nu)-\frac{5}{42}P_2(\lambda) , \tag{4.18}$$

$$\mathcal{B}_4^{(4)} = -\frac{20}{77}P_2(\nu) - \frac{27}{77}P_2(\lambda) - \frac{4}{11}P_4(\lambda) - \frac{10}{11}P_2(\lambda)P_2(\nu) - \frac{1}{11}, \quad (4.19)$$

$$\mathcal{B}_6^{(4)} = \frac{7}{33}P_2(\nu) + \frac{25}{66}P_2(\lambda) + \frac{4}{11}P_4(\lambda) + + \frac{10}{11}P_2(\lambda)P_2(\nu) + \frac{1}{11} + P_4(\lambda)P_2(\nu)\,; \quad (4.20)$$

satisfying a general relation of the form

$$\mathcal{B}_{j-2}^{(j)} + \mathcal{B}_j^{(j)} + \mathcal{B}_{j+2}^{(j)} = P_j(\lambda)P_2(\nu)\,. \quad (4.21)$$

In consequence, a full-dress expression for the radius-vector r' of a tidally-distorted configuration rotating in synchronism with its revolution about an axis perpendicular to the orbital plane should—correctly to quantities of second order in surficial distortion—contain the terms[4]

$$r' = a\left\{1 + \ldots + \sum_{j=0}^{4} g_j(a)P_j(\lambda') + \sum_{j=0}^{4} h_j(a)P_j(\lambda')P_2(\nu')\right\} \quad (4.22)$$

additional to those arising from pure rotation or tides. The latter have already been investigated in previous parts of our work; but the explicit form of the mixed terms with amplitudes g_j and h_j remain yet to be established.

In order to do so, let—in accordance with Equations (1.17) and (1.18)—the interior and exterior potential of our distorted configuration be expressible as

$$(2n+1)U = 4\pi G \sum_{n=0}^{\infty} r^n U_n \quad (4.23)$$

and

$$(2n+1)V = 4\pi G \sum_{n=0}^{\infty} r^{-n-1} V_n\,, \quad (4.24)$$

respectively, where U_n and V_n continue to be given by Equations (1.24)–(1.26) for any value of n. Since, moreover, by use of the generalized orthogonality theorem (4.4),

$$\int_0^{\pi}\int_0^{2\pi}\left\{\sum_{j=0}^{4} g_jP_j(\lambda') + \sum_{j=0}^{4} h_jP_j(\lambda')P_2(\nu')\right\}P_n(\cos\gamma)\sin\theta'\,d\theta'\,d\phi' = = \frac{4\pi}{2n+1}\left\{g_nP_n(\lambda) + \sum_{j=0}^{4} h_j\mathcal{B}_n^{(j)}(\lambda,\nu)\right\}, \quad (4.25)$$

[4] The expansion of r' on the r.h.s. of Equation (4.22) has been carried out in terms of the zonal harmonic $P_2(\nu)$, rather than the sectorial harmonic $P_2^2(\nu)$, of the colatitude. Since, however, $P_2^2 = 2(1 - P_2)$, the coefficients f_j and h_j of different harmonics used in our previous work (Kopal, 1960) are related with g_j and h_j used presently by $f_j = g_j + h_j$ and $h_j = -\frac{1}{2}h_j$.

the contributions to U_n and V_n arising from the *interaction* between rotation and tides (and, therefore, *additive* to those due to pure rotation and tides) will be given by the equations

$$U_0 = \int_a^{a_1} \rho \frac{\partial}{\partial a}(a^2 g_0) da + \mathcal{B}_0^{(2)}(\lambda,\nu) \int_a^{a_1} \rho \frac{\partial}{\partial a}(a^2 h_2) da + \dots ;, \quad (4.26)$$

$$3U_1 = P_1(\lambda) \int_a^{a_1} \rho \frac{\partial}{\partial a}(a g_1) da + \mathcal{B}_1^{(1)}(\lambda,\nu) \int_a^{a_1} \rho \frac{\partial}{\partial a}(a h_1) da + + \mathcal{B}_1^{(3)}(\lambda,\nu) \int_a^{a_1} \rho \frac{\partial}{\partial a}(a h_3) da + \dots , \quad (4.27)$$

$$5U_2 = P_2(\lambda) \int_a^{a_1} \rho \frac{\partial}{\partial a}(g_2) da + \mathcal{B}_2^{(0)}(\lambda,\nu) \int_a^{a_1} \rho \frac{\partial}{\partial a}(h_0) da + + \mathcal{B}_2^{(2)}(\lambda,\nu) \int_a^{a_1} \rho \frac{\partial}{\partial a}(h_2) da + + \mathcal{B}_2^{(4)}(\lambda,\nu) \int_a^{a_1} \rho \frac{\partial}{\partial a}(h_4) da + \dots , \quad (4.28)$$

$$7U_3 = P_3(\lambda) \int_a^{a_1} \rho \frac{\partial}{\partial a}\left(\frac{g_3}{a}\right) da + \mathcal{B}_3^{(1)}(\lambda,\nu) \int_a^{a_1} \rho \frac{\partial}{\partial a}\left(\frac{h_3}{a}\right) da + + \mathcal{B}_3^{(3)}(\lambda,\nu) \int_a^{a_1} \rho \frac{\partial}{\partial a}\left(\frac{h_3}{a}\right) da + \dots , \quad (4.29)$$

$$9U_4 = P_4(\lambda) \int_a^{a_1} \rho \frac{\partial}{\partial a}\left(\frac{g_4}{a^2}\right) da + \mathcal{B}_4^{(2)}(\lambda,\nu) \int_a^{a_1} \rho \frac{\partial}{\partial a}\left(\frac{h_2}{a^2}\right) da + + \mathcal{B}_4^{(4)}(\lambda,\nu) \int_a^{a_1} \rho \frac{\partial}{\partial a}\left(\frac{h_4}{a^2}\right) da + \dots , \quad (4.30)$$

$$11U_5 = \mathcal{B}_5^{(4)}(\lambda,\nu) \int a^{a_1} \rho \frac{\partial}{\partial a}\left(\frac{h_3}{a^3}\right) da + \dots , \quad (4.31)$$

$$13U_6 = \mathcal{B}_6^{(4)}(\lambda,\nu) \int_a^{a_1} \rho \frac{\partial}{\partial a}\left(\frac{h_4}{a^4}\right) da + \dots ; \quad (4.32)$$

and

$$V_0 = \int_0^a \rho \frac{\partial}{\partial a}(a^3 f_0) da + \mathcal{B}_0^{(2)}(\lambda,\nu) \int_0^a \rho \frac{\partial}{\partial a}(a^3 h_2) da + \dots , \quad (4.33)$$

$$3V_1 = P_1(\lambda) \int_0^a \rho \frac{\partial}{\partial a}(a^4 g_1) da + \mathcal{B}_1^{(1)}(\lambda,\nu) \int_0^a \rho \frac{\partial}{\partial a}(a^4 g_1) da + + \mathcal{B}_1^{(3)}(\lambda,\nu) \int_0^a \rho \frac{\partial}{\partial a}(a^4 h_3) da + \dots , \quad (4.34)$$

$$5V_2 = P_2(\lambda) \int_0^a \rho \frac{\partial}{\partial a}(a^5 g_2) da + \mathcal{B}_2^{(0)}(\lambda,\nu) \int_0^a \rho \frac{\partial}{\partial a}(a^5 h_0) da + \mathcal{B}_2^{(2)}(\lambda,\nu) \int_0^a \rho \frac{\partial}{\partial a}(a^5 h_2) da + \mathcal{B}_2^{(4)} \rho \frac{\partial}{\partial a}(a^5 h_4) da + \dots , \quad (4.35)$$

$$\begin{aligned} 7V_3 &= P_3(\lambda)\int_0^a \rho\frac{\partial}{\partial a}(a^6 g_3)da + \mathcal{B}_3^{(1)}(\lambda,\nu)\int_0^a \rho\frac{\partial}{\partial a}(a^6 h_1)da + \\ &+ \mathcal{B}_3^{(3)}(\lambda,\nu)\int_0^a \rho\frac{\partial}{\partial a}(a^6 h_3)da + \dots , \end{aligned} \tag{4.36}$$

$$\begin{aligned} 9V_4 &= P_4(\lambda)\int_0^a \rho\frac{\partial}{\partial a}(a^7 g_4)da + \mathcal{B}_4^{(2)}(\lambda,\nu)\int_0^a \rho\frac{\partial}{\partial a}(a^7 h_4)da + \\ &+ \mathcal{B}_4^{(4)}(\lambda,\nu)\int_0^a \rho\frac{\partial}{\partial a}(a^7 h_4)da + \dots , \end{aligned} \tag{4.37}$$

$$11V_5 = +\mathcal{B}_5^{(3)}(\lambda,\nu)\int_0^a \rho\frac{\partial}{\partial a}(a^8 h_3)da + \dots , \tag{4.38}$$

$$13V_6 = +\mathcal{B}_6^{(4)}(\lambda,\nu)\int_0^a \rho\frac{\partial}{\partial a}(a^9 h_4)da + \dots ; \tag{4.39}$$

where — in addition to the $\mathcal{B}_n^{(j)}$'s given by Equations (4.17)–(4.20) earlier in this section,

$$\mathcal{B}_2^{(0)} = P_2(\nu) \tag{4.40}$$

and

$$\mathcal{B}_1^{(0)} = -\frac{1}{5}P_1(\lambda) , \tag{4.41}$$

$$\mathcal{B}_3^{(1)} = \frac{1}{5}P_1(\lambda) + P_1(\lambda)P_2(\nu) . \tag{4.42}$$

If we factor out common harmonics in the expressions (4.26)–(4.39) for U_n and V_n, the latter can be rewritten more concisely in the form

$$U_n = \sum_{j,k} P_j(\nu)P_k(\lambda)\int_a^{a_1} \rho\frac{\partial}{\partial a}\{a^{2-n}\Phi_{j,k}^{(n)}\}da \tag{4.43}$$

and

$$V_n = \sum_{j,k} P_j(\nu)P_k(\lambda)\int_0^a \rho\frac{\partial}{\partial a}\{a^{n+3}\Phi_{j,k}^{(n)}\}da \tag{4.44}$$

for $j = 0$, 2 and $k = 0(1)4$, where

$$\Phi_{0,n}^{(n)} = G_n, \quad n = 0(1)4 ; \tag{4.45}$$

$$\Phi_{2,0}^{(2)} = H_0 , \tag{4.46}$$

$$\Phi_{2,1}^{(3)} = 5\Phi_{0,1}^{(3)} = H_1 , \tag{4.47}$$

$$\Phi_{2,2}^{(4)} = \frac{7}{2}\Phi_{0,2}^{(4)} = 10\Phi_{0,0}^{(4)} = H_2 , \tag{4.48}$$

$$\Phi_{2,3}^{(5)} = 3\Phi_{0,3}^{(5)} = \frac{42}{11}\Phi_{0,1}^{(5)} = H_3 , \tag{4.49}$$

$$\Phi_{2,4}^{(6)} = \frac{11}{4}\Phi_{0,4}^{(6)} = \frac{11}{10}\Phi_{2,2}^{(6)} = \frac{66}{25}\Phi_{0,2}^{(6)} = \frac{33}{7}\Phi_{2,0}^{(6)} = H_4 , \tag{4.50}$$

in which

$$G_0 = g_0 - \frac{1}{10}h_2 \tag{4.51}$$

$$G_1 = g_1 - \frac{1}{5}h_1 - \frac{9}{70}h_3 , \tag{4.52}$$

$$G_2 = g_2 - \frac{2}{7}h_2 - \frac{5}{42}h_4 , \tag{4.53}$$

$$G_3 = g_3 - \frac{1}{3}h_3 ;, \tag{4.54}$$

$$G_4 = g_4 - \frac{4}{11}h_4 ; \tag{4.55}$$

and

$$H_0 = h_0 - \frac{2}{7}h_2 + \frac{1}{21}h_4 , \tag{4.56}$$

$$H_1 = h_1 - \frac{2}{3}h_3 , \tag{4.57}$$

$$H_2 = h_2 - \frac{10}{11}h_4 , \tag{4.58}$$

$$H_3 = h_3 , \tag{4.59}$$

$$H_4 = h_4 . \tag{4.60}$$

With the potential expansions for U and V now explicitly formulated by Equations (4.23)–(4.24) and (4.43)–(4.44) in terms of amplitudes g_j and h_j of the 'mixed' tides, we are in a position to set up the constraints which these amplitudes are to obey if the total potential

$$\Psi(r', \lambda, \nu) = U(r', \lambda, \nu) + V(r', \lambda \nu) + V'(r', \lambda, \nu) \tag{4.61}$$

is to remain constant over a distorted surface of radius r'. As pointed out in section 1, this constancy will be ensured if a sum of the coefficients of all harmonics of equal orders on the r.h.s. of the foregoing Equation (4.61) will be made to add up to zero for any value of $j > 0$; thus rendering $\Psi(r', \lambda, \nu)$ independent of angular variables θ and ϕ. The conditions so obtained for $j > 0$ should then be sufficient completely to specify the new functions g_j and h_j introduced on the r.h.s. of the expansion (4.22) for r'; and our aim, in what follows, will be to establish their explicit form.

In doing so, we should observe that whereas the expected interaction terms in the first power of r' are stated on the r.h.s. of (4.22), in the binomial expansion for $r'^m (m > 0)$ the complete interaction terms will be given by

$$\begin{aligned} r'^m = a^m \Bigg\{ 1 + \ldots + m(m-1) \sum_{j=2}^{4} \hbar_2 f_2 P_2(\nu) P_j(\lambda) + \\ + m \sum_{j=0}^{4} g_j P_j(\lambda) + m \sum_{j=0}^{4} h_j P_2(\nu) P_j(\lambda) + \ldots \Bigg\} \end{aligned} \tag{4.62}$$

correctly to quantities of second order, where $\hbar_2$ and f_j represent the first-order effects of rotational and tidal distortion, respectively.

With the aid of the foregoing expansion, we are now in a position to proceed with the formulation of the necessary conditions on g_j and h_j to ensure the constancy of the total potential $\Psi(r', \lambda, \nu)$ over a level surface. In order to so, let us expand $\Psi(r', \lambda, \nu)$ in a series of the form

$$\Psi(r', \lambda, \nu) = \sum_{i,j} \alpha_{i,j}(a)\gamma_j^{(i)}(\lambda, \nu) , \tag{4.63}$$

and seek to obtain the explicit form of the equations

$$\alpha_{i,j}(a) = 0 , \tag{4.64}$$

analogous to (4.46), for all types of surface harmonics $Y_j^{(i)}(\theta, \phi)$ involved in r'.

Turning last to the disturbing potential $V'(r')$, we note that the structure of the one of tidal origin—as given by Equation (3.11)—is such that its terms varying as $r'^j P_j(\lambda)$ contain mixed terms of the form $jc_j a^j \hbar_2 P_2(\nu)P_j(\lambda)$ for $j =$ 2, 3 and 4 of second order. In addition, the centrifugal potential arising from axial rotation—as represented by Equation (1.39)—will contribute mixed terms of the form $\frac{2}{3}\omega^2 a^2 f_j P_j(\lambda)$ and $\frac{2}{3}\omega^2 a^2 f_j P_2(\nu)P_j(\lambda)$, likewise for $j = 2$, 3, and 4. Therefore, the total disturbing potential of the rotational as well as tidal origin will contain interaction terms of the form

$$\begin{aligned} V_t^{(2)}(r', \lambda, \nu) &= \frac{2}{3}\omega^2 a^2 \sum_{j=2}^{4} f_j P_j(\lambda) + \\ &+ \sum_{j=2}^{4} \left\{ jc_j a^j \hbar_2 - \frac{2}{3}\omega^2 a^2 f_j \right\} P_2(\nu)P_j(\lambda) ; \end{aligned} \tag{4.65}$$

and if so, the conditions (4.64) required for the constancy of the total potential $\Psi(r', \theta, \phi)$ over a level surface $r'(a, \theta, \phi)$ will assume the form

$$\begin{aligned} G_j \int_0^a \rho a^2 \, da &- \frac{1}{(2j+1)a^j} \int_0^a \rho \frac{\partial}{\partial a}(a^{j+3} G_j) da - \\ &- \frac{a^{j+1}}{2j+1} \int_a^{a_1} \rho \frac{\partial}{\partial a}(a^{2-j} G_j) da = \mathcal{G}_j + \frac{\omega^2 a^2}{6\pi G} f_j , \end{aligned} \tag{4.66}$$

$$\begin{aligned} H_j \int_0^a \rho a^2 \, da &- \frac{1}{(2j+1)a^{j+2}} \int_0^a \rho \frac{\partial}{\partial a}(a^{j+5} H_j) da - \\ &- \frac{a^{j+3}}{2j+5} \int_a^{a_1} \rho \frac{\partial}{\partial a}(a^{-j} H_j) da \\ &= \mathcal{H}_j + \frac{jc_j a^{j+1}}{4\pi G} \hbar_j - \frac{\omega^2 a^3}{6\pi G} f_j \end{aligned} \tag{4.67}$$

for $j = 0(1)4$, where

$$\begin{aligned}\mathcal{G}_j &= \frac{2(2j+1)}{j+2}\mathbf{g}_j \int_0^a \rho a^2\, da + \frac{1}{a^j}\int_0^a \rho\frac{\partial}{\partial a}(a^{j+3}\mathbf{g}_j)da - \\ &- \frac{j-1}{j+2}a^{j+1}\int_0^{a_1} \rho\frac{\partial}{\partial a}(a^{2-j}\mathbf{g}_j)da \end{aligned} \tag{4.68}$$

and

$$\begin{aligned}\mathcal{H}_j &= \frac{2(2+5)}{j+4}\mathbf{h}_j \int_o^a \rho a^2\, da + \frac{1}{a^{j+2}}\int_0^a \rho\frac{\partial}{\partial a}(a^{j+5}\mathbf{h}_j)da - \\ &- \frac{j+1}{j+4}a^{j+3}\int_a^{a_1} \rho\frac{\partial}{\partial a}(a^{-j}\mathbf{h}_j)da\,, \end{aligned} \tag{4.69}$$

where

$$\mathbf{g}_0 = -\frac{1}{5}\hbar_2 f_2\,, \tag{4.70}$$

$$\mathbf{g}_1 = -\frac{9}{70}\hbar_2 f_3\,, \tag{4.71}$$

$$\mathbf{g}_2 = -\frac{8}{35}\hbar_2 f_2 - \frac{2}{21}\hbar_2 f_4\,, \tag{4.72}$$

$$\mathbf{g}_3 = -\frac{5}{21}\hbar_2 f_3\,, \tag{4.73}$$

$$\mathbf{g}_4 = -\frac{8}{33}\hbar_2 f_4\,; \tag{4.74}$$

and

$$\mathbf{h}_0 = -\frac{8}{35}\hbar_2 f_2 + \frac{4}{105}\hbar_2 f_4\,, \tag{4.75}$$

$$\mathbf{h}_1 = -\frac{10}{21}\hbar_2 f_3\,, \tag{4.76}$$

$$\mathbf{h}_2 = \frac{2}{3}\hbar_2 f_2 - \frac{20}{33}\hbar f_4\,, \tag{4.77}$$

$$\mathbf{h}_3 = \frac{7}{11}\hbar_2 f_3\,, \tag{4.78}$$

$$\mathbf{h}_4 = \frac{8}{13}\hbar_2 f_4\,. \tag{4.79}$$

It may also be noted that the tidal amplitudes f_j on the r.h.s. of Equations (4.66) and (4.67) for $j = 0$ and 1 are (unlike those for $j = 2$, 3 and 4) quantities of second order—a fact which justifies the neglect of their products with ω^2 in our present approximation. The same is, moreover, true of c_1 as given by Equation (3.13)—which justifies again a disregard, on the r.h.s. of Equations (4.66) and (4.67), of its products with $\hbar_2$.

The foregoing Equations (4.66) and (4.67) will play the same role in the study of the effects produced by interaction between rotation and tides in close binary systems as do Equations (3.49) for purely tidal distortion; and Equations (4.68) with (4.69) are then analogous to (3.50). Their subsequent treatment can, consequently, follow similar lines. In order to follow them, multiply both sides of Equations (4.66) by a^j, and differentiate with respect to a; the result will be the relation

$$\frac{\partial}{\partial a}(a^j G_j)\int_0^a \rho a^2\,da \;-\; a^{2j}\int_a^{a_1} \rho\frac{\partial}{\partial a}(a^{2-j}G_j)da = \\ = \frac{\partial}{\partial a}(a^j \mathcal{G}_j) + \frac{\omega^2}{6\pi G}\frac{\partial}{\partial a}(a^{j+3}f_j)\,; \qquad (4.80)$$

while if (4.67) is multiplied by a^{j+2} and differentiated with respect to a, we find that

$$\frac{\partial}{\partial a}(a^{j+2}H_h)\int_0^a \rho a^2\,da \;-\; a^{2(j+2)}\int_a^{a_1} \rho\frac{\partial}{\partial a}(a^{-j}H_j)da = \\ = \frac{\partial}{\partial a}(a^{j+2}\mathcal{H}_j) + \frac{jc_j}{4\pi G}\frac{\partial}{\partial a}(a^{2j+3}\hbar_2) - \\ - \frac{\omega^2}{6\pi G}\frac{\partial}{\partial a}(a^{j+5}f_j)\,. \qquad (4.81)$$

Next, divide the foregoing Equations (4.80) and (4.81) by a^{2j} and $a^{2(j+2)}$, respectively; differentiate once more with respect to a; and divide subsequently by $a^{-(j+2)}F_0$ and $a^{-(j+4)}F_0$. The outcome of these operations will disclose that

$$a^2G_j'' + 6D(aG_j + G_j) - j(j+1)G_j = \mathcal{A}_j + \frac{3\omega^2}{\pi G\overline{q}}(1-D)(f_j + af_j') \qquad (4.82)$$

and

$$a^2H_j'' \;+\; 6D(aH_j' + H_j) - (j+2)(j+3)H_j = \\ = \;\mathcal{B}_j \;+\frac{jc_ja^{j+1}}{Gm(a)}\{2(j+1)a\hbar_2 + (j^2-3j-6)\hbar_2 - 6D(a\hbar_2' + \hbar_2)\}\,(4.83)$$

by use of Equation (3.47) which is bound to be satisfied by both f_j and $\hbar_2$; where primes continue to denote differentiation with respect to a; $m(a)$ represents the mass of the distorted configuration interior to a; and $D \equiv \rho/\overline{\rho}$ continues to be given by Equation (3.52). Moreover, the quantities

$$\mathcal{A}_j\int_0^a \rho a^2\,da = a^2\mathcal{G}_j'' - j(j+1)\mathcal{G}_j \qquad (4.84)$$

and

$$\mathcal{B}_j\int_0^a \rho a^2\,da = a^2\mathcal{H}_j'' - (j+2)(j+3)\mathcal{H}_j \qquad (4.85)$$

occurring on the r.h.s. of Equations (4.66) and (4.67) can be obtained by repeated differentiation of the expressions (4.68) and (4.69) in the form

$$\mathcal{A}_j \;=\; \frac{2(2j+1)}{j+2}\{a^2\mathrm{g}_j'' - j(j+1)\mathrm{g}_j\} +$$

$$+\frac{3(2j+1)}{j+2}\{a^2\mathrm{g}_j''+7a\mathrm{g}_j'-(j^2+j+8)\mathrm{g}_j\}\,D+$$
$$+\frac{3(2j+1)}{j+2}\{a\mathrm{g}_j'+6\mathrm{g}_j\}(aD'+eD^2)=\mathcal{B}_{j-2}(\mathrm{h}_{j-2})\,, \qquad (4.86)$$

in which advantage has been taken of the relation

$$a^2\frac{\rho'}{\overline{\rho}}=aD'+3D(D-1) \qquad (4.87)$$

obtained by differentiation of (3.52).

The functions $\mathcal{G}_j$ and $\mathcal{H}_j$ as given by Equations (4.70)–(4.79) are linear combinations of the products $\hbar_2 f_j$. Successive differentiation of the latter discloses, moreover, that

$$a\frac{\partial}{\partial a}(\hbar_2 f_j)=(\zeta_2+\eta_j)\hbar_2 f_j \qquad (4.88)$$

and

$$a^2\frac{\partial^2}{\partial a^2}(\hbar_2 f_j)=\{2\zeta_2\eta_j-6D(\zeta_2+\eta_2+2)+j(j+1)+6\}\hbar_2 f_j \qquad (4.89)$$

in terms of the logarithmic derivatives

$$\zeta_2\equiv a\,\hbar_2'/\hbar_2 \quad\text{and}\quad \eta_j\equiv af_j'/f_j \qquad (4.90)$$

if advantage is again taken of the fact that, correctly to quantities of first order, ζ_2 as well as η_j satisfy the differential Equation (3.61).

With the functions $\mathcal{A}_j$ and $\mathcal{B}_j$ thus explicitly formulated, Equations (4.82) and (4.83) represent second-order nonhomogeneous linear differential equations for G_j and H_j—analogous to Equation (3.51) for the purely tidal problem. In comparing the latter with (4.82)–(4.83) we may note that, whereas its nonhomogeneous terms $T_j(a)$ as given by Equations (3.53)–(3.59) depend on the internal structure of the respective configuration only through the first powers of the ratio $D\equiv\rho/\overline{\rho}$, those on the right-hand sides of Equations (4.82) and (4.83) governing the interaction between rotation and tides involve terms factored not only by D, but also D^2 and aD.

Like in the case of purely tidal distortion discussed in Section IV.3, the particular solutions of Equations (4.82) and (4.83) representing the amplitudes of 'mixed' tides are subject to specific boundary conditions imposed at both ends of the interval of integration. At the origin ($a=0$), both $G_j(a)$ as well as $H_j(a)$ should be minimum—the necessary conditions of which require that

$$G'(0)=H'(0)=0 \text{ for } j=0(1)4\,. \qquad (4.91)$$

At the other end—on the surface $a=a_1$ defined by $\rho(a_1)=0$—Equations (4.80) and (4.81) reduce to

$$\left\{\frac{\partial}{\partial a}(a^jG_j)\right\}_{a_1}=\frac{4\pi}{m_1}\left\{\frac{\partial}{\partial a}\left(a^j\mathcal{G}_j+\frac{\omega^2 f_j}{6\pi G}a^{j+3}\right)\right\}_{a_1} \qquad (4.92)$$

and

$$\left\{ \frac{\partial}{\partial a}(a^{j+2}H_j) \right\}_{a_1} = \frac{4\pi}{m_1} \left\{ \frac{\partial}{\partial a} \left(a^{j+2}\mathcal{H}_j - \frac{\omega^2 f_j}{6\pi G} a^{j+5} + \frac{c_j \hbar_2}{4\pi G} a^{2j+3} \right) \right\}_{a_1} \tag{4.93}$$

where m_1 denotes the total mass of the distorted configuration, and where

$$\begin{aligned} \left\{ \frac{\partial}{\partial a}[a\mathcal{G}_j(\mathbf{g}_j)] \right\}_{a_1} &= \frac{(2j+1)m_1}{2(j+2)\pi} \{a\mathbf{g}'_j + \mathbf{g}_j\}_{a_1} - \\ - \frac{j-1}{a_1^j} \int_0^{a_1} \rho \frac{\partial}{\partial a}(a^{j+3}\mathbf{g}_j) da &= \left\{ \frac{\partial}{\partial a}[a\mathbf{h}_{j-2}(\mathbf{h}_{j-2})] \right\}_{a_1} . \end{aligned} \tag{4.94}$$

It may be noted that whereas the conditions (4.91) at $a = 0$ are homogeneous in the dependent variables, those given by Equations (4.92) and (4.93) representing linear relations between G_j and G'_j or H_j and H'_j at $a = a_1$ are nonhomogeneous and, as such, impose specific amplitudes on the respective tides.

Moreover, once the particular solutions of (4.82) and (4.83) subject to the boundary conditions (4.91) and (4.92)–(4.94) have been constructed (by numerical or other methods), the amplitudes g_j and h_j of the 'mixed' equilibrium terms arising from interaction between rotation and tides will—by an inversion of Equations (4.51)–(4.60) be given by the relations

$$g_0 = G_0 + \frac{1}{10}H_2 + \frac{1}{11}H_4 , \tag{4.95}$$

$$g_1 = G_1 + \frac{1}{5}H_1 + \frac{11}{42}H_3 , \tag{4.96}$$

$$g_2 = G_2 + \frac{2}{7}H_2 + \frac{25}{66}H_4 , \tag{4.97}$$

$$g_3 = G_3 + \frac{1}{3}H_3 , \tag{4.98}$$

$$g_4 = G_4 + \frac{4}{11}H_4 ; \tag{4.99}$$

and

$$h_0 = H_0 + \frac{2}{7}H_2 + \frac{7}{33}H_4 , \tag{4.100}$$

$$h_1 = H_1 + \frac{2}{3}H_3 , \tag{4.101}$$

$$h_2 = H_2 + \frac{10}{11}H_4 , \tag{4.102}$$

$$h_3 = H_3 , \tag{4.103}$$

$$h_4 = H_4 . \tag{4.104}$$

In conclusion of the present section on the interaction effects between rotation and tides, it should be noted that the total mass m_1 of our distorted configuration, defined by Equation (1.26) for $n = 0$ as

$$m(a_1) = \frac{1}{3}\int_0^{a_1} \rho \frac{\partial}{\partial a}\left\{\int_0^{\pi}\int_0^{2\pi} (r')^3 \sin\theta'\, d\theta'\, d\phi'\right\} da \tag{4.105}$$

requires for its constancy (i.e., independence of distortion) that the coefficients of zero-order harmonics should, within the scheme of our approximation, be related with those of higher harmonics by the equation

$$\begin{aligned} f_0 + \hbar_2 + g_0 &= -\frac{1}{5}f_2^2 - \frac{1}{7}f_2^3 - \dots \\ &\quad - \frac{1}{5}\hbar_2^2 + \frac{1}{5}\hbar_2 f_2 + \dots + \frac{1}{10}h_2 + \dots, \end{aligned} \tag{4.106}$$

where the terms of purely tidal origin on the r.h.s. have already been established by Equation (3.33); while the 'mixed' terms go back to that factored by $\mathcal{B}_0^{(2)}$ on the r.h.s. of Equation (4.33).

IV.5 Effects of Internal Structure

In the preceding Sections 2-4 of this chapter we established that the form of the rotationally and (or) tidally distorted self-gravitating fluids can be described by expansions of the form (2.1), (3.16) or (4.22), in which the amplitudes $f_j(a)$ or $g_j(a)$ and $h_j(a)$ of the respective harmonic are defined as particular solutions of Equations (2.33)–(2.35), (3.51)–(3.61) or (4.82)–(4.87) subject to appropriate boundary conditions. Correctly to quantities of first order in surficial distortion, all amplitudes f_j have been found to satisfy the linear second-order differential equation

$$a^2\frac{\partial^2 f_j}{\partial a^2} + 6\frac{\rho}{\overline{\rho}}\left(a\frac{\partial f_j}{\partial a} + f_j\right) - j(j+1)f_j = 0\,, \tag{5.1}$$

where

$$\overline{\rho} = \frac{3}{a^3}\int_0^a \rho a^2\, da \tag{5.2}$$

denotes the mean density of the respective configuration interior to the radius a. *It is through this equation that the equilibrium form (or potential) of a distorted body depends on its internal structure.*

Moreover, the form of the boundary conditions at $a = a_1$ discloses *that the surface values of $f_j(a_1)$ depend on the internal structure of the distorted configuration only through the logarithmic derivative*

$$\frac{a}{f_j}\frac{df_j}{da} \equiv \eta_j(a)\,, \tag{5.3}$$

introduced already through Equations (2.36), (3.60) or (4.90) to simplify our previous results. If we insert (5.3) in (5.1) we find—in accordance with Equations (3.52) and (3.61)—that Equation (5.1) transforms into

$$a\frac{d\eta_j}{da} + 6D(\eta_j + 1) + \eta_j(\eta_j - 1) = j(j+1)\,, \tag{5.4}$$

which is of first order, but second degree, in the dependent variable a.

In the contemporary literature Equation (5.1) is usually referred to as *Clairaut's equation*—in honour of the distinguished 18th century French mathematician who first deduced its form for the case of $j = 2$; while Equation (5.4) often carries the name of ***Radau's equation***. As is frequently the case in the annals of science, such summary labels are to be regarded as matters of convenience rather than historical truth. For Equation (5.1) with arbitrary j did not appear till in the works of Laplace some 80 years after Clairaut's time; while Radau's equation (5.4) was already known to Clairaut for $j = 2$. As it came, however, to play—very much later—a central part in Radau's investigations (Radau, 1885a,b) in the second half of the 19th century, it seems appropriate to associate (5.4) henceforth with his name—if alone to distinguish it by name from its second-order form (5.1).

The initial condition which a solution of Radau's equation (5.4) must satisfy at the origin—and which by itself will specify uniquely the requisite particular solution for any $\rho(r)$—is easy to establish. For finite values of $\rho(0)$ it follows that $\rho/\overline{\rho} \equiv D = 1$ as $a \to 0$; and if so, the structure of Equation (5.4) makes it evident that

$$\eta_j(0) = j - 2 \, . \tag{5.5}$$

Moreover, if our configuration were homogeneous ($\rho =$ constant) and, accordingly, $D = 1$ for $0 \leq a \leq a_1$, we find that $\eta_j(a)$ remains constant and equal to

$$\eta_j(a) = j - 2 \tag{5.6}$$

throughout the interior.

If, on the other hand, the entire mass of our configuration were condensed at its centre—i.e., if $D = \infty$ for $a = 0$ and zero for $a > 0$—the solution of (5.4) becomes

$$\eta_j(a) = j + 1, \quad a > 0 \, . \tag{5.7}$$

If, conversely, the whole mass of our configuration were confined to an infinitesimally thin shell—so that $D = \infty$ for $0 \leq a \leq a_1$ and ∞ at a_1 Equation (5.4) yields

$$\eta_j(a) = -1 \, . \tag{5.8}$$

Therefore—if we presume no knowledge of the density distribution in the interior of our configuration—the absolute limits of the solution of (5.4) at $a = a_1$ are

$$-1 \leq \eta_j(a_1) \leq j + 1 \, ; \tag{5.9}$$

while if the internal density $\rho(a)$ is supposed nowhere to increase outwards (i.e., for $\rho'(a) \leq 0$), the foregoing inequality (5.9) becomes restricted to

$$j - 2 \leq \eta_j(a_1) \leq j + 1 \, . \tag{5.10}$$

It may also be noted that, whatever the structure,

$$\eta_j(a) < \eta_{j+1}(a) \, . \tag{5.11}$$

The existence of the inequality (5.10) was first pointed out (for $j = 2$) by Clairaut (1743); and that of (5.9), by Kopal (1941). The latter discloses that, for any given external field of force, *the surficial distortion of a configuration whose mass is confined to an infinitesimally thin surface shell would—to the first order in small quantities—be twice as large as if the configuration were homogeneous, and five times as large as that appropriate for a mass-point model.*

It may be added that, for a continuous distribution of density in the interior, the inequality (5.10) may be refined further in the following manner. In order to do so, let us return to Equation (3.47) which, on integration by parts of its left-hand side, assumes the form

$$a^j f_j\{[j+1-\eta_j(a)]\int_0^a \rho a^2\, da - \rho a^3\} = -\int_0^a \frac{d\rho}{da} f_j\, a^{j+3}\, da\,. \tag{5.12}$$

Since, however, it follows from (5.2) and (3.52) that

$$\int_0^a \rho a^2\, da = \frac{1}{3}a^3\overline{\rho} \text{ and } \rho = D\overline{\rho}\,, \tag{5.13}$$

on insertion of these expressions in the left-hand side of (5.12) this latter equation can be rewritten as

$$a^{j+3} f_j\{j+1-3D-\eta_j(a)\}\overline{\rho} = 3\int_0^a \left(-\frac{d\rho}{da}\right) f_j\, a^{j+3}\, da\,. \tag{5.14}$$

Now let us assume next that $\rho(a)$ is a diminishing function of a. If so, however, both sides of (5.15) must be non-negative[5]; and this can obviously be true only if

$$\eta_j(a) \leq j+1-3D(a)\,, \tag{5.15}$$

where the quantity $D(a)$ introduced by Equation (3.52) is, in turn, constrained to obey the inequality

$$0 \leq D(a) \leq 1\,. \tag{5.16}$$

The left-hand side of (5.15) is necessarily true if $d\overline{\rho}/da$ is negative; the right-hand side, because ρ is positive. Therefore, any variation of $\eta_j(a)$ throughout the interior must be comprised within the range

$$j-2 \leq \eta_j(a) \leq j+1-3D(a)\,, \tag{5.17}$$

which constitutes a refinement of our previous inequality (5.10).

How does the function $\eta_j(a)$ behave within this permitted interval? In considering a manifold of the solutions of Radau's equation (5.4) in the $(a-\eta)$-plane,

[5] It may be noted that the positivity of both sides of (5.14)—and, therefore the validity of (5.15)—does not necessarily require $d\rho/da$ to remain negative throughout the interior. This latter condition is sufficient, but not necessary. The validity of (5.15) would not, for instance, be impaired by an occurrence of positive density gradients in the interior—provided that these are not sufficient to alter the sign of the integral on the r.h.s. of (5.14).

we may note the Equation (5.4) defines the first derivative η'_j in terms of a and η_j uniquely everywhere except at the origin. Along the axis $a = 0$ the function $\eta'_j(a)$ ceases to be holomorphic, and (5.4) can be satisfied by η_j assuming the alternative values of $j-2$ or $-j-3$. At the points $(0, j-2)$ and $(0, -j-3)$ the integral curves obeying Equation (5.4) are thus not uniquely defined. It can, however, be shown (cf. Poincaré, 1903) that, of the total manifold of such curves passing through an arbitrary point of the $(a-\eta)$-plane, only *one* will pass through $(0, j-2)$; all others passing through $(0, -j-3)$. Therefore, the initial condition (5.5) is sufficient to ensure the uniqueness of the particular solution of the (nonlinear) Equation (5.4) which is of interest to us in this connection.

Let us inquire next as to the behaviour of such solutions in the $(\eta-\eta')$-plane. For the limiting model of a homogeneous configuration ($D = 1$) Equation (5.4) reduces to the parabola

$$a\eta' = j(j+1) - (\eta+2)(\eta+3)\,; \tag{5.18}$$

while for a centrally-condensed model ($D = 0$ for $a > 0$), Radau's equation reduces likewise to

$$a\eta' = j(j+1) - \eta(\eta-1)\,. \tag{5.19}$$

On the other hand, for values of $D(a)$ for which (5.15) becomes an equality,

$$a\eta' = (j+1)(j-2) - (2j-1)\eta + \eta^2\,. \tag{5.20}$$

Any point of the solution of Equation (5.4) for a configuration intermediate in structure between a homogeneous and a mass-point model must, in the $(\eta-\eta')$-plane, lie in a region between the foregoing parabolae: in fact, for such models as render the integral on the r.h.s. of Equation (5.14) positive, the permitted region is comprised between the parabolae (5.18) and (5.19) intersecting at the points $j-2$ and $j+1$. These curves have been plotted on the accompanying Figure IV.3; and delimit an area in which $\eta'_j(a)$ can be both positive and negative.

In order to investigate further a possible behaviour of $\eta_j(a)$ in this area, let us differentiate Equation (5.4) with respect to a, to find that

$$a\eta'' + \frac{d}{da}\{\eta(\eta+6) - 6(1-D)(1+\eta)\} = 0\,. \tag{5.21}$$

Accordingly, $\eta''(a) = 0$ for $a \neq 0$ only if

$$\eta(\eta+6) - 6(1-D)(1+\eta) = \text{constant}\,; \tag{5.22}$$

and the value of this constant can be obtained from the initial conditions $D(0) = 1$ and $\eta(0) = j = 2$ to be equal to $(j-2)(j+4)$. If so, however, an insertion of (5.22) in (5.4) discloses that

$$a\eta' = \eta - j + 2\,. \tag{5.23}$$

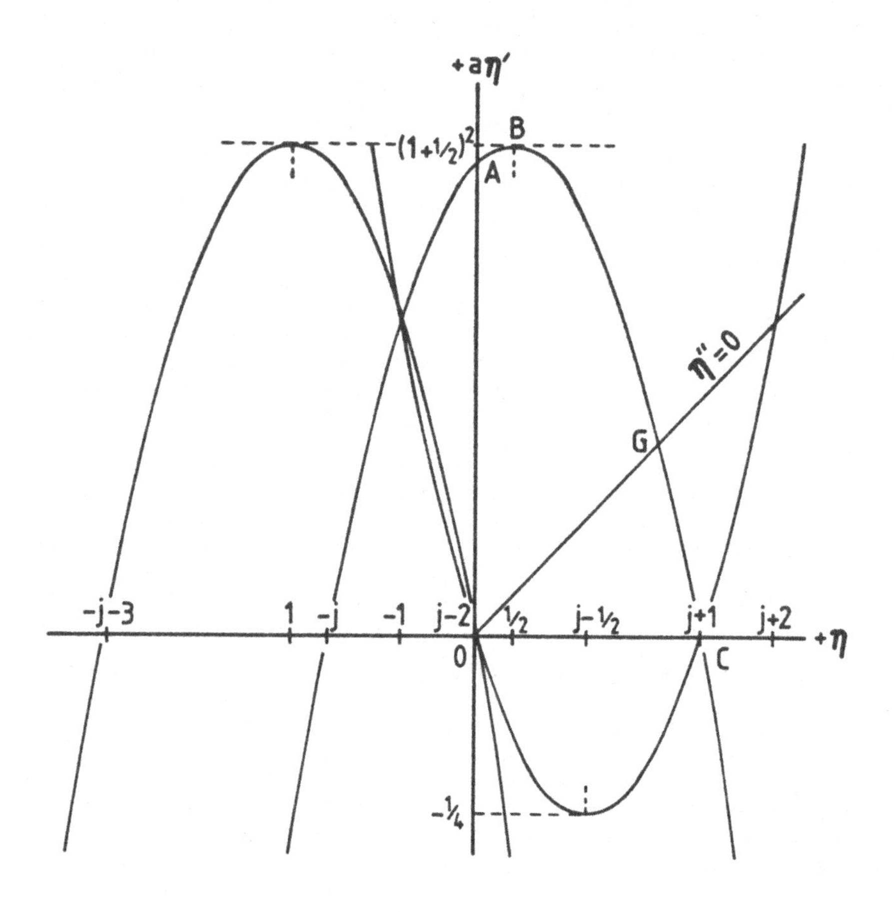

Figure IV.3:

The only solution of (5.23) consistent with (5.4) is η = const and D = const, requiring $\eta' = D' = 0$. Therefore, the locus $\eta'' = \eta' = 0$ of inflection points will be represented on Figure IV.3 by a straight line passing through the point $\eta = j - 2$, and inclined by 45° to the locus $\eta' = 0$ of the extrema which coincides with the η-axis.

Apart from the limiting cases (5.6) or (5.7) representing the solutions of (5.4) for the homogeneous or mass-point models, Clairaut's or Radau's equations admit also of other closed solutions for certain less extreme internal density distributions—adopted, however, for the sake of mathematical tractability rather than their physical reasonableness; and for an account of them the reader is referred, e.g., to sections II-2 of Kopal (1959, 1960), or to Lanzano (1973). Some—particularly those corresponding to weak internal density concentration—are of geophysical (or planetological) rather than astrophysical interest, and an account of their more detailed properties need not detain us in this place. The density profiles of the stars (with the exception of white dwarfs) are characterized by a generally high degree of central condensation; and a construction of the corresponding particular solutions of Radau's equation (5.4) subject to the initial condition (5.5) can, in general, proceed only by numerical integration. Extensive integrations carried out for the polytropic family of models by Brooker and Olle (1955) are now of only mainly historical significance—except, perhaps, for the case of polytropic index $n = 1.5$ which should approximate closely the structure of (non-relativistic) white dwarfs. The most comprehensive integrations for a wide range of actual stellar models corresponding to different evolutionary stages we owe to Petty (1973); and their outcome bears out indeed the general properties of the η-functions discussed earlier in this section.

For relatively *high degree* of *central condensation* of the distorted configuration —i.e., if the ratio $\rho/\overline{\rho} \equiv D(a) << 1$ throughout most part of the range $0 \leq a \leq a_1$—approximate solutions of Clairaut's equation (5.1) can be constructed in the following manner. Let $y(a)$ define a new function related with the amplitudes f_j by the equation

$$y(a) = \frac{1}{3}a^3\overline{\rho}\, f_j(a)\,, \tag{5.24}$$

where $\overline{\rho}$ continues to be defined by Equation (5.2), so that

$$\eta_j(a) = \frac{a}{y}\frac{dy}{da} - 3\frac{\rho}{\overline{\rho}}\,. \tag{5.25}$$

Let, moreover, the surface value of

$$\eta_j(a_1) = \lambda a_1\,, \tag{5.26}$$

where λ represents the desired value of a parameter specifying the effects of internal structure on the external shape of the respective configuration. Its determination is obviously tantamount (cf. Kopal, 1953) to a search for the characteristic parameters λ of a Sturm-Liouville problem, consisting of the equation

$$L[y] = \mathbf{r}(a)y\,, \tag{5.27}$$

subject to the homogeneous boundary conditions

$$y(0) = 0 \text{ and } y'(a_1) = \lambda\, y(a_1)\,, \tag{5.28}$$

where

$$\mathbf{r}(a) \equiv \frac{3}{a\overline{\rho}}\frac{d\rho}{da} = \frac{3}{a^2\overline{\rho}}\frac{d}{da}\left(a^2\frac{d\overline{\rho}}{da}\right) \tag{5.29}$$

and L stands for the operator

$$L \equiv \frac{d^2}{da^2} - \frac{j(j+1)}{a^2}\,. \tag{5.30}$$

The operator L is evidently self-adjoint. Its Green's function appropriate for the given boundary conditions will, therefore, be symmetrical and of the form

$$G(a,\alpha) = -\frac{a^{j+1}}{2j+1}\{\alpha^{-j} + C_j\alpha^{j+1}\}\ ,\ \alpha \geq a\ , \tag{5.31}$$

in which we have abbreviated

$$a_1^{2j+1}C_j = \frac{j+\lambda a_1}{j-\lambda a_1+1}\ . \tag{5.32}$$

Our foregoing boundary-value problem can, accordingly, be rewritten in the form of the single integral equation

$$\begin{aligned}(2j+1)\,y(a) \;=\;& -a^{j+1}C_j\int_0^{a_1} \mathrm{r}(\alpha)y(\alpha)\alpha^{j+1}\,d\alpha - \\ & - a^{-j}\int_0^{a} \mathrm{r}(\alpha)y(\alpha)\alpha^{j+1}\,d\alpha - a^{j+1}\int_a^{a_1} \mathrm{r}(\alpha)y(\alpha)\alpha^{-j}\,d\alpha. \end{aligned}\tag{5.33}$$

Now in the preceding sections of this chapter we found that, for homogeneous configurations $(\rho/\overline{\rho} = D = 1), \eta(a) = j-2$; while for a mass-point model ($D = 0$ for $a > 0$), $\eta(a) = j+1$. In the former case, therefore, $\lambda a_1 = j-2$ and, in the latter, $\lambda a_1 = j+1$ rendering the constant C_j as defined by the Equation (5.33) infinite. In actual stars this limit can never be attained; for the central condensation of no existing configuration can be infinite. If it, however, happens to be high (such as is likely to be in many stars), the constants C_j may become very large. The second and third terms on the right-hand side of our integral Equation (5.33) will then be small in comparison with the first; for this latter alone is magnified by multiplication with C_j. To the order of accuracy to which their disparity is such as to render the second and third terms ignorable, Equation (5.33) should become essentially equivalent to

$$(2j+1)y(a) = -a^{j+1}C_j\int_0^{a_1} \mathrm{r}(\alpha)y(\alpha)\alpha^{j+1}d\alpha\ ; \tag{5.34}$$

and since the limits of the integral on the right-hand side now are constant, it implies that

$$y(a) = Aa^{j+1}\ , \tag{5.35}$$

where A is a constant — or, by (3.38)–(3.44)

$$f_j(a) = \frac{c_j a^{j+1}}{4\pi G F_0}\ , \tag{5.36}$$

where $4\pi F_0 \equiv m(a)$ continues to be given by Equations (2.17) or (3.45), and the constants c_j are given by Equations (1.42) or (3.12)–(3.15), depending on whether the distortion is caused by the rotation or tides.

The reader may, furthermore, note from Equations (2.27)–(2.29) or (3.38)–(3.44) that—to the first order in small quantities—we have

$$F_0 f_j(a) = \frac{c_j a^{j+1}}{4\pi G} + \frac{a^{j+1}E_j + a^{-j}F_j}{2j+1}, \tag{5.37}$$

where $j = 2, 3, 4$, for configurations of *any* structure. If, therefore, in accordance with Equation (5.35), the first term on the r.h.s. of (5.36) represents a good approximation to the actual solution for a pronounced degree of central condensation, it follows that the second term on the r.h.s. of (5.37) should generally be small. In such a case, however, the idea suggests itself to construct a solution of Equation (5.37) by an *iterative process*, which can generate successive approximations to the solution of our problem in a systematic manner.

Let, in what follows, Y_n stand for the n-th approximation to the desired function $Y \equiv f_j(a)$. In order to generate such approximations, let us replace Equation (5.37) symbolically by a *quadrature* formula of the form

$$Y_n = \frac{c_j a^{j+1}}{Gm(a)} + J[Y_{n-1}], \tag{5.38}$$

where (for sufficiently high central condensation) the second term on the r.h.s. should be small in comparison with the first to justify the use in it of an approximation of lower order.

Let, moreover, the absolute upper bound of Y be denoted by M. If so, then

$$|J[Y]| < MJ[1]; \tag{5.39}$$

and as (by the Mean Value Theorem)

$$J[Y] < \frac{3}{2j+1}, \tag{5.40}$$

it follows that

$$|J[Y]| < \frac{3M}{2j+1}. \tag{5.41}$$

Now let

$$|J[Y_n - Y_{n-1}]| < \frac{3M_n}{2j+1}, \tag{5.42}$$

where M_n stands for the upper bound of the individual Y_n's in the interval $(0, a_1)$. If so, however, then Equation (5.38) yields

$$|Y_{n+1} - Y_n| < \frac{3M_n}{2j+1} < \left(\frac{3}{2j+1}\right)^n M_1 \tag{5.43}$$

and, therefore, the sequence

$$x = Y_0 + (Y_1 - Y_0) + (Y_2 - Y_1) + \dots \tag{5.44}$$

converges absolutely and uniformly throughout the interval $(0, a_1)$ to a definite limit x.

Does this limit actually satisfy Clairaut's equation? In order to demonstrate that this is indeed the case, let us rewrite Equation (5.38) as

$$x - J[x] - \frac{ca^{j+1}}{\int_0^a \rho a^2 da} = x - Y_{n+1} - J[x - Y_n] \,. \tag{5.45}$$

Now if ϵ_n denotes the absolute upper bound of the differences $x - Y_n$ in the interval $(0, a_1)$, Equation (5.41) reveals that

$$|J[x - Y_n]| < \frac{3\epsilon_n}{2j+1} \,, \tag{5.46}$$

so that

$$x - J[x] - \frac{c_j a^{j+1}}{Gm(a)} < \epsilon_{n+1} + \frac{3\epsilon_n}{2j+1} \,, \tag{5.47}$$

which can be true only if the left-hand side of this inequality vanishes—i.e., if $x \equiv Y$.

The sequence of the Y_n's in terms of which we have defined x depends on the form of the adopted starting function Y_0. However, the limit x is clearly independent of it; for if an arbitrary function X_0 is used to generate another sequence of the X_n's through the mill of Equation (5.38), and if—analogously with Equation (5.43),

$$|Y_n - X_n| < \left\{ \frac{3}{2j+1} \right\}^n N \,, \tag{5.48}$$

where N represents the absolute upper bound of $J[X]$, it follows that (as $3/(2j+1)$ is less than one for $j > 1$) the difference $Y_n - X_n$ can be diminished arbitrarily by choosing a sufficiently large value of n—no matter how large N may be.

In order to demonstrate this procedure in a more concrete case, let us return to Equation (5.27) for y, and replace the latter by the logarithmic derivative

$$\mu z = \frac{1}{y}\frac{dy}{da} \,, \tag{5.49}$$

where

$$\mu^2 = j(j+1) \,. \tag{5.50}$$

If so, (5.27) transforms into a non-homogeneous Riccati equation

$$\frac{1}{\mu}\frac{dz}{da} + z^2 = \frac{1}{a^2} + \frac{\mathrm{r}(a)}{\mu^2} \tag{5.51}$$

of first order but second degree.

Since the constant parameter $\mu >> 1$, the possibility suggests itself (cf. Kopal and Lanzano, 1973) to seek the solution of (5.51) for $a > 0$ in the form of an expansion

$$\mathfrak{z} = \sum_{j=0}^{\infty} \mathcal{L}_j(a)\mu^{-j} \tag{5.52}$$

in inverse powers of μ. If we insert this expansion in (5.51) and set the coefficients of successive powers of μ equal to zero, we find that

$$\mathcal{L}_0 = a^{-1} , \tag{5.53}$$

$$\mathcal{L}_0' + 2\mathcal{L}_0\mathcal{L}_1 = 0 , \tag{5.54}$$

$$\mathcal{L}_1' + 2\mathcal{L}_0\mathcal{L}_2 + \mathcal{L}_1^2 = \mathbf{r}(a) ; \tag{5.55}$$

while, for $j > 2$,

$$\mathcal{L}_{j-1}' + 2\mathcal{L}_0\mathcal{L}_j + \sum_{k=1}^{j-1} \mathcal{L}_k\mathcal{L}_{j-k} = 0 . \tag{5.56}$$

The foregoing equations constitute a set of recursion relations by which individual terms $\mathcal{L}_j$ on the r.h.s. of Equation (5.52) can be successively generated. The only condition for the validity of such a procedure is the requirement that the function $\mathbf{r}(a)$ be analytic within $0 < a \leq a_1$. Moreover, once this has been done, the function $\eta_j(a)$ can by a combination of Equations (5.24), (5.25) and (5.49) be expressed as

$$\eta_j(a) = a\mu\mathcal{L} - 3\frac{\rho}{\overline{\rho}} = \sum_{j=0}^{\infty} \frac{a\mathcal{L}_j(a)}{\mu^{j-1}} - 3\frac{\rho}{\overline{\rho}} \tag{5.57}$$

reducing, on the surface $(a = a_1)$, to

$$\begin{aligned} \eta_j(a_1) &= a_1\mu \sum_{j=0}^{\infty} \mathcal{L}_j(a_1)\mu^{-j} = \\ &= \mu + \frac{1}{2} + \frac{1}{8\mu} + \frac{3}{2\mu}\left(\frac{a\rho'}{\overline{\rho}}\right)_{a_1} + a_1\mu\sum_{j=3}^{\infty} \mathcal{L}_j(a)\mu^{-j} \end{aligned} \tag{5.58}$$

by (5.53)–(5.55) while, for $j > 3$, the $\mathcal{L}_j$'s can be generated by use of (5.56).

For a centrally condensed model $\rho'(a_1) = 0$. The reader is invited to verify that, in such a case, the first three terms on the r.h.s. of (5.58), approximating the exact expression $\eta_j(a_1) = j + 1$, err by excess of only 52, 19 and 9 units of the fifth decimal place—a quality of approximation which is unlikely to change much for models of finite density concentration as long as the latter remains high. The structure of the expansion on the r.h.s. of Equation (5.58) demonstrates convincingly the extent to which *the values of $\eta_j(a_1)$ depend primarily on the surface density gradient $\rho'(a_1)$ of the respective configuration.*

Before concluding the present chapter, we wish still to return to one task left unfinished in Section IV.3: namely, to specify the constants k_j in the expansion (3.2) for the exterior potential $V(r)$ in terms of the internal structure of the respective configuration. In order to do so, let us note that this potential is of the form (3.28), where, by (3.48),

$$F_j(a_1) = a_1^j f_j(a_1)\{j + 1 - \eta_j(a_1)\} f_0(a_1) . \tag{5.59}$$

Since, however, $F_0(a_1) \equiv m_1/4\pi$ and, by (3.63),

$$f_j(a_1) = \frac{(2j+1)a_1^{j+1}c_j}{G(j+\eta_j)m_1} \tag{5.60}$$

to the first order in small quantities, it follows that

$$F_j(a_1) = \frac{(2j+1)(j+1-\eta_j)c_j}{4\pi G(j+\eta_j)} a_1^{2j+1} . \tag{5.61}$$

If, lastly, we insert for c_j from (3.15), we find (3.3) to be indeed an expansion of the form (3.28) if we set

$$k_j = \frac{1+j-\eta_j(a_1)}{2[j+\eta_j(a_1)]} . \tag{5.62}$$

The foregoing coefficients define the constants k_j in the expansion of the exterior potential of a distorted configuration in terms of its internal structure; and as such they are going to play an important role in Chapter VI concerned with the perturbations of the orbital elements in close binary systems. For the present we may note that if the internal density concentration were infinite—and, in accordance with (5.7), $\eta_j(a) = j+1$—all k_j's will become zero, and the potential expansion (3.28) will reduce (as it should) to its leading term appropriate for a mass point.

For a finite degree of mass concentration, the surface values of $\eta_j(a_1)$ for any particular stellar model must in general be obtained by a numerical integration of Equations (5.1)–(5.4). If, however, this concentration is high enough to enable us to approximate the function $y(a)$ by (5.35), the fact that the integral Equation (5.34) is homogeneous in y discloses that—within this scheme of approximation—the value of C_j involved in it should be given by the equation

$$C_j^{-1} = 2a_1^{2j+1}k_j = -\frac{3}{2j+1}\int_0^{a_1}\left(\frac{1}{\bar{\rho}}\frac{d\rho}{da}\right)a^{2j+1}da . \tag{5.63}$$

A partial integration of the expression on its right-hand side reveals that

$$\begin{aligned}\int_0^{a_1}\left(\frac{1}{\bar{\rho}}\frac{d\rho}{da}\right)a^{2j+1}da &= -\int_0^{a_1}\rho\frac{d}{da}\left(\frac{a^{2j+1}}{\bar{\rho}}\right)da \\ &= -\int_0^{a_1}\frac{\rho}{\bar{\rho}}\left\{2j+1-\frac{a}{\bar{\rho}}\frac{d\bar{\rho}}{da}\right\}a^{2j}da ;\end{aligned} \tag{5.64}$$

and as, by (5.2),

$$\frac{1}{\bar{\rho}}\frac{d\bar{\rho}}{da} = \frac{3}{a}\left\{\frac{\rho}{\bar{\rho}}-1\right\} , \tag{5.65}$$

if follows that

$$\int_0^{a_1}\frac{1}{\bar{\rho}}\frac{d\rho}{da}a^{2j+1}da = 3\int_0^{a_1}D^2\,a^{2j}da - 2(j+1)\int_0^{a_1}D\,a^{2j}da , \tag{5.66}$$

where, as before, $D \equiv \rho/\overline{\rho}$. For centrally-condensed configurations $D << 1$ and, consequently, $D^2 << D$. In such cases, the first integral on the right-hand side of the foregoing equation is likely to be very small in comparison with the second and may, therefore, be neglected. Doing so we arrive at the equation

$$a_1^{2j+1}k_j = \frac{3(j+2)}{2j+1}\int_0^{a_1} Da^{2j}\,da\,, \tag{5.67}$$

permitting us to approximate closely the numerical values of the constants k_j for centrally-condensed configurations by simple quadratures of the products $a^{2j}D$ throughout the interior.

Therefore, outside the distorted body ($r > a_1$), where the interior potential $U(a)$ becomes identically zero, the potential arising from the mass will—to the first order in surficial distortion—reduce to

$$\Omega \equiv V(r) = \frac{Gm_1}{r}\left\{1 + \sum_{j=2}^{4}\frac{j+1-\eta_j}{2j+1}\frac{a_1^j}{r^j}Y_j(a_1)\right\} \tag{5.68}$$

where, by Equation (3.47) particularized for $a = a_1$, it follows that

$$Y_j(a_1) = \frac{2j+1}{j+\eta_j(a_1)}\frac{a_1^{j+1}}{Gm_1}c_jP_j(\theta,\phi)\,, \tag{5.69}$$

in agreement with (5.60). Inserting (5.69) in (5.68), we find that

$$\begin{aligned} V(r) &= \frac{GM_1}{r} + \sum_{j=2}^{4}\frac{j+1-\eta_j}{j+\eta_j}\frac{a_1^{2j+1}}{r^{j+1}}c_jP_j(\theta,\phi) \\ &= \frac{Gm_1}{r} + \sum_{j=2}^{4}2k_j\frac{a_1^{2j+1}}{r^{j+1}}c_jP_j(\theta,\phi)\,, \end{aligned} \tag{5.70}$$

in agreement with Equation (3.2), where the constants k_j continue to be given by (5.62).

Lastly, it should be stressed that a determination of these constants as outlined in this section is correct only to quantities of first order in surficial distortion of the respective stars. For an extension of their precision to quantities of second order the reader is referred to sections III.2 and 3 of Kopal (1960).

IV.6 Bibliographical Notes

The methods expounded in this chapter for a specification of the gravitational potential of self-gravitating fluid configuration distorted by axial rotation have their source in Clairaut's book on the *Théorie de la Figure de la Terre etc.* (Paris, 1743) already quoted on p. 3; and was extended to include the effects of the tides by Legendre (1793) or Laplace (1825). It was to meet the need of such a theory which led both these distinguished authors to develop what became subsequently known under the name of spherical harmonics—just as, towards the end of that century, Henri Poincaré (1885) was led to develop a theory of ellipsoidal harmonics to study the secular stability of pear-shaped figures arising from a distortion of the Jacobian ellipsoids. It may be added that a systematic work on the solution of Kepler's equation (relating the mean and eccentric anomaly of celestial bodies) led Bessel in 1824 to introduce another family of special functions bearing his name, that represented a third permanent gift to applied mathematics of the 19th century arising from astronomical problems.[6]

The cornerstone of the mathematical foundations of Clairaut's theory—the Dirichlet Theorem on the uniqueness of the potential—goes back to a memoir by G. P. Lejeune-Dirichlet (1805–1859) which appeared in 1850 under a rather involved title, *Über einen neuen Ausdruck zur Bestimmung der Dichtigkeit einer unendlich dünnen Kugelschale, wenn der Werth des Potentials derselben in jedem Punkte ihrer Oberfläche gegeben ist.* Dirichlet's name has been given to this theorem by Riemann; but—at least according to Courant and Hilbert (1931)—*"diese Personalbezeichnung entspricht keineswegs den historischen Tatsachen"* (op. cit., p.153 footnote).

We may add that although, mathematically, Dirichlet was concerned with his theorem in three-dimensional space, his arguments admit readily of an extension to an arbitrary number of dimensions. For a more modern (and more easily accessible) form of this theorem see Poincaré (1899).

Throughout the 19th century, applications of Clairaut's theory to astronomical problems had planetary rather than stellar systems in mind; and as most of them were directed towards geophysics, a study of first-order departures from spherical form was deemed sufficient to compare theory with observations. The first investigators who addressed themselves to a systematic extension of Clairaut's theory to terms of second order in rotational distortion were Darwin (1900) and de Sitter (1924); and to terms of third order, by Lanzano (1962). All these investigators limited themselves, however, to a derivation of higher-order Clairaut equations (2.27)–(2.29) in their *integral* form — and omitted the last step of the procedure to deduce Clairaut's equations (2.33)–(2.35) in their *differential* form subject to the boundary conditions (2.39)–(2.60). This latter task was carried out by the author of this book correctly to terms of the second order (cf. Kopal, 1960), and extended to terms of third order in 1973. It is this latter source which has been followed in Section IV.2 of the present book; although its further extension of the entire theory to terms of the fourth order was subsequently carried out by Kopal and Kamala Mahanta (1974).

An essential idea of Clairaut's approach to our problem, followed in Section IV.1, has been to expand the potential arising from self-attraction in spherical (in fact, zonal) harmonics of increasing order. A consistent theory of the rotational distortion, expansible in terms of sectorial (and also of Hansen-Tisserand coefficients; cf. Hagihara, 1972) rather than zonal harmonics has recently been developed by El-Shaarawy (1974) correctly to quantities of third order in surficial distortion. To a first-order approximation all these approaches lead to identical results; and these do not commence to differ till in quantities of second (and higher) orders. Since, however, these are no simpler than the results previously obtained by use of the zonal harmonics, they have not been reproduced in this book.

A first-order theory of the tidal distortion expounded in Section IV.3 is so severely classical that nothing of significance has been added to it since the days of Laplace. Its

[6] The first solution of Kepler's equation in terms of Bessel's coefficients were, to be sure, constructed already by Lagrange (1770); but Bessel was the first to express such functions in integral form.

systematic extension to terms of second order was not undertaken till more recently by the present writer (cf. Kopal, 1960), and extended to third-order terms by Rahimi-Ardabili (1979). The cross-terms between rotation and the tides were likewise investigated by Kopal (1960, 1974); their present version as given in Section IV.4 follows the latter source.

For particular solutions of the Clairaut or Radau equations given in Section IV.5 in the case of limiting density distributions, cf., e.g., Tisserand (1891); though that corresponding to a shell model does not seem to have been noted before Kopal (1941b). The inequality (5.10) is likewise classical, and implicit (for $j = 2$) already in Clairaut's work of 1743; though its rigorous proof had to await Poincaré (1902). The inequality (5.17) is due to Kopal (1941b); though (for $j = 2$) its existence was noted already by Callandreau (1888); cf. also Véronnet (1912). A discussion of the manifold of the solutions of Radau's equation in the $a - \eta$ and $\eta - \eta'$ planes, as given in Section IV.5, represents a generalization (for an arbitrary value of j) of a treatment of the subject previously given for $j = 0$ by Poincaré (1902); and (for any j) by Kopal (1959).

A formulation (5.70)–(5.30) of our problem as one of the Sturm-Liouville type is due to Kopal (1953); later, Kopal and Lanzano (1973) discovered other substitutions leading to equations of self-adjoint form. The convergence argument for an iterative solution of Equation (5.38) is due to Liapounov (1904), who discussed also the cases arising if the internal density-distribution function $\rho(a)$ happens to be discontinuous.

For further applications of these theories to problems arising in double-star astronomy, cf. Kopal (1960), or Kopal and Lanzano (1973). For a separate discussion of the Roche model as a limiting case of polytropic gas spheres as polytropic index $n \rightarrow 5$ cf. Chandrasekhar (1933); or, more recently, Caimmi (1980).

Chapter V

CLAIRAUT COORDINATES

The principal aim of the preceding chapter—which was to outline a systematic procedure for the construction of the radius-vector $r(a, \theta, \phi)$ of a self-gravitating fluid configuration of arbitrary structure in hydrostatic equilibrium in an arbitrary field of force—antedates, in fact, the Roche model described in Chapter II by more than a hundred years; though the latter was not recognized as a particular case of the Clairaut model till in the past half a century. In order to deal effectively with the problems arising in connection with the Roche model, a special system of curvilinear coordinates has been introduced in Chapter III; and the aim of the present chapter will be to extend the same treatment to the Clairaut model as well.

In order to outline the essential steps to be followed to this end, let—in accordance with the procedure developed in Chapter IV—the radius $r(a, \theta, \phi)$ connecting the centre of mass of a self-gravitating body in hydrostatic equilibrium with an arbitrary point of its surface be expressed as

$$r(a, \theta, \phi) = a\left\{1 + \sum_{i,j} Y_j^i(a; \theta, \phi)\right\} \tag{0.1}$$

where (as in Equation (1.31) of Chapter IV) the solid harmonics

$$Y_j^i(a; \theta, \phi) \equiv f_j(a)\, P_j(\theta, \phi)\,, \tag{0.2}$$

in which the Clairaut radial variable a is connected with the total potential Ψ by Equations (1.44)–(1.45) of Chapter IV for $j = 0$ — an equation which in terms of the scaled Roche potential ξ reduces (cf. Equation (2.3) of Chapter II) to the simple form

$$a = \frac{1}{\xi - q}\,, \tag{0.3}$$

with $q \equiv m_2/m_1$; vanishing at the centre of mass m_1.

By virtue of the uniqueness of the potential (cf. Section IV.1), only one surface of the form (0.1) can pass through a point characterized by a certain value of a; while the angular part of the solid harmonics $Y_j^i(a; \theta, \phi)$ depends on the symmetry of the external field of force causing distortion; and (as we have learned in Chapter IV) the whole essence of Clairaut's theory rests on the mechanism by which the amplitudes $f_j(a)$ of the individual solid harmonics (0.2) can be ascertained for self-gravitating configurations of any structure.

V.1 Metric Transformation

The Cartesian coordinates x, y, z at any point of an equipotential surface are related with the coordinates a, θ, ϕ by means of the equations

$$\left.\begin{aligned} x &= r(a,\theta,\phi)\cos\phi\sin\theta &\equiv r\lambda\,, \\ y &= r(a,\theta,\phi)\sin\phi\sin\theta &\equiv r\mu\,, \\ z &= r(a,\theta,\phi)\cos\theta &\equiv r\nu\,, \end{aligned}\right\} \tag{1.1}$$

where, for the specific form of $r(a, \theta, \phi)$ deduced from Clairaut's theory, the trio a, θ, ϕ will hereafter be referred to as the *Clairaut coordinates*; of which extensive use will be made in the sequel.

Like in the case of the Roche coordinates ξ, η, ζ, the Clairaut coordinates a, θ, ϕ are related with the Cartesian coordinates x, y, z by a metric transformation of the form

$$\begin{aligned} (dx)^2+(dy)^2+(dz)^2 &= g_{11}(da)^2+g_{22}(d\theta)^2+g_{33}(d\phi)^2+ \\ &\quad + 2g_{12}da\,d\theta+2g_{13}da\,d\phi+2g_{23}d\theta\,d\phi\,, \end{aligned} \tag{1.2}$$

specified by the metric coefficients

$$g_{11} = (x_a)^2+(y_a)^2+(z_a)^2\,, \tag{1.3}$$

$$g_{22} = (x_\theta)^2+(y_\theta)^2+(z_\theta)^2\,, \tag{1.4}$$

$$g_{33} = (x_\phi)^2+(y_\phi)^2+(z_\phi)^2\,; \tag{1.5}$$

and

$$g_{12} = x_a x_\theta + y_a y_\theta + z_a z_\theta = g_{21}\,, \tag{1.6}$$

$$g_{13} = x_a x_\phi + y_a y_\phi + z_a z_\phi = g_{31}\,, \tag{1.7}$$

$$g_{23} = x_\theta x_\phi + y_\theta y_\phi + z_\theta z_\phi = g_{32}\,; \tag{1.8}$$

where the subscripts a, θ, ϕ of x, y, z stand, as before, for the *partial* derivatives of the rectangular with respect to the Clairaut coordinates.

By a differentiation of Equations (1.1)–(1.2) for r, the G_{ij}'s assume the more explicit forms

$$g_{11} = (r_a)^2\,, \tag{1.9}$$

$$g_{22} = r^2+(r_\theta)^2\,, \tag{1.10}$$

$$g_{33} = (r\sin\theta)^2+(r_\phi)^2 \tag{1.11}$$

and

$$g_{12} = r_a r_\theta = g_{21}\,, \tag{1.12}$$

$$g_{13} = r_a r_\phi = g_{31}\,, \tag{1.13}$$

$$g_{23} = r_\theta r_\phi = g_{32}\,, \tag{1.14}$$

in which the radius-vector $r(a, \theta, \phi)$ can be established by the procedures developed in the preceding Chapter IV and can, therefore, be regarded hereafter as known. Moreover, the determinant of the metric coefficients G_{ij} proves to be equal to

$$||g_{ij}|| \equiv g \equiv (r^2 r_a \sin\theta)^2 , \tag{1.15}$$

a result to be frequently used later on.

In comparing the Roche coordinates ξ, η, ζ introduced in Chapter III with the Clairaut coordinates a, θ, ϕ developed presently we may note that the Roche radial coordinate ξ has been given to us in a closed form represented by Equation (1.6) of Chapter II, but its associated angular variables η and ζ must then be obtained from the system of partial differential equations (1.23)–(1.25) of Chapter III. In contrast, the Clairaut angular variables θ and ϕ have been identified with those of spherical polar coordinates, but their radial variable a is related with the Roche potential ξ by means of the simple equation (0.3).

Next, we wish to transform into Clairaut coordinates a, θ, ϕ certain differential operators which we shall encounter later in this chapter. The partial derivatives of the Clairaut coordinates a, θ, ϕ with respect to the rectangular coordinates x, y, z transform (by analogy to Equation (2.15) of Chapter III) in accordance with the scheme

$$\left\{ \begin{array}{c} \dfrac{\partial}{\partial x} \\ \dfrac{\partial}{\partial y} \\ \dfrac{\partial}{\partial z} \end{array} \right\} = \left\{ \begin{array}{ccc} a_x & \theta_x & \phi_x \\ a_y & \theta_y & \phi_y \\ a_z & \theta_z & \phi_z \end{array} \right\} \left\{ \begin{array}{c} \dfrac{\partial}{\partial a} \\ \dfrac{\partial}{\partial \theta} \\ \dfrac{\partial}{\partial \phi} \end{array} \right\} , \tag{1.16}$$

where the reciprocal partial derivatives $a_{x,y,z}$, $\theta_{x,y,z}$, $\phi_{x,y,z}$ are (by analogy with Equations (1.8)–(1.10) of Chapter III) given by the equations

$$\begin{aligned} Ja_x &= y_\theta z_\phi - y_\phi z_\theta = \\ &= r^2 \cos\phi \sin^2\theta - r r_\theta \cos\phi \sin\theta \cos\theta + r r_\phi \sin\phi , \end{aligned} \tag{1.17}$$

$$\begin{aligned} Ja_y &= x_\phi z_\theta - x_\theta z_\phi = \\ &= r^2 \sin\phi \sin^2\theta - r r_\theta \sin\phi \sin\theta \cos\theta - r r_\phi \cos\phi , \end{aligned} \tag{1.18}$$

$$\begin{aligned} Ja_z &= x_\theta y_\phi - x_\phi y_\theta = \\ &= r^2 \sin\theta \cos\theta + r r_\theta \sin^2\theta ; \end{aligned} \tag{1.19}$$

$$J\theta_x = y_\phi z_a - y_a z_\phi = r r_a \cos\phi \sin\theta \cos\theta , \tag{1.20}$$

$$J\theta_y = x_a z_\phi - y_a z_\phi = r r_a \sin\phi \sin\theta \cos\theta , \tag{1.21}$$

$$J\theta_z = x_\phi y_a - x_a y_\phi = -r r_a \sin^2\theta ; \tag{1.22}$$

and

$$J\phi_x = y_a z_\theta - y_\theta z_a = -r r_a \sin\phi , \tag{1.23}$$

$$J\phi_y = x_\theta z_a - x_a z_\theta = r r_a \cos\phi , \tag{1.24}$$

$$J\phi_z = 0 ; \tag{1.25}$$

where the Jacobian

$$J \equiv \begin{vmatrix} x_a & y_a & z_a \\ x_\theta & y_\theta & z_\theta \\ x_\phi & y_\phi & z_\phi \end{vmatrix} = r^2 r_a \sin\theta \equiv \sqrt{g} \tag{1.26}$$

is equal to the square-root of the determinant $||g_{ij}||$ of the metric coefficients as given by Equation (1.15).

From the foregoing equations (1.17)–(1.25) it follows that

$$\begin{aligned} a_x^2 + a_y^2 + a_z^2 &= \frac{1}{r_a^2}\left\{1 + \left(\frac{r_\theta}{r}\right)^2 + \left(\frac{r_\phi}{r\sin\theta}\right)\right\} \\ &= (r_a \cos\beta)^{-2} \equiv g^{11}, \end{aligned} \tag{1.27}$$

$$\theta_x^2 + \theta_y^2 + \theta_z^2 = r^{-2} \equiv g^{22}, \tag{1.28}$$

$$\phi_x^2 + \phi_y^2 + \phi_z^2 = (r\sin\theta)^{-2} \equiv g^{33}, \tag{1.29}$$

while

$$a_x\theta_x + a_y\theta_y + a_z\theta_z = -r_\theta/r^2 r_a \equiv g^{21}, \tag{1.30}$$

$$a_x\phi_x + a_y\phi_y + a_z\phi_z = -r_\phi/r^2 r_a \sin^2\theta \equiv g^{31}, \tag{1.31}$$

$$\theta_x\phi_x + \theta_y\phi_y + \theta_z\phi_z = 0 \equiv g^{32}; \tag{1.32}$$

the $g^{ij} \equiv g^{ji}$ being cofactors of g_{ij} in the determinant $||g_{ij}||$, divided by g; and β stands for the angle between the radius-vector of direction cosines λ, μ, ν and a normal to the surface a = constant. Lastly, the reader should have no difficulty to verify that

$$\lambda a_x + \mu a_y + \nu a_z = r_a^{-1}, \tag{1.33}$$

$$\lambda\theta_x + \mu\theta_y + \nu\theta_z = 0, \tag{1.34}$$

$$\lambda\phi_x + \mu\phi_y + \nu\phi_z = 0. \tag{1.35}$$

If (as before) the direction cosines of a curve perpendicular to the equipotential a = constant are denoted by ℓ_1, m_1, n_1, it follows that

$$\ell_1, m_1, n_1 = \frac{a_{x,y,z}}{\sqrt{a_x^2 + a_y^2 + a_z^2}} \equiv \frac{a_{x,y,z}}{\sqrt{g^{11}}}, \tag{1.36}$$

whence

$$\cos\beta = \lambda\ell_1 + \mu m_1 + \nu m_1. \tag{1.37}$$

The direction cosines of the normals to the cones θ = constants and meridional planes ϕ = constants are similarly given by

$$\ell_2, m_2, n_2 = \frac{\theta_{x,y,z}}{\sqrt{\theta_x^2 + \theta_y^2 + \theta_z^2}} \equiv \frac{\theta_{x,y,z}}{\sqrt{g^{22}}} \tag{1.38}$$

and

$$\ell_3,\, m_3,\, n_3 \;=\; \frac{\phi_{x,y,z}}{\sqrt{\phi_x^2+\phi_y^2+\phi_z^2}} \equiv \frac{\phi_{x,y,z}}{\sqrt{g^{33}}} \tag{1.39}$$

disclose that

$$\ell_1\ell_2 + m_1m_2 + n_1n_2 \;=\; \frac{g^{12}}{\sqrt{g^{11}g^{22}}}\,, \tag{1.40}$$

$$\ell_1\ell_3 + m_1m_3 + n_1n_3 \;=\; \frac{g^{13}}{\sqrt{g^{11}g^{33}}}\,, \tag{1.41}$$

$$\ell_2\ell_3 + m_2m_3 + n_2n_3 \;=\; 0\,. \tag{1.42}$$

This latter equation merely bears out the fact that the intersections of the cones θ = constant and ϕ = constant are orthogonal. The intersections of the latter with the equipotentials a = constants will, however, be orthogonal only if $g^{12} = g^{13} = 0$. In the case of purely rotational distortion which is symmetrical in the meridional planes, $r_\phi = 0$—rendering, by (1.31), $g^{31} \equiv g^{13} = 0$. However, even so $g^{12} \neq 0$ and, therefore, the corresponding system of Clairaut coordinates proves to be *non-orthogonal* (and the same is true for the case of tidal distortion as well).

In the case of rotational distortion ($d\phi = 0$), the metric transformation (1.2) for Clairaut coordinates reduces to

$$ds^2 \;=\; g_{11}da^2 + g_{22}d\theta^2 + 2g_{12}da\,d\theta\,; \tag{1.43}$$

but as long as the coefficient $g_{12} \equiv r_a r_\theta \neq 0$, the Clairaut coordinates a, θ in the meridional planes ϕ = constant remain non-orthogonal.

Their orthogonality can, however, be restored by introducing a new angular coordinate ζ, such that

$$\theta \;=\; \Theta(a,\zeta)\,. \tag{1.44}$$

If so,

$$d\theta \;=\; \Theta_a\,da + \Theta_\zeta\,d\zeta\,; \tag{1.45}$$

which on insertion in (1.43) permits us to rewrite the latter as

$$\begin{aligned} ds^2 \;=\;& \{g_{11} + 2g_{12}\Theta_a + g_{22}\Theta_a^2\}\,da^2 + \\ & + \{g_{22}\Theta_\zeta^2\}\,d\zeta^2 + 2\{g_{22}\Theta_a + g_{12}\}\,\Theta_\zeta\,da\,d\zeta\,. \end{aligned} \tag{1.46}$$

Now—unlike the Clairaut coordinates a, θ—the new coordinates a, ζ can be orthogonalized provided that the coefficient of $da\,d\zeta$ on the right-hand side of Equation (1.46) can be made to vanish—i.e., ζ is chosen so that

$$g_{22}\Theta_a + g_{12} \;=\; 0\,, \tag{1.47}$$

a requirement which reduces the metric element (1.46) to

$$ds^2 \;=\; \{g_{11} + g_{12}\Theta_a\}\,da^2 + g_{22}\Theta_\zeta^2\,. \tag{1.48}$$

The radial coordinate a continues to be defined by Equation (1.29) of Chapter IV; but how to specify the angular coordinate ζ to be orthogonal to it? By virtue of Equations (1.45) and (1.47), the curves ζ = constant (i.e., $d\zeta = 0$) will be defined by the ordinary differential equation

$$\frac{d\Theta}{da} = \Theta_a = -\frac{g_{12}}{g_{22}}, \tag{1.49}$$

which can be integrated to yield

$$\Theta = \zeta - \int_0^a \frac{g_{12}}{g_{22}}\, da = \zeta - \int_0^a \frac{r_a r_\theta}{r^2 + r_\theta^2}\, da\,; \tag{1.50}$$

for the rotating Roche model; the radius-vector r or its derivatives can be expanded in ascending powers of a by use of Equations (0.1)–(0.3); and ζ represents the constant of integration.

Since—in accordance with Equation (1.44)—$\theta \equiv \Theta(a, \zeta)$, the solution of (1.49) can obviously proceed only by sucessive approximations. If we ignore terms of the order of a^3 in (0.1), the zero-order approximation to Θ will be $\theta_0 = \zeta$; and, therefore, inclusive of terms of the order or a^3,

$$\theta_1 = \zeta - \frac{1}{3}a^3 \sin\zeta \cos\zeta\,, \tag{1.51}$$

yielding

$$\cos\theta_1 = \cos\zeta \left\{1 + \frac{1}{3}a^3 \sin^2\zeta + \ldots\right\}. \tag{1.52}$$

If we insert this in Equation (1.49), the second approximation to its solution (including terms of the order of a^6) will be of the form

$$\cos\theta_2 = \cos\zeta \left\{a + \frac{1}{3}a^3 \sin^2\zeta + \frac{1}{18}a^6(1 + \frac{9}{2}\sin^2\theta)\sin^2\theta + \ldots\right\}, \tag{1.53}$$

the inversion of which yields

$$\cos\zeta = \cos\theta \left\{1 - \frac{1}{3}a^3 \sin^2\theta - \frac{1}{9}a^6 \sin^2\theta(1 + \frac{9}{2}\sin^2\theta) + \ldots\right\}, \tag{1.54}$$

etc.

The reader may note that this last equation turns out to be identical (apart from notations) with Equation (1.76) of Chapter III defining the Roche angular coordinate ζ; therefore, the orthogonalization of Clairaut coordinates leads back to the Roche coordinates as was to have been expected on general grounds.

A: Differential Operators; Clairaut Harmonics

In order to establish the form of the Eulerian equations of hydrodynamics in Clairaut coordinates introduced in this chapter, it remains still to rewrite such operators in their terms as we shall encounter in this book. To begin with, let us define the velocity components in the direction of increasing Clairaut coordinates by

$$\dot{a} \equiv U, \quad r\dot{\theta} \equiv V, \quad (r \sin \theta)\dot{\phi} \equiv W\,. \tag{1.55}$$

On the other hand, the velocity components u, v, w in rectangular coordinates can be expressed (by a similar differentiation of Equations (1.1)) in terms of Clairaut's velocity components U, V, W by

$$\begin{aligned} u &= (r_a \cos \phi \sin \theta)\, U + \{(r_\theta/r) \sin \theta + \cos \theta\}\, V \cos \phi + \\ &\quad + \{(r_\phi/r) \cos \phi - \sin \phi\}\, W\,, \end{aligned} \tag{1.56}$$

$$\begin{aligned} v &= (r_a \sin \phi \sin \theta)\, U + \{(r_\phi/r) \sin \theta + \cos \theta\}\, V \sin \phi + \\ &\quad + \{(r_\phi/r) \sin \phi + \cos \phi\}\, W\,, \end{aligned} \tag{1.57}$$

$$\begin{aligned} w &= (r_a \cos \theta)\, U + \{(r_\theta/r) \cos \theta - \sin \theta\}\, V + \\ &\quad + \{(r_\phi/r) \cot \theta\}\, W\,, \end{aligned} \tag{1.58}$$

If so, by insertion from the foregoing Equations (1.56)–(1.58) in Equation (2.1) of Chapter III and a resort to (1.16) we find that the Lagrangian time-derivative

$$\frac{D}{Dt} \equiv \frac{\partial}{\partial t} + U\frac{\partial}{\partial a} + \frac{V}{r}\frac{\partial}{\partial \theta} + \frac{W}{r \sin \theta}\frac{\partial}{\partial \phi}\,; \tag{1.59}$$

while the divergence Δ of the velocity vector in Clairaut's coordinates will assume the form

$$\begin{aligned} \Delta &\equiv \frac{1}{r^2 r_a}\frac{\partial}{\partial a}(U r^2 r_a) + \frac{1}{r^2 r_a \sin \theta}\frac{\partial}{\partial \theta}(V r r_a \sin \theta) + \\ &\quad + \frac{1}{r^2 r_a \sin \theta}\frac{\partial}{\partial \phi}(W r r_a)\,; \end{aligned} \tag{1.60}$$

In order to proceed further, let us keep in mind that—unlike the Roche coordinates of Chapter III which are by definition orthogonal—Clairaut's coordinates do not necessarily share with them the same property; and, as a result, the gradient or Laplacian operators may be quite different when written down in their terms. In fact, the gradient ∇ of a scalar function ψ expressed in terms of the Clairaut coordinates $q^i \equiv a, \theta, \phi$ can be generally written as

$$\nabla \equiv \mathbf{e}_j\, g^{ij} \frac{\partial}{\partial q^i} \tag{1.61}$$

where the $\mathbf{e}_j$'s (j = 1,2,3) are the base vectors in the direction of increasing coordinates q^i — in our case,

$$e_1^2 = \frac{1}{g^{11} r_a^2}\,, \quad e_2^2 = \frac{1}{g^{11} g^{22}}\,, \quad e_3^2 = \frac{1}{g^{11} g^{33}}\,. \tag{1.62}$$

Summing up the above expression (1.61) by its dummy indices, we arrive at the more explicit expression for the gradient in Clairaut coordinates a, θ, ϕ to be

$$\begin{aligned} \nabla \equiv \quad & \mathbf{i}\,\frac{\cos\beta}{g^{11}}\left\{g^{11}\frac{\partial}{\partial a}+g^{12}\frac{\partial}{\partial\theta}+g^{13}\frac{\partial}{\partial\phi}\right\}+ \\ & +\frac{\mathbf{j}}{\sqrt{g^{11}g^{33}}}\left\{g^{21}\frac{\partial}{\partial a}+g^{22}\frac{\partial}{\partial\theta}\right\}+ \\ & +\frac{\mathbf{k}}{\sqrt{g^{11}g^{33}}}\left\{g^{31}\frac{\partial}{\partial a}+g^{33}\frac{\partial}{\partial\phi}\right\}, \end{aligned} \tag{1.63}$$

where $\mathbf{i}, \mathbf{j}, \mathbf{k}$ are unit vectors in the direction of a, θ, ϕ. If the coordinates q^i were orthogonal, $g^{ij}=0$ for $i \neq j$; and the right-hand side of the preceding equation would reduce to its diagonal elements—in agreement with Equation (2.20) of Chapter III.

Let us apply Equation (1.63) first to a function $\psi \equiv \psi(a)$—such as would be the case if ψ were to represent the total potential $\Psi_0(a)$, the equilibrium pressure $P_0(a)$, or the density $\rho_0(a)$. If so, then

$$(\nabla\psi)^2 = \frac{1}{g^{11}}\left\{\frac{1}{r_a^2}+\frac{(g^{21})^2}{g^{22}}+\frac{(g^{31})^2}{g^{33}}\right\}\left(\frac{\partial\psi}{\partial a}\right)^2, \tag{1.64}$$

which on insertion for the g^{ij}'s and $\cos\beta$ reduces to

$$\nabla\psi = \frac{d\psi}{da}, \tag{1.65}$$

in agreement with (1.27). However, for $\psi \equiv \psi(a, \theta, \phi)$ this will not be the case; and the full-dress Equation (1.64) must be used to evaluate grad ψ; the divergence of the velocity vector $\mathbf{V}$ will likewise be given by

$$\operatorname{div}\mathbf{V} = \frac{1}{\sqrt{g}}\frac{\partial}{\partial q^i}\left\{\sqrt{g}\,V^i\right\}, \tag{1.66}$$

where V^i stands for the i-th component of the vector $\mathbf{V}$, and g by Equation (1.15).

If, in particular, $\mathbf{V} \equiv \nabla\psi$, $V^i = g^{ij}(\partial\psi/\partial q^i)$, where ψ stands for any arbitrary scalar function, the Laplacian

$$\nabla^2\psi \equiv \text{div.grad}\psi = \frac{1}{\sqrt{g}}\frac{\partial}{\partial q^i}\left\{\sqrt{g}\,g^{ij}\frac{\partial\psi}{\partial q^j}\right\}; \tag{1.67}$$

of which Equation (2.22) of Chapter III represents a particular case, obtaining if the curvilinear coordinates q^i are orthogonal (i.e., $g_{ii} \equiv h_i$ and $g_{ij}=0$ for $i \neq j$).

Summing up the r.h.s. of Equation (1.67) by their dummy indices we find the expression for the Laplacian ∇^2 to be given, more explicitly, as

$$\nabla^2 \equiv \frac{1}{r^2 r_a}\frac{\partial}{\partial a}\left\{\frac{r^2}{(g^{11})^{3/2}}\left[g^{11}\frac{\partial}{\partial a}+g^{12}\frac{\partial}{\partial\theta}+g^{13}\frac{\partial}{\partial\phi}\right]\right\}+$$

$$+ \frac{1}{r^2 r_a \sin\theta} \frac{\partial}{\partial\theta} \left\{ \frac{r r_a \sin\theta}{\sqrt{g^{11} g^{22}}} \left[g^{21} \frac{\partial}{\partial a} + g^{22} \frac{\partial}{\partial\theta} \right] \right\} +$$

$$+ \frac{1}{r^2 r_a \sin\theta} \frac{\partial}{\partial\phi} \left\{ \frac{r r_a}{\sqrt{g^{11} g^{33}}} \left[g^{31} \frac{\partial}{\partial a} + g^{33} \frac{\partial}{\partial\phi} \right] \right\} ; \tag{1.68}$$

or, on insertion for the g^{ij}'s from (1.27)-(1.32) we find that

$$\begin{aligned} r^2 r_a \nabla^2 \equiv \; & \frac{\partial}{\partial a} \left\{ \frac{r^2}{\sqrt{g^{11}}} \frac{\partial}{\partial a} - \frac{\cos\beta}{g^{11}} \left[r_\theta \frac{\partial}{\partial\theta} + \frac{r_\phi}{\sin^2\theta} \frac{\partial}{\partial\phi} \right] \right\} + \\ & + \frac{1}{\sin\theta} \frac{\partial}{\partial\theta} \left\{ \frac{\sin\theta}{\sqrt{g^{11}}} \left[r_a \frac{\partial}{\partial\theta} - r_\theta \frac{\partial}{\partial a} \right] \right\} + \\ & + \frac{1}{\sin^2\theta} \frac{\partial}{\partial\phi} \left\{ \frac{1}{\sqrt{g^{11}}} \left\{ r_a \frac{\partial}{\partial\phi} - r_\phi \frac{\partial}{\partial a} \right] \right\} , \end{aligned} \tag{1.69}$$

which for spherical configurations (when $r_a = 1, r_\theta = r_\phi = 0$; $\cos\beta = 1$ and $g^{11} = 1$) reduces to the well-known Laplacian operator in spherical polar coordinates.

The Clairaut coordinates a, θ, ϕ introduced in this chapter should play a similar role in the studies of close binary systems whose components are approximable by Clairaut's theory of the preceding chapter as the use of spherical polars in connection with ellipsoidal models for their components, or ellipsoidal coordinates in connection with Poincaré's pear-shaped figures (cf. Poincaré, 1885). Each of these systems of coordinates is associated with a family of special funtions representing separable solutions of Laplace's equation

$$\nabla^2 \Phi = 0 , \tag{1.70}$$

in which the Laplacian operator ∇^2 has been rewritten in terms of the respective coordinates. In Clairaut's coordinates the explicit form of this operator has already been given by Equations (1.68) or (1.69).

If, moreover, the solution of (1.70) can be expanded in ascending powers of a of the form

$$\Phi(a, \theta, \phi) = \sum_{n=0}^{\infty} a^n f_n(\theta, \phi) , \tag{1.71}$$

the functions $f_n(\theta, \phi)$ of the angular variables θ and ϕ can be regarded as "Clairaut harmonics"—analogous with spherical or ellipsoidal harmonics associated with simpler geometrical models. A systematic study of such functions associated with more realistic models of the components of close binary systems is, however, so far only in its infancy (see references at the end of this chapter); and offers ample scope for further work.

V.2 Internal Structure

Let us return next to Section IV.2 of Chapter IV and its Equation (1.6): dividing both sides by the differential element da, we see that the condition of hydrostatic equilibrium of a fluid configuration of internal pressure P and density ρ can be expressed by the vector equation

$$\mathrm{grad}\, P = \rho\, \mathrm{grad}\, \Psi , \tag{2.1}$$

where both P and ρ as well as Ψ are functions of a. In rectangular coordinates, the equilibrium condition can be split up into three *partial* differential equations; with boundary conditions imposed on a free surface which (for distorted bodies) requires three spatial coordinates for its specification. If, however, this free surface happens to be an equipotential, the vector equation (2.1) can be reduced to a single *ordinary* differential equation of the form

$$\frac{dP}{da} = \rho \frac{d\Psi}{da} , \tag{2.2}$$

with boundary conditions imposed at $a = a_1$ such that $\rho(a_1) = 0$. To solve such problems represents a simple task; and its significance is enhanced by the fact that the respective solution applies *regardless of the distortion of the equipotential surface* $a =$ constant *in other spatial coordinates.*

This can indeed be done by an obvious extension of the strategy followed in Section IV.1, in which the vanishing of the coefficients $\alpha_j(a)$ of the Neumann expansion (1.44) of Chapter IV for $j > 0$ has already been used to specify the amplitudes $f_j(a)$ of distortion impressed by the external field of force. For $j = 0$, however, by Equations (1.44)–(1.45) of Chapter IV it follows that

$$\alpha_0(a) \equiv \Psi(a) = \frac{1}{4\pi} \int_0^{\pi} \int_0^{2\pi} \Psi(r', \theta, \phi) \sin\theta \, d\theta \, d\phi , \tag{2.3}$$

in which the function $\Psi(r', \theta, \phi)$ continues to be given by Equation (1.42) and r', by the expansion (1.29) of the same chapter.

Unlike the $\alpha_j(a)$'s—the forms of which were already established in Chapter IV for the rotational as well as tidal distortion and used to specify the corresponding amplitudes $f_j(a)$—the function $\alpha_0(a)$ as defined by (2.2) above has not yet been evaluated. In what follows we shall proceed to do so for the case of rotational distortion; and an extension of the result to tidal distortion can be left again as an exercise for the interested reader.

A: Rotational Problem

In order to establish the explicit form of the potential $\Psi(a)$ as defined by Equation (2.2), our first task will be to evaluate the coefficients E_j and F_j on the right-hand side of Equation (1.42) of Chapter IV in the same way as we did it in that chapter: doing so now to quantities of the *fourth* order in surficial distortion,

$$\begin{aligned} E_0(a) = & \int_a^{a_1} \rho \frac{\partial}{\partial a} \left\{ a^2 \left[\frac{1}{2} + f_0 + \frac{1}{2} f_0^2 + \frac{1}{10} f_2^2 + \right.\right. \\ & \left.\left. + \frac{1}{18} f_4^2 \right] \right\} da \end{aligned} \tag{2.4}$$

and

$$
\begin{aligned}
F_0(a) &= \int_0^a \rho \frac{\partial}{\partial a}\left\{a^3\left[\frac{1}{3} + f_0 + f_0^2 + \frac{1}{5}f_2^2 + \frac{2}{105}f_2^3 + \right.\right. \\
&\left.\left. + \frac{1}{9}f_4^2 + \frac{2}{35}f_2^2 f_4\right]\right\} da \,. \qquad (2.5)
\end{aligned}
$$

Since, moreover, $4\pi F_0(a)$ represents the mass of our configuration interior to a which must be independent of distortion, the foregoing equation must reduce to

$$
4\pi F_0(a) = \int_0^a \rho\, a^2 da \,; \qquad (2.6)
$$

and this will be true only if (to the same degree of accuracy) the zero-order solid harmonic will be specified by

$$
f_0 = -\frac{1}{5}f_2^2 - \frac{2}{105}f_2^3 - \frac{1}{25}f_2^4 - \frac{1}{9}f_4^2 - \frac{2}{35}f_2^2 f_4 \qquad (2.7)
$$

to safeguard the constancy of mass.

For $j > 0$, the explicit values of the coefficients E_j and F_j are already known to us from Equations (2.13)–(2.20) of Chapter IV: i.e.,

$$
\begin{aligned}
E_2(a) &= \int_a^{a_1} \rho \frac{\partial}{\partial a}\left\{f_2 - \frac{1}{7}f_2^2 + \frac{12}{35}f_2^3 - \frac{2}{7}f_2 f_4\right\} da \,, \qquad (2.8) \\
E_4(a) &= \int_a^{a_1} \rho \frac{\partial}{\partial a}\left\{\frac{1}{a^2}\left[f_4 - \frac{27}{35}f_2^2\right]\right\} da \,; \qquad (2.9) \\
F_2(a) &= \int_0^a \rho \frac{\partial}{\partial a}\left\{a^5\left[f_2 + \frac{4}{7}f_2^2 + \frac{2}{35}f_2^3 + \frac{8}{7}f_2 f_4\right]\right\} da \,, \qquad (2.10) \\
F_4(a) &= \int_0^a \rho \frac{\partial}{\partial a}\left\{a^7\left[f_4 + \frac{54}{35}f_2^2\right]\right\} da \,; \qquad (2.11)
\end{aligned}
$$

while F_0 continues to be given by (2.6).

Equations (2.8)–(2.11) permit us to evaluate these integrals explicitly in terms of the f_j's. The actual procedure for doing so is rather complicated (for its details cf. Kopal, 1973; Kopal and Kamala Mahanta, 1974; cf. also sec. IV-2 of Kopal, 1978), but the outcome is, fortunately, simple; for, consistent with our scheme of approximation,

$$
\begin{aligned}
a^3 E_2(a) &= \left\{(\eta_2 + 2)f_2 - \frac{2}{7}(\eta_2^2 + 3\eta_2 + 6)f_2^2 + \right. \\
&+ \frac{4}{35}(2\eta_2^3 + 15\eta_2^2 + 30\eta_2 + 38)f_2^3 - \\
&\left. - \frac{2}{7}(2\eta_2\eta_4 + 3\eta_2 + 3\eta_4 + 26)f_2 f_4\right\} F_0 + \qquad (2.12) \\
&\frac{\omega^2 a^3}{12\pi G}\left\{5 - 2(\eta_2 + 5)f_2 + \frac{2}{7}(2\eta_2^2 + 6\eta_2 + 15)f_2^2\right\} \,, \\
a^5 E_4(a) &= \left\{(\eta_4 + 4)f_4 - \frac{18}{35}(\eta_2^2 + 5\eta_2 + 6)f_2^2\right\} F_0 \qquad (2.13)
\end{aligned}
$$

and

$$\begin{aligned} F_2(a) &= -a^2\left\{(\eta_2-3)f_2-\frac{2}{7}(\eta_2^2-2\eta_2+6)f_2^2+\right.\\ &+\frac{23}{(}4\eta_2^3+10\eta_2^2+5\eta_2+51)f_2^3-\\ &\left.-\frac{4}{7}(\eta_2\eta_4-\eta_2-\eta_4+13)f_2f_4\right\}F_0+ \\ &+\frac{2}{3}\left(\frac{\omega^2a^5}{4\pi G}\right)\left\{1-\frac{2}{7}(\eta_2-2)f_2\right\}f_2\eta_2\,, \end{aligned} \tag{2.14}$$

$$F_4(a) = a^2\left\{(\eta_4-5)f_4-\frac{18}{35}(\eta_2^2-4\eta_2+6)f_2^2\right\}F_0\,, \tag{2.15}$$

etc., so that, correctly to the fourth order in surficial distortion,

$$\begin{aligned} \alpha_0(a) &= E_0+\frac{2}{25}a^2E_2\left\{f_2+\frac{1}{7}f_4-\frac{1}{5}f_2^3+\frac{2}{7}f_2f_4\right\}+\\ &+\frac{4}{81}a^4E_4\left\{f_4+\frac{27}{35}f_2^2\right\}+\\ &+\frac{F_0}{a}\left\{1+\frac{2}{5}f_2^2-\frac{4}{105}f_2^3+\frac{8}{175}f_2^4+\frac{2}{9}f_4^2-\frac{4}{35}f_2f_4\right\}+\\ &+\frac{3F_2}{25a^3}\left\{-f_2+\frac{4}{7}f_2^2-\frac{78}{35}f_2^3+\frac{8}{7}f_2f_4\right\}+\\ &+\frac{5F_4}{81a^5}\left\{-f_4+\frac{54}{35}f_2^2\right\}+\\ &+\frac{1}{3}\omega^2a^2\left\{1-\frac{2}{5}f_2-\frac{9}{35}f_2^2+\frac{22}{525}f_2^3-\frac{4}{35}f_2f_4\right\}\,, \end{aligned} \tag{2.16}$$

where

$$E_0(a) = \int_a^{a_1}\rho\frac{\partial}{\partial a}\left\{a^2\left[\frac{1}{2}-\frac{1}{10}f_2^2-\frac{2}{105}f_2^3-\frac{1}{50}f_2^4-\frac{1}{18}f_4^2-\frac{2}{35}f_2^2f_4\right]\right\}da\,. \tag{2.17}$$

After insertion in (2.4) for f_0 from (2.7); and, consistent with our degree of accuracy, we similarly find that

$$\begin{aligned} \alpha_0(a) &= E_0+\frac{F_0}{a}\left\{1+\frac{1}{5}(\eta_2+1)f_2^2-\right.\\ &-\frac{2}{105}(3\eta_2^2+3\eta_2+8)f_2^3+\frac{1}{9}(\eta_4-1)f_4^2-\\ &-\frac{2}{35}(2\eta_2\eta_4+(\eta_2+1)^2+\eta_4+14)f_2^2f_4+\\ &\left.+\frac{2}{175}(4\eta_2^3+22\eta_2^2+19\eta_2+16)f_2^4\right\}+\\ &+\frac{\omega^2a^2}{12\pi G}\left\{1-\frac{1}{5}(2\eta_2+5)f_2^2+\frac{4}{105}(3\eta_2^2+3\eta_2+5)f_2^3\right\}\,; \end{aligned} \tag{2.18}$$

which, by (2.3) and on differentiation with respect to a discloses that

$$\begin{aligned}
\frac{d\Psi}{da} = & -\frac{Gm(a)}{a^2}\Big\{1-\frac{1}{5}\Big[(\eta_2+1)^2+4\Big]f_2^2+ \\
& +\frac{2}{105}\Big[3(\eta_2+1)^3-6(\eta_2+1)^2+9(\eta_2+1)+4\Big]f_2^3- \\
& -\frac{2}{175}\Big[4(\eta_2+1)^4-10(\eta_2+1)^3-77(\eta_2+1)^2+24(\eta_2+1)-9\Big]f_2^4- \\
& -\frac{1}{9}\Big[(\eta_4+1)^2+20\Big]f_4^2+ \qquad (2.19)\\
& +\frac{2}{35}\left[(\eta_2+1)(\eta_2+17)+(\eta_2-1)(3\eta_2+5)\eta_4\right]f_2^2f_4\Big\}+ \\
& +\frac{2}{3}\omega^2 a\Big\{1-\frac{1}{5}(\eta_2^2+2\eta_2+5)f_2^2+\frac{2}{105}(3\eta_2^3+3\eta_2^2+6\eta_2+10)f_2^3\Big\},
\end{aligned}$$

in which any derivative of Ψ higher than the first can be obtained from (2.19) by successive differentiation; for these will be needed to relate the density $\rho(a)$ with the potential $\Psi(a)$ by Poisson's equation $4\pi G\rho = -\nabla^2\Psi$.

As a result, both Equations (2.2) and (2.19) can be solved for $P(\rho)$ and $\rho(a)$ by sucessive approximations for configurations of any shape or form. To be sure, Equations (2.2) and (2.19) do not yet specify the problem uniquely; for a third relation between P, ρ and Ψ must be sought to render its solution unique. As is well known, such a missing relation can be constructed on the basis of the principle of conservation of energy; and its consequences may assume a different form—algebraic if the principal process of energy transport is convection, or differential if energy is transported mainly by radiation; and both may on occasion participate in the final outcome. Such participation depends, however, on specific characteristics (mass, chemical composition) of each respective star; and an investigation of their consequences is outside the scope of this book.

In conclusion of this section, the reader may be interested in the specific form which Equation (2.2) assumes in the limiting case of the Roche model discussed in Chapter II. Should the gravitational attraction of the component of mass m_1 be reduced to a central point, the coefficients $E_j(a) = 0$ for $j > 0$; while Equations (2.13) and (2.14) of this section reduce to

$$4\pi G\, F_2(a) = 3Gm_1a^2\left(f_2+\frac{4}{7}f_2^2+\ldots\right) \qquad (2.20)$$

and

$$4\pi G\, f_4(a) = 3Gm_1a^4\left(f_4+\frac{54}{35}f_2^2+\ldots\right) \qquad (2.21)$$

where

$$f_2(a) = -\frac{1}{3}\left(\frac{\omega^2}{Gm_1}\right)a^3 - \frac{4}{7}\left(\frac{\omega^2}{Gm_1}\right)^2 a^6+\ldots, \qquad (2.22)$$

$$f_4(a) = -\frac{6}{35}\left(\frac{\omega^2}{Gm_1}\right)^2 a^6+\ldots \qquad (2.23)$$

etc.

The next step towards obtaining the desired relation between the total potential Ψ and the Clairaut radial coordinate should be a resort to Equations (1.44)–(1.45) of Chapter IV in which we insert for r' from Equation (2.1) of that chapter, and evaluate $\alpha_0(a)$ by integration of (1.45). Should we do this, however, we would obtain expressions seemingly at variance with Equation (2.11) of Chapter II which, for $q = 0$ reduces to the hypergeometric expansion given by Equation (1.67) of Chapter III which, however, expresses r' as a series of terms factored by successive *sectorial* harmonics

$$(1-\nu^2)^j \equiv (1.3.5\ldots 2j+1)^{-1} P_{2j}^{2j}(\nu)\,, \tag{2.24}$$

in place of the *zonal* harmonics $P_{2j}(\nu)$ occurring on the right-hand side of Equation (2.1) of Chapter IV. Indeed,the Roche parameter r_0 defined by Equation (2.3) of Chapter II and the Clairaut radial coordinate a introduced by the expansion (1.29) of Chapter IV are *not* identical; the latter being equal to r_0 times the cube-root of the expression in curly brackets on the right-hand side of Equation (2.18), Chapter II—yielding

$$a = r_0\left\{1 + \frac{1}{3}r_0^3 + \frac{19}{45}r_0^6 + \frac{2137}{2835}r_0^9 + \ldots\right\} \tag{2.25}$$

or its inverse

$$r_0 = a\left\{1 - \frac{1}{3}a^3 + \frac{1}{45}a^6 - \frac{58}{2835}a^9 + \ldots\right\}. \tag{2.26}$$

With the aid of the foregoing results the reader should experience no difficulty that the limiting case of Equations (2.2) appropriate for the Roche model reduces indeed to the closed expression given by Equation (0.3) earlier in this chapter—QED.

V.3 Vibrational Stability

In the preceding section of this chapter our aim has been to simplify a mathematical description of the *equilibrium* models of stellar structure by use of the Clairaut coordinates, which made it possible to express the fundamental equations of internal structure of self-gravitating configurations in terms of *ordinary* differential equations—regardless of the extent to which rotation or tides can deform them from spherical shape. The aim of the present concluding section of Chapter V should be to establish the *variational form* of such equations, to enable us to analyze such equilibrium configurations for their *vibrational stability*, as long as their departures from equilibrium forms are small enough for their squares and cross-products to be neglected.

In order to do so, let

$$\mathbf{r} = \mathbf{i}x' + \mathbf{j}y' + \mathbf{k}z' \tag{3.1}$$

stand for the displacement vector of any mass element referred to the rectangular *body* axes $X'Y'Z'$ (cf.Section II.3A), *rotating* in space with the vector angular velocity

$$\vec{\omega} = \mathbf{i}\omega_x + \mathbf{j}\omega_y + \mathbf{k}\omega_z \; ; , \tag{3.2}$$

where $\omega_{x,y,z}$ are the angular velocity components about the inertial (space) axes XYZ; and let the velocity vector $\mathbf{V}$ of motion in rotating coordinates be expressible as

$$\mathbf{V} = \mathbf{i}u' + \mathbf{j}v' + \mathbf{k}w' \, , \tag{3.3}$$

where

$$u' \equiv \dot{x}', \quad v' \equiv \dot{y}', \quad w' \equiv \dot{z}' \, , \tag{3.4}$$

and in which $\mathbf{i}$, $\mathbf{j}$, $\mathbf{k}$ stand for the unit vectors in the direction of the body-axes $X'Y'Z'$.

Let, moreover, P, ρ and Ψ denote (as before) the pressure, density and total potential (arising from the mass as well as from tidal attraction and centrifugal force). If so, the conservative form of the Eulerian equations of motion (in which dissipative terms arising from possible viscosity or interaction between mass and radiation have been ignored) can be written (cf., e.g., p.285 of Ledoux, 1958) as

$$\rho \frac{D}{Dt}(\mathbf{V} + \vec{\omega} \times \mathbf{r}) + 2\rho(\vec{\omega} \times \mathbf{V}) = \rho \, \mathrm{grad}\, \Psi - \mathrm{grad}\, P \, , \tag{3.5}$$

in which the vector product

$$\vec{\omega} \times \mathbf{V} = \mathbf{i}(w'\omega_y - v'\omega_z) + \mathbf{j}(u'\omega_z - w'\omega_x) + \mathbf{k}(v'\omega_x - u'\omega_y) \tag{3.6}$$

stands for the effects of Coriolis force; while the Lagrangian time-derivative of the product $\vec{\omega} \times \mathbf{r}$ can be expressed as

$$\begin{aligned} \frac{D}{Dt}(\vec{\omega} \times \mathbf{r}) &= -\mathbf{r} \times \frac{D\vec{\omega}}{Dt} = \mathbf{i}\left(z'\frac{d\omega_y}{Dt} - y'\frac{D\omega_z}{Dt}\right) + \\ &+ \mathbf{j}\left(x'\frac{D\omega_z}{Dt} - z'\frac{D\omega_x}{Dt}\right) + \mathbf{k}\left(y'\frac{D\omega_x}{Dt} - x'\frac{D\omega_y}{Dt}\right) . \end{aligned} \tag{3.7}$$

When dissipative forces are ignored, the vector equation (3.5) safeguards the conservation of *momentum* of the respective dynamical system; and can be decomposed in three scalar equations involving six dependent variables — i.e., three velocity components u', v', w'; the pressure P, density ρ, and total potential Ψ. Therefore, in order to render the problem determinate, three additional equations relating our six dependent variables must be sought. These are, in turn, provided by the equation of continuity

$$\frac{D\rho}{Dt} + \rho \, \mathrm{div}\, \mathbf{V} = 0 \tag{3.8}$$

safeguarding the conservation of *mass*, Poisson's equation

$$\nabla^2 \Psi = -4\pi G \rho \tag{3.9}$$

relating the density ρ with the potential Ψ; and, lastly the equation for the conservation of *energy*, the general formulation of which presents a problem of its own. If, however, the motions can be regarded as adiabatic, this equation reduces—as is well known—to the simple form

$$\frac{DP}{Dt} = c^2\frac{d\rho}{Dt}, \quad c^2 = \gamma\frac{P}{\rho} \tag{3.10}$$

where γ denotes the ratio of specific heats of stellar matter.

In a conservative framework (and for adiabatic changes of state) Equations (3.5)–(3.10) represent the exact formulation of our problem—without restriction on the magnitude or variation of any one of its dependent variables. The system is, however, also non-linear in them; and its general solution encounters unsurmountable difficulties. In order to lessen these, we propose first to limit our problem to the case of *uni-axial* rotation—i.e., to set in Equation (3.2)

$$\omega_x = \omega_y = 0 \text{ and } \omega_z = \text{constant (say)}, \tag{3.11}$$

an assumption which reduces the expression (3.6) for the Coriolis force to

$$\vec{\omega} \times \mathbf{V} = (-\mathbf{i}v' + \mathbf{j}u')\omega_z \tag{3.12}$$

in agreement with Equation (2.26) of Chapter III. A removal of the restrictions (3.11) will be the principal subject of the subsequent Chapter VI; but for the remainder of the present chapter we shall regard $\dot{\omega}_z = 0$ and $\omega_z \equiv$ constant (say, ω); which renders the r.h.s. of Equation (3.7) to vanish identically.

Secondly, let us — in what follows — *linearize* our system of equations by assuming that the velocity components u', v', w' of motion are small enough for their squares and cross-products to be negligible; and that so are also the changes P', ρ', Ψ' in pressure, density and potential caused by such motions. In more specific terms, let us hereafter assume that

$$P(r, \theta, \phi) = P_0(a) + P'(a, \theta, \phi; t), \tag{3.13}$$
$$\rho(r, \theta, \phi) = \rho_0(a) + \rho'(a, \theta, \phi; t), \tag{3.14}$$
$$\Psi(r, \theta, \phi) = \Psi_0(a) + \Psi'(a, \theta, \phi; t); \tag{3.15}$$

where the quantities P_0, ρ_0 and Ψ_0 remain constant over equipotential surfaces characterized by the Clairaut coordinate a. Moreover, if the angular velocity ω_z of axial rotation become identical with the Keplerian angular velocity ω_K *and* the rotating system of coordinates $X'Y'Z'$ become identical with the doubly-primed "revolving" coordinates $X''Y''Z''$, it follows that

$$\Psi'(a, \theta, \phi; t) \equiv \Psi'(a) \tag{3.16}$$

as well; for the conditions under which this may cease to be true, see Section II-3A.

Let us set out now to linearize the right-hand side of the vector equation (3.5) in terms of P', ρ' and grad Ψ'. By virtue of the fact that (cf.Equation 2.1) grad $P_0 = \rho_0$ grad Ψ_0 this can be written out as

$$\begin{aligned} \rho \operatorname{grad} \Psi - \operatorname{grad} P &= \mathbf{i}\left\{\rho_0 \frac{\partial \Psi'}{\partial x} + \rho' \frac{\partial \Psi_0}{\partial x} - \frac{\partial P'}{\partial x}\right\} + \\ &+ \mathbf{j}\left\{\rho_0 \frac{\partial \Psi'}{\partial y} + \rho' \frac{\partial \Psi_0}{\partial y} - \frac{\partial P'}{\partial y}\right\} + \\ &+ \mathbf{k}\left\{\rho_0 \frac{\partial \Psi'}{\partial z} + \rho' \frac{\partial \Psi_0}{\partial z} - \frac{\partial P'}{\partial z}\right\}, \end{aligned} \tag{3.17}$$

and the operators $\partial/\partial x$, $\partial/\partial y$, $\partial/\partial z$ can be expressed in terms of the Clairaut coordinates by means of Equation (1.16).

Next, let us convert the rectangular velocity components u', v', w' into the Clairaut components U', V', W', as defined by (1.55), by means of Equations (1.56)–(1.58). To this end, let us introduce an auxiliary square matrix $\{L\}$, consisting of the elements of Clairaut angular variables θ, ϕ and defined by

$$\left\{\begin{array}{lll} \cos\phi \sin\theta & \sin\phi \sin\theta & \cos\theta \\ \cos\phi \cos\theta & \sin\phi\cos\theta & -\sin\theta \\ -\sin\phi & \cos\phi & 0 \end{array}\right\} \equiv \{L\}. \tag{3.18}$$

With its aid, the product

$$\{L\}\left\{\begin{array}{c} u' \\ v' \\ w' \end{array}\right\} = \left\{\begin{array}{c} r_a U' \quad \frac{r_\theta}{r} V' \quad \frac{r_\phi}{r \sin\theta} W' \\ V' \\ W' \end{array}\right\}, \tag{3.19}$$

where r_a, r_θ, and r_ϕ stand for partial derivatives of the radius-vector $r(a, \theta, \phi)$ of the equipotential a = constant, as represented by an expansion on the r.h.s. of Equation (1.47) of Chapter IV.

Moreover, by operating with the matrix $\{L\}$ on the individual columns of the square matrix on the r.h.s. of Equation (1.16), we find that

$$\{L\}\left\{\begin{array}{c} a_x \\ a_y \\ a_z \end{array}\right\} = \left\{\begin{array}{c} \frac{1}{r_a} \\ -\frac{1}{r_a}\left(\frac{r_\theta}{r}\right) \\ -\frac{1}{r_a}\left(\frac{r_\phi}{r\sin\theta}\right) \end{array}\right\}, \tag{3.20}$$

$$\{L\}\left\{\begin{array}{c} \theta_x \\ \theta_y \\ \theta_z \end{array}\right\} = \left\{\begin{array}{c} 0 \\ \frac{1}{r} \\ 0 \end{array}\right\}, \tag{3.21}$$

$$\{L\}\left\{\begin{array}{c}\phi_x\\ \phi_y\\ \phi_z\end{array}\right\}=\left\{\begin{array}{c}0\\ 0\\ \dfrac{1}{r\sin\theta}\end{array}\right\}; \tag{3.22}$$

while the Coriolis terms

$$\{L\}\left\{\begin{array}{c}-v'\\ +u'\\ 0\end{array}\right\}=\left\{\begin{array}{l}-W\sin\theta\\ -W\cos\theta\\ \left[r_aU+\dfrac{r_\theta}{r}V+\dfrac{r_\phi}{r\sin\theta}W\right]\sin\theta+V\cos\theta\end{array}\right\}. \tag{3.23}$$

Moreover, with the aid of the foregoing Equations (3.18)–(3.22), it follows from (1.16) that

$$\{L\}\left\{\begin{array}{c}\dfrac{\partial}{\partial x}\\ \dfrac{\partial}{\partial y}\\ \dfrac{\partial}{\partial z}\end{array}\right\}=\left\{\begin{array}{lll}\dfrac{1}{r_a}\dfrac{\partial}{\partial a} & & \\ -\dfrac{1}{r_a}\left(\dfrac{r_\theta}{r}\right)\dfrac{\partial}{\partial a} & +\dfrac{1}{r}\dfrac{\partial}{\partial\theta} & \\ -\dfrac{1}{r_a}\left(\dfrac{r_\phi}{r\sin\theta}\right)\dfrac{\partial}{\partial a} & & +\dfrac{1}{r\sin\theta}\dfrac{\partial}{\partial\phi}\end{array}\right\}, \tag{3.24}$$

and, accordingly, the linearized scalar components of the vector equation (3.5) will assume the forms

$$\begin{aligned}\frac{\partial}{\partial t}\left\{r_aU+\left(\frac{r_\theta}{r}\right)V+\left(\frac{r_\phi}{r\sin\theta}W\right)\right\}-2\omega_K W\sin\theta &=\\ =\frac{1}{r_a}\left\{\frac{\partial\Psi'}{\partial a}+\frac{\rho'}{\rho_0}\frac{\partial\Psi_0}{\partial a}-\frac{1}{\rho_0}\frac{\partial P'}{\partial a}\right\}&,\end{aligned} \tag{3.25}$$

$$\begin{aligned}\frac{\partial V}{\partial t}-2\omega_K W\cos\theta &= \frac{1}{r}\left\{\frac{\partial\Psi'}{\partial\theta}-\frac{1}{\rho_0}\frac{\partial P'}{\partial\theta}\right\}-\\ &-\frac{r_\theta}{rr_a}\left\{\frac{\partial\Psi'}{\partial a}+\frac{\rho'}{\rho_0}\frac{\partial\Psi_0}{\partial a}-\frac{1}{\rho_0}\frac{\partial P'}{\partial a}\right\},\end{aligned} \tag{3.26}$$

$$\begin{aligned}\frac{\partial W}{\partial t}+2\omega_K\left\{\left[r_aU+\frac{r_\theta}{r}V+\frac{r_\phi W}{r\sin\theta}\right]\sin\theta+V\cos\theta\right\} &=\\ =\frac{1}{r\sin\theta}\left\{\frac{\partial\Psi'}{\partial\phi}-\frac{1}{\rho_0}\frac{\partial P'}{\partial\phi}\right\}-&\\ -\frac{1}{r_a}\left(\frac{r_\phi}{r\sin\theta}\right)\left\{\frac{\partial\Psi'}{\partial a}+\frac{\rho'}{\rho_0}\frac{\partial\Psi_0}{\partial a}-\frac{1}{\rho_0}\frac{\partial P'}{\partial a}\right\}&,\end{aligned} \tag{3.27}$$

in which the primed velocity components have merely been replaced (there should be no danger of confusion) by the unprimed ones.

Moreover, a similar linearization of the equation (3.8) of continuity discloses that the density changes ρ' are bound to satisfy the relation

$$\frac{\partial \rho'}{\partial t} + U\frac{\partial \rho_0}{\partial a} = -\rho_0 \operatorname{div} \mathbf{V} \equiv -\rho_0 \Delta , \tag{3.28}$$

where the divergence Δ of the velocity vector $\mathbf{V}$ continues to be given by Equation (1.60); and the linearized pressure changes P'—if adiabatic—are likewise constrained to satisfy the relation

$$\frac{\partial P'}{\partial t} + U\frac{\partial P_0}{\partial a} = c^2\left\{\frac{\partial \rho'}{\partial a} + U\frac{\partial \rho_0}{\partial a}\right\} = \rho_0(g_0 U - c^2\Delta) , \tag{3.29}$$

where g_0 stands for the gravitational acceleration in the equilibrium state; and $c^2 = \gamma(P_0/\rho_0)$, the equilibrium speed of sound.

Lastly, since the Poisson equation (3.9) is linear in its dependent variables and $\nabla^2\Psi_0 = -4\pi G\rho_0$, it can be replaced in terms of the perturbed (primed) variables by

$$\nabla^2\Psi' = -4\pi G\rho' \tag{3.30}$$

exactly; and this, together with (3.28) and (3.29), completes with Equations (3.25)–(3.27) a self-contained set of equations to specify U, V, W together with P', ρ' and Ψ'.

A: The Stability of Roche Equipotentials: Rotational Problem

In conclusion of the present chapter, let us apply the results obtained in the preceding part of this section to investigate in Clairaut coordinates the stability of motion arising if a self-gravitating configuration is disturbed from its state of equilibrium. In order to render such a problem solvable in a closed form, let us assume that the configuration in question exhibits so high an internal density concentration that not only its geometrical, but also dynamical properties can be sufficiently approximated by the Roche Model described in Chapter II; for if so, we are entitled to set in Equations (3.25)–(3.27) of this section

$$\rho' = P' = 0 . \tag{3.31}$$

If, moreover, the angular velocity ω of axial rotation can be identified with the Keplerian angular velocity ω_K, then (in accordance with Equation (3.16)) $\Psi'(a, \theta, \phi; t) \equiv \Psi'(a)$, while

$$\frac{\partial \Psi'}{\partial \theta} = \frac{\partial \Psi'}{\partial \phi} = 0 ; \tag{3.32}$$

and if so, the linearized Equations (3.25)–(3.27) reduce to

$$\frac{\partial}{\partial t}\left\{r_a U + \frac{r_\theta}{r}V + \frac{r_\phi}{r \sin\theta}W\right\} - 2\omega_K W \sin\theta = \frac{1}{r_a}\frac{\partial \Psi'}{\partial a} , \tag{3.33}$$

$$\frac{\partial V}{\partial t} - 2\omega_K W \cos\theta = -\frac{1}{r_a}\left(\frac{r_\theta}{r}\right)\frac{\partial \Psi'}{\partial a}, \tag{3.34}$$

$$\frac{\partial W}{\partial t} + 2\omega_K \left\{ \left[r_a U + \frac{r_\theta}{r} V + \frac{r_\phi}{r \sin\theta} W \right] \sin\theta + V \cos\theta \right\} - \\ = -\frac{1}{r_a}\left(\frac{r_\phi}{r \sin\theta}\right)\frac{\partial \Psi'}{\partial a}, \tag{3.35}$$

where $r_{a,\theta,\phi}$ denote, as before, partial derivatives of the radius-vector $r(a, \theta, \phi)$ of an equipotential surface with respect to a, θ and ϕ. The reader should, moreover, keep in mind that whereas, in deriving Equations (3.25)–(3.27), we assumed their dependent variables U, V, W as well as the perturbations P', ρ', Ψ' caused by the motion to be small enough for their squares and cross-products to be ignorable, *no such assumption has been made about the amplitudes $f_j(a)$ in the expansion of $r(a, \theta, \phi)$ on the right-hand side of Eqauations (0.1) and (0.2), or their derivatives.* In other words, while departures from equipotential surfaces caused by motion will be regarded as small, no restriction has been imposed on the departures of such surfaces from spherical form (other than that they remain *closed* about their respective centres of mass.

In order to particularize the foregoing equations (3.33)–(3.35) for the case of *rotational distortion*, let us adopt the mass m_1 of the rotating configuration as the unit of mass; R as the unit of length; and the unit of time be chosen so that the gravitational constant $G = 1$ (so that $\omega_K^2 \equiv Gm_1/R^3 = 1$ define the unit of frequency). If so, it follows from Equation (1.6) of Chapter II that the (normalized) potential ξ of our rotating configuration can be rewritten (for $m_2 = q = 0$) as[1]

$$\xi = \frac{1}{r} + \frac{r^2}{2}\sin^2\theta \equiv \frac{1}{a}, \tag{3.36}$$

the radius-vector r of which is defined as a solution of the cubic equation

$$r^3 \sin^2\theta - 2(r/a) + 2 = 0, \tag{3.37}$$

satisfied (cf. Equation (1.67) of Section III.1B) by

$$\frac{r}{a} = \frac{\sin\frac{1}{3}\sin^{-1} x}{\frac{1}{3}x} = {}_2F_1\left(\frac{1}{3}, \frac{2}{3}; \frac{3}{2}; x^2\right) \tag{3.38}$$

i.e., expressible as an ordinary hypergeometric series of the type ${}_2F_1$, of the argument

$$x^2 = \left(\frac{3}{2}a\right)^3 \sin^2\theta. \tag{3.39}$$

[1] In Chapter II we used for a the notation r_0; and the two (cf. Equations 2.25 and 2.26) are, strictly speaking, not identical. The difference between the two is, however, only formal; and in what follows we shall replace r_0 of Chapter II with a introduced in Chapter IV for the sake of consistency.

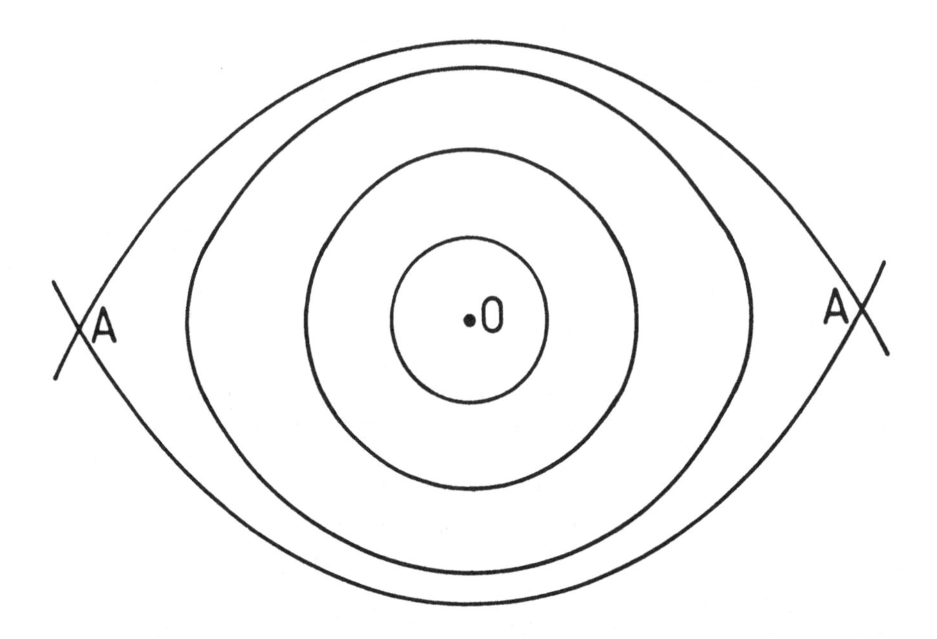

Figure V.1:

The necessary condition for our configuration to remain closed is the requirement that our hypergeometric series converges—i.e., that $x \leq 1$. At the equator ($\theta = 90°$) this requires that

$$r = 1, \; \xi = \frac{3}{2}, \; a = \frac{2}{3}, \tag{3.40}$$

corresponding to the "Roche Limit" (cf. Section II.2B; marked by a heavy line on Figure II.1), when Equation (3.38) reduces to

$$r = 2\frac{\sin \frac{1}{3}\theta}{\sin \theta}; \tag{3.41}$$

and because of the rotational symmetry about the Z'-axis,

$$r_\phi = 0 . \tag{3.42}$$

If so, Equations (3.33)–(3.35) of this section reduce further to

$$\frac{\partial}{\partial t}\left\{r_a U + \frac{r_\theta}{r} V\right\} - 2W \sin \theta = \frac{\xi_a}{r_a}, \tag{3.43}$$

$$\frac{\partial V}{\partial t} - 2W \cos \theta = -\frac{r_\theta}{r}\left(\frac{\xi_a}{r_a}\right), \tag{3.44}$$

$$\frac{\partial W}{\partial t} + 2\left\{\left[r_a U + \frac{r_\theta}{r} V\right] \sin \theta + V \cos \theta\right\} = 0; \tag{3.45}$$

where, by (3.36)

$$\xi_a \equiv \frac{\partial \xi}{\partial a} = -\frac{1}{a^2} \,. \tag{3.46}$$

Before we can proceed to solve Equations (3.43)–(3.45) subject to appropriate initial conditions, we have to evaluate the coefficients of the unknowns; and to this end we have to return to Equation (3.37). By differentiating the latter partially with respect to a and θ, we find that

$$r_a = \frac{r^2}{a(3a-2r)} = \frac{r^2}{a^2(1-r^3\sin^2\theta)} \tag{3.47}$$

and

$$r_\theta = (ar)^2\, r_a \sin\theta \cos\theta \tag{3.48}$$

or

$$\frac{r_\theta}{r} = \frac{r^3 \sin\theta \cos\theta}{1-r^3\sin^2\theta} \tag{3.49}$$

for $a < 2/3$; so that, at the equator ($\theta = 90°$) of the Roche limit ($a = 2/3$ and $r = 1$), $r_a = \infty$[2]; while the expression (3.48), proportional to the product $r_\theta(2/3, 90°) = 1/\sqrt{3}$.

Moreover, for $a \leq 2/3$

$$\frac{1}{r_a}\left\{\cos\theta + \frac{r_\theta}{r}\sin\theta\right\} = \left[\frac{a}{r}\right]^2 \cos\theta \tag{3.50}$$

while

$$\frac{1}{r_a}\left\{\sin\theta - \frac{r_\theta}{r}\cos\theta\right\} = \left\{\left[\frac{a}{r}\right]^2 - a^2 r\right\}\sin\theta \,, \tag{3.51}$$

and Equations (3.43)–(3.44) can be rewritten as

$$\frac{\partial U}{\partial t} + a^2 r \sin\theta\cos\theta \frac{\partial V}{\partial t} - 2\frac{\sin\theta}{r_a} W = -\frac{1}{(ar_a)^2} \,, \tag{3.52}$$

$$\frac{\partial V}{\partial t} - 2W\cos\theta = r\sin\theta\cos\theta \tag{3.53}$$

and

$$\frac{\partial W}{\partial t} + 2r_a\left\{U\sin\theta + \left[\frac{a}{r}\right]^2 V\cos\theta\right\} = 0 \,. \tag{3.54}$$

---

[2] The proof is supplied by a differentiation of Equation (3.38) with respect to a, disclosing that

$$r_a = {}_2F_1\left(\frac{1}{3}, \frac{2}{3}; \frac{3}{2}; x^2\right) + \left(\frac{2}{3}x\right)^2 {}_2F_1\left(\frac{4}{3}, \frac{5}{3}; \frac{5}{2}; x^2\right) \,;$$

where x^2 continues to be given by Equation (3.39). For $x^2 = 1$ (which obtains at the Lagrangian points of the restricted problem of three bodies), the second hypergeometric series on the right-hand side of the above equation for r_a (since 4/3 + 5/3 - 5/2 > 0) diverges.

In order to separate the dependent variables in (3.52)–(3.54), differentiate these equations repeatedly with respect to the time. In doing so we can take advantage of the fact that only the velocity components U, V, W in these equations depend explicitly on the time; their coefficients are specified by the *equilibrium* properties of our model and, as such, are, by definition, time-independent. Accordingly, the outcome of the *partial* differentiation with respect to t discloses that

$$\frac{\partial^2 U}{\partial t^2} + a^2 r \sin\theta \cos\theta \frac{\partial^2 V}{\partial t^2} = 2\frac{\sin\theta}{r_a}\frac{\partial W}{\partial t}, \tag{3.55}$$

$$\frac{\partial^3 U}{\partial t^2} + a^2 r \sin\theta \cos\theta \frac{\partial^3 V}{\partial t^3} = 2\frac{\sin\theta}{r_a}\frac{\partial W}{\partial t^2}; \tag{3.56}$$

$$\frac{\partial^2 V}{\partial t^2} = 2\cos\theta \frac{\partial W}{\partial t}, \tag{3.57}$$

$$\frac{\partial^3 V}{\partial t^3} = 2\cos\theta \frac{\partial^2 W}{\partial t^2} \tag{3.58}$$

and

$$\frac{\partial^2 W}{\partial t^2} + 2r_a \left\{ \frac{\partial U}{\partial t}\sin\theta + \left[\frac{a}{r}\right]^2 \frac{\partial V}{\partial t} \right\} = 0. \tag{3.59}$$

Inserting in the latter for $\partial U/\partial t$ and $\partial V/\partial t$ from (3.43) and (3.45) in terms of W, and for r_a from (3.47), we find that

$$\frac{\partial^2 W}{\partial t^2} + 4W = \frac{2}{r^2}(1 - r^3)\sin\theta. \tag{3.60}$$

Next, combining (3.58) with (3.53) and (3.60) we obtain

$$\frac{\partial^3 V}{\partial t^3} + 4\frac{\partial V}{\partial t} = \left[\frac{2}{r}\right]^2 \sin\theta\cos\theta; \tag{3.61}$$

while a combination of (3.56) with (3.60) and (3.61) together with (3.52)–(3.53) and elimination of r_a by means of (3.47) discloses that

$$\frac{\partial^3 U}{\partial t^3} + 4\frac{\partial U}{\partial t} + \left\{\frac{2a}{r^2}\cos\theta\right\}^2 = 0. \tag{3.62}$$

Equations (3.60)–(3.62) are all linear in their dependent variables with constant coefficients, but non-homogeneous; and their exact closed solutions are of the form

$$\frac{\partial U}{\partial t} = c_1 \sin(2t + c_1) - \left[\frac{a\cos\theta}{r^2}\right]^2, \tag{3.63}$$

$$\frac{\partial V}{\partial t} = c_2 \sin(2t + c_2) + \frac{\sin\theta\cos\theta}{r^2}, \tag{3.64}$$

$$W = c_3 \sin(2t + c_3) + \frac{(1 - r^3)\sin\theta}{2r^2}. \tag{3.65}$$

Moreover, by integrating Equations (3.63)–(3.64) once more with respect to the time we arrive at the equations

$$U = \bar{c}_1 - \frac{c_1}{2}\cos(2t + c_1) - \left[\frac{a \cos\theta}{2r^2}\right]^2 t \tag{3.66}$$

and

$$V = \bar{c}_2 - \frac{c_2}{2}\cos(2t + c_2) + \left[\frac{\sin\theta \cos\theta}{2r^2}\right] t\,, \tag{3.67}$$

which together with (3.65) represent the closed solution (complete primitive) of Equations (3.60)–(3.62), containing eight integration constants $c_{1,2,3}$–$C_{1,2,3}$ and $\bar{c}_{1,2}$, the values of which must still be specified from given initial conditions.

In order to do so, let us depart from the assumption that, at the time t, the mass-particle is at rest on the equilibrium surface given (cf. Equation 3.36) by

$$\xi = \frac{1}{a}\,, \tag{3.68}$$

where

$$U_0 = V_0 = W_0 = 0\,; \tag{3.69}$$

this will be the case if

$$C_{1,2}\cos c_{1,2} = 2\overline{C}_{1,2} \tag{3.70}$$

and

$$C_3 \sin c_3 = \frac{(1 - r^3)\sin\theta}{2r^2}\,. \tag{3.71}$$

Next, by (3.45),

$$\left(\frac{\partial W}{\partial t}\right)_0 = 0\,, \tag{3.72}$$

which by (3.60) implies tht

$$C_3 \cos c_3 = 0\,. \tag{3.73}$$

Similarly, by (3.53) and (3.64),

$$\left(\frac{\partial V}{\partial t}\right)_0 = r\sin\theta\cos\theta = C_2 \sin c_2 + \frac{\sin\theta\cos\theta}{r^2} \tag{3.74}$$

yielding

$$C_2 \sin c_2 = -\frac{1 - r^3}{r^2}\sin\theta\cos\theta\,; \tag{3.75}$$

while from (3.52) and (3.63) it follows that

$$\begin{aligned} \left(\frac{\partial U}{\partial t}\right)_0 &= -\left\{\frac{a}{r^2}(1 - r^3\sin^2\theta)\right\}^2 - a^2 r^2 \sin^2\theta = \\ &= C_1 \sin c_1 - \left(\frac{a\cos\theta}{r^2}\right)^2\,, \end{aligned} \tag{3.76}$$

yielding

$$C_1 \sin c_1 = -\left\{\frac{a}{r^2}(1-r^3)\sin\theta\right\}^2 ; \tag{3.77}$$

From (3.55)–(3.63) and (3.56)–(3.64) it follows, however, that

$$\left(\frac{\partial^2 U}{\partial t^2}\right)_0 = 2C_1 \cos c_1 = 0 \tag{3.78}$$

and

$$\left(\frac{\partial^2 V}{\partial t^2}\right)_0 = 2C_2 \cos c_2 = 0 \tag{3.79}$$

from which it transpires that

$$c_1 = c_2 = \frac{\pi}{2} . \tag{3.80}$$

If so, however, from (3.70) we see that

$$\overline{C}_1 = \overline{C}_2 = 0 . \tag{3.81}$$

On the other hand, from (3.66) and (3.67) we have

$$C_1 = -\left\{\frac{a}{r^2}(1-r^3)\sin\theta\right\}^2 , \tag{3.82}$$

$$C_2 = -\frac{1-r^3}{r^2}\sin\theta\cos\theta ; \tag{3.83}$$

and, lastly, from (3.71)and (3.73),

$$c_3 = \frac{\pi}{2} \text{ and } c_3 = \frac{(1-r^3)\sin\theta}{2r^2} . \tag{3.84}$$

Inserting for the constants $c_{1,2,3}, C_{1,2,3}$ and $\overline{C}_{1,2}$ from (3.80)–(3.84) in Equations (3.63)–(3.65) for U, V and W, we obtain

$$U = -\left\{\frac{a(1-r^3)}{r^2}\right\}^2 \left\{\sin^2\theta \sin t \cos t + \left[\frac{\cos\theta}{1-r^3}\right]^2 t\right\} , \tag{3.85}$$

$$V = -\frac{1-r^3}{r^2}\sin\theta\cos\theta\left\{\sin t \cos t - \frac{t}{1-r^3}\right\} , \tag{3.86}$$

$$W = \frac{1-r^3}{r^2}\sin\theta \sin^2 t , \tag{3.87}$$

as the final result of this section (the correctness of which can be readily verified by insertion in Equations (3.52)–(3.54).

Of the three velocity components U, V, W given by Equations (3.85)–(3.87), only W—varying as $\sin^2 t = \frac{1}{2}(1-\cos 2t)$—is a purely periodic function of time;

and any disturbance exciting it results in stable harmonic oscillations. The other two components U and V consist, however, of two time-dependent parts. The first, representing a complementary function of Equations (3.61) and (3.62), varies as $\sin t \cos t \equiv \frac{1}{2}\sin 2t$, is likewise periodic in t, but their particular integrals are not; and these render the corresponding oscillations unstable. The secular terms (varying as t) arise, moreover, as soon as the equilibrium form of the Roche model departs from a sphere (i.e., $\omega > 0$); the latter represents, therefore, the only figure of equilibrium around which the Roche model can oscillate harmonically in the course of time.

B: Stability of Roche Equipotentials: Double-Star Problem

In any effort to extend our inquiry on the stability of the Roche equipotentials to double-star configurations, and its treatment by use of the Clairaut coordinates, the following considerations should be kept in mind. First, that within the scheme of our approximation, the Equations (3.33)–(3.35) continue to hold in the present case as well; but we are no longer entitled to set, as before, $r_\phi = 0$ in their coefficients. And—unlike in the case of a purely rotational problem—neither the radius-vector $r(a, \theta, \phi)$, nor its derivatives with respect to a, θ or ϕ can any longer be expressed in a closed form. In the purely rotational case, this was possible (see Equation (3.38)) because the function r was defined as a solution of the cubic equation (3.37); its degree is increased from 3 to 8 if the combined effects of rotation and tides are to be simultaneously taken into account. Lastly, if $m_2 > 0$ (and, therefore, $q \equiv m_2/m_1 > 0$), the (normalized) Keplerian angular velocity ω_K^2 in (3.33)–(3.35) is to be replaced by $1 + q$ in place of unity.

To be sure, a solution of the 8-th degree Equation (2.37) of Chapter II can be constructed by successive approximation; we already did so in Section II.2A, with the outcome represented by Equation (2.21) of that chapter, in which we now only have to replace r_0 by a. Differentiating this latter equation partially with respect to a, θ and ϕ we find that, to the first order in small quantities,

$$\begin{aligned} r_a &= 1 + 4na^3(1-\nu^2) + qa^3\{P_2^2(\nu)\cos 2\phi - 2P_2(\nu)\} - \\ &- \frac{5}{4}qa^4\left\{\frac{1}{6}P_3^3(\nu)\cos 3\phi + P_3^1(\nu)\cos\phi\right\} + \\ &+ qa^5\left\{\frac{1}{32}P_4^4(\nu)\cos 4\phi - \frac{1}{6}P_4^2(\nu)\cos 2\phi + \frac{9}{4}P_4(\nu)\right\} + \dots , \qquad (3.88) \end{aligned}$$

$$\begin{aligned} \frac{r_\theta}{r} &= -\frac{1}{6}na^3P_1^1(\nu) - qa^3\{P_2^2(\nu)\cos\phi - \frac{1}{4}a[P_3^2(\nu)\cos 2\phi - 6P_3(\nu)] + \\ &+ \frac{1}{24}a^2[P_4^3(\nu)\cos 3\phi - 18P_4^1(\nu)\cos\phi] + \dots\}\cos\phi \qquad (3.89) \end{aligned}$$

and

$$\frac{r_\phi}{r} = -\frac{1}{2}qa^3P_2^2(\nu)\sin 2\phi - \frac{1}{8}qa^4[P_3^3(\nu)\sin 3\phi - P_3^1(\nu)\sin\phi] -$$

$$- \frac{1}{48} q a^5 [P_4^4(\nu) \sin 4\phi - 4P_4^2(\nu) \sin 2\phi] - \dots , \tag{3.90}$$

where the P_n^m's stand for associated Legendre polynomials of the argument $\nu \equiv \cos\theta$, of order m and degree n; and in their terms all coefficients in Equations (3.33)–(3.35) can be specified.

In order to solve these equations, let us avail ourselves of the same technique as in the preceding sub-section, and differentiate them partially with respect to the time. Inserting, moreover, for $\partial U/\partial t$ and $\partial V/\partial t$ from (3.33) and (3.34) in (3.35), this latter equation will assume the form

$$\frac{\partial^2 W}{\partial t^2} + 4\omega^2 W = \frac{2\omega\xi_a}{r r_a} \frac{\partial}{\partial\theta}(r \cos\theta) . \tag{3.91}$$

Next, differentiating (3.34) twice with respect to t and inserting from (3.35) we arrive at

$$\frac{\partial^3 V}{\partial t^3} + 4\omega^2 \frac{\partial V}{\partial t} = -\frac{2\omega^2}{r^2} \left[\frac{\xi_a}{r_a} \right] . \tag{3.92}$$

Lastly, by a combination of (2.3)–(2.5) with the preceding Equations (3.1) and (3.2) we establish that

$$\frac{\partial^3 U}{\partial t^3} + 4\omega^2 \frac{\partial U}{\partial t} = \frac{4\omega^2 \xi_a}{(r r_a)^2} \left\{ \frac{\partial}{\partial\theta}(r \sin\theta) \right\}^2 , \tag{3.93}$$

where

$$\xi = \frac{1}{a} + q, \quad \xi_a = -\frac{1}{a^2} \tag{3.94}$$

and

$$\omega = 1 + q . \tag{3.95}$$

Equations (3.91)–(3.93) represent a fundamental set of differential equations of our problem—analogous to Equations (3.43)–(3.45) of the purely rotational case. The reader may note that, although the inclusion of tidal effects renders the partial derivative r_ϕ different from zero, Equations (3.91)–(3.93) do not contain r_ϕ explicitly; the tidal effects are contained in them only indirectly through r, r_a and r_θ.

However, the partial derivative r_ϕ does occur explicitly in the *initial conditions* of our problem at $t = 0$. By analogy with the rotational problem treated in Section V.3A, we depart from the equilibrium form of the Roche model, where

$$U_0 = V_0 = W_0 = 0 . \tag{3.96}$$

Therefore, by (3.33)–(3.35),

$$\left(\frac{\partial u}{\partial t}\right)_0 = \left(\frac{\xi_a}{r_a^2}\right) \left\{ 1 + \left(\frac{r_\theta}{r}\right)^2 + \left(\frac{r_\phi}{r \sin\theta}\right) \right\} = \frac{g^{11}}{a^2} \tag{3.97}$$

and

$$\left(\frac{\partial V}{\partial t}\right)_0 = -\frac{r_\theta}{r}\left(\frac{\xi_a}{r_a}\right) = \frac{r_\theta}{a^2 r r_a} = -\left(\frac{g^{12}}{a^2}\right) r\,, \tag{3.98}$$

$$\left(\frac{\partial W}{\partial t}\right)_0 = -\frac{r_\phi}{r\,\sin\,\theta}\left(\frac{\xi_a}{r_a}\right) = \frac{r_\phi}{a^2 r r_a \sin\,\theta} = -\left(\frac{g^{13}}{a^2}\right) r\,\sin\,\theta, \tag{3.99}$$

where the metric coefficients g^{ij} occurring on the right-hand sides have already been specified by Equations (1.27) and (1.30)–(1.31).

Moreover, from the same equations it follows that

$$\left(\frac{\partial^2 U}{\partial t^2}\right)_0 = 0 \tag{3.100}$$

and

$$\left(\frac{\partial^2 V}{\partial t^2}\right)_0 = -\frac{2\omega r_\phi}{r\,\sin\,\theta}\left(\frac{\xi_a}{r_a}\right)\cos\,\theta = -\omega\left(\frac{g^{13}}{a^2}\right) r\,\sin\,2\theta\,; \tag{3.101}$$

which, together with the preceding Equations (3.96)–(3.99), is sufficient to specify the particular solutions of the simultaneous system (3.91)–(3.93) of eighth order in t, representing the solution of our problem.

Equations (3.91)–(3.93) are linear but non-homogeneous, with coefficients (specified by the equilibrium properties of the respective configuration) independent of the time. Therefore, their complete primitives will be represented by a sum of their complementary functions, of the general form $\sin(2\omega t + c)$, augmented by the particular integrals specified by their right-hand sides. Those of (3.91) and (3.92) are of second order in the first time-derivatives of U and V; and, accordingly, their complete primitives can be expressed as

$$\frac{\partial U}{\partial t} = C_1 \sin(2\omega t + c_1) - (a r r_a)^{-2}(r_\theta \sin\,\theta + r\,\cos\,\theta)^2 \tag{3.102}$$

and

$$\frac{\partial V}{\partial t} = C_2 \sin(2\omega t + c_2) + (a^2 r r_a)^{-1}(r_\theta \sin\,\theta + r\,\cos\,\theta)\,, \tag{3.103}$$

while (3.91) yields similarly

$$W = C_3 \sin(2\omega t + c_3) - (2\omega a^2 r r_a)^{-1}(r_\theta \cos\,\theta - r\,\sin\,\theta)\,; \tag{3.104}$$

in which $C_{1,2,3}$ and $c_{1,2,3}$ stand for six integration constants of our problem.

Moreover, Equations (3.102) and (3.103) can be integrated by simple quadratures, to disclose that

$$U = -\frac{C_1}{2\omega}\cos(2\omega t + c_1) - (r r_a a)^{-2}(r_\theta \sin\,\theta + r\,\cos\,\theta)^2 t + \overline{C}_1\,, \tag{3.105}$$

and

$$V = -\frac{C_2}{2\omega}\cos(2\omega t + c_2) + (a^2 r_a^2)^{-1}(r\sin\theta)(r_\theta \sin\theta + r\cos\theta)t + \overline{C}_2\,; \quad (3.106)$$

where $\overline{C}_{1,2}$ are two additional integration constants; augmenting their complete set to eight.

In order to specify all these constants, we fall back on the initial conditions (3.96)–(3.101) of our problem. Equations (3.105) and (3.106) readily disclose that

$$C_{1,2}\cos c_{1,2} = 2\omega\overline{C}_{1,2}\,. \quad (3.107)$$

Furthermore, Equation (3.104) supplies the relation

$$C_3 \sin c_3 = (2\omega a^2 r r_a)^{-1}\{r_\theta \cos\theta - r\sin\theta\}\,; \quad (3.108)$$

while, in accordance with (3.99) and a time-derivative of (3.104),

$$C_3 \cos c_3 = \frac{r_\phi}{2\omega a^2 r r_a}\sin\theta\,. \quad (3.109)$$

A combination of the foregoing two equations yields

$$\begin{aligned} C_3 &= \frac{1}{2\omega a^2 r_s}\left\{\left[\frac{1}{r}\frac{\partial}{\partial\theta}(r\cos\theta)\right]^2 + \left[\frac{r_\phi}{r\sin\theta}\right]^2\right\}^{\frac{1}{2}} = \\ &= \frac{r_\phi \sec c_3}{2\omega a^2 r r_a \sin\theta}\,, \end{aligned} \quad (3.110)$$

where

$$\tan c_3 = \frac{\sin\theta}{r_\phi}\frac{\partial}{\partial\theta}(r\cos\theta)\,. \quad (3.111)$$

Turning next to a determination of $C_{1,2}$ and $c_{1,2}$, from Equations (3.97) and (3.102) for $t = 0$ we find that

$$C_1 \sin c_1 = -(a r_a)^{-2}\left\{\left[\frac{1}{r}\frac{\partial}{\partial\theta}(r\cos\theta)\right]^2 + \left[\frac{r_\phi}{r\sin\theta}\right]^2\right\}\,; \quad (3.112)$$

while (3.98) and (3.102) similarly yield

$$C_2 \sin c_2 = \frac{\cos\theta}{a^2 r r_a}\frac{\partial}{\partial\theta}(r\cos\theta)\,. \quad (3.113)$$

On the other hand, the initial conditions (3.100) and (3.101) combined with the time-derivatives of (3.102) and (3.103) furnish us with the relations

$$C_1 \cos c_1 = 0 \quad (3.114)$$

and

$$C_2 \cos c_2 = \frac{1}{a^2 r_a}\left(\frac{r_\phi}{r\cos\theta}\right)\cos\theta\,. \quad (3.115)$$

A combination of (3.112) and (3.113) with (3.114) and (3.115) discloses that

$$C_1 = -\frac{1}{a^2 r_a^2}\left\{\left[\frac{1}{r}\frac{\partial}{\partial\theta}(r\cos\theta)\right]^2 + \left[\frac{r_\phi}{r\sin\theta}\right]^2\right\} = $$

$$= -\left\{\frac{r_\phi}{r\sin\theta}\,\frac{\sec c_3}{a r_a}\right\}^2, \tag{3.116}$$

$$c_1 = \frac{\pi}{2} \tag{3.117}$$

and

$$C_2 = \frac{\cos\theta}{a^2 r_a}\left\{\left[\frac{1}{r}\frac{\partial}{\partial\theta}(r\cos\theta)\right]^2 + \left[\frac{r_\phi}{r\sin\theta}\right]^2\right\}^{\frac{1}{2}} = $$

$$= \frac{r_\phi \cot\theta}{a^2 r r_a \cos c_2}, \tag{3.118}$$

where

$$\tan c_2 = \frac{\sin\theta}{r_\phi}\frac{\partial}{\partial\theta}(r\cos\theta) = \tan c_3\,; \tag{3.119}$$

so that

$$c_2 = c_3 \pm n\pi\,, \tag{3.120}$$

where n is an integer (including zero).

The remaining integration constants of our solution to be determined are the barred values of $\overline{C}_{1,2}$ as defined by Equation (3.107). In view of (3.117) it follows immediately that

$$\overline{C}_1 = 0\,; \tag{3.121}$$

while by (3.118) and (3.119) we have

$$\overline{C}_2 = \frac{r_\phi \cot\theta}{2\omega a^2 r r_a}. \tag{3.122}$$

If, eventually, we insert the values of all integration constants previously established in this section in the complete primitives (3.104)–(3.106), the respective particular solutions of our problem assume the forms

$$2\omega(ar_a)^2 U = -\left\{\sec^2\beta - \left[\frac{1}{r}\frac{\partial}{\partial\theta}(r\cos\theta)\right]^2\right\}\sin 2\omega t - $$

$$- \left[\frac{1}{r}\frac{\partial}{\partial\theta}(r\sin\theta)\right]^2 2\omega t\,, \tag{3.123}$$

$$2\omega a^2 r_a V = 2\cos\theta\left[\frac{r_\phi}{r\sin\theta}\right]\sin^2\omega t + $$

$$+ \left[\frac{1}{r^2}\frac{\partial}{\partial\theta}(r\cos\theta)^2\right]\sin\omega t\cos\omega t + $$

$$+ \left[\frac{1}{r^2}\frac{\partial}{\partial\theta}(r\sin\theta)^2\right]\omega t \tag{3.124}$$

and

$$2\omega a^2 r_a W = \left[\frac{r_\phi}{r \sin\theta}\right] \sin 2\omega t - \left[\frac{2}{r}\frac{\partial}{\partial\theta}(r \cos\theta)\right] \sin^2 \omega t \,. \qquad (3.125)$$

As the reader can easily verify, the foregoing particular solutions of our problem satisfy indeed Equations (3.33)–(3.35) as well as the initial conditions (3.96)–(3.101). Moreover, if we set in Equations (3.123)–(3.125) $q = 0$ ($\omega = 1$), and $r_\phi = 0$, insert for $r_{a,\theta}$ from Equations (3.50) and (3.51), we find that Equations (3.123)–(3.125) reduce indeed to Equations (3.85)–(3.87) valid for the purely rotational case.

Equations (3.123)–(3.125) represent, therefore, the final outcome of our analysis. Its most important feature is the presence (as in the case of a purely rotational problem) in (3.123) and (3.125) of secular terms varying as ωt. However, before we accept these as indicators of secular instability of the respective configurations, we should remember that these were derived with the tacit assumption that the (normalized) values of the velocity components U, V, W—set at zero for $t = 0$— remain small enough (in comparison with unity) for their squares and cross-products to be ignorable; for this was the assumption on which Equations (3.43)–(3.45) were derived in Section V.3A as a price of their linearization. Should this not be the case (i.e., for large values of ωt), the velocity componments U and V need not remain small indefinitely; and a retention of quadratic terms in (3.43)–(3.45) may possibly (though not necessarily) restore stability. However, as long as such non-linear analysis has not been actually performed it is impossible to anticipate the outcome.

In conclusion, let us mention that an investigation of the motion of mass particles in the proximity of Roche equipotentials simulating a close binary is, in effect, equivalent to a study of three-dimensional motions of the restricted problem of three bodies in the proximity of its surfaces of zero velocity.

As is well known, in the immediate proximity of such a surface, the velocity components U, V, W can be expanded in Taylor series of the form

$$U = U_0 + \left(\frac{\partial U}{\partial t}\right)_0 t + \left(\frac{\partial^2 U}{\partial t^2}\right)_0 \frac{t^2}{2} + \left(\frac{\partial^3 U}{\partial t^3}\right)_0 \frac{t^3}{6} + \ldots, \qquad (3.126)$$

$$V = V_0 + \left(\frac{\partial V}{\partial t}\right)_0 t + \left(\frac{\partial^2 V}{\partial t^2}\right)_0 \frac{t^2}{2} + \left(\frac{\partial^3 V}{\partial t^3}\right)_0 \frac{t^3}{6} + \ldots, \qquad (3.127)$$

$$W = W_0 + \left(\frac{\partial W}{\partial t}\right)_0 t + \left(\frac{\partial^2 W}{\partial t^2}\right)_0 \frac{t^2}{2} + \left(\frac{\partial^3 W}{\partial t^3}\right)_0 \frac{t^3}{6} + \ldots, \qquad (3.128)$$

which in view of the differential equations (3.91)–(3.93) and of their initial conditions (3.96)–(3.101) assume the more explicit forms

$$U = \left(\frac{\sec\beta}{r_a}\right)^2 \left(\frac{\xi_a}{r_a}\right) t - \frac{2}{3}\frac{\omega^2}{r_a}\left\{\left[\frac{1}{r}\frac{\partial}{\partial\theta}(r\cos\theta)\right]^2 + \right.$$

$$+ \left[\frac{r_\phi}{r \sin\theta}\right]^2 \Bigg\} \left(\frac{\xi_a}{r_a}\right) t^3 + \dots, \tag{3.129}$$

$$V = -\frac{r_\theta}{r}\left(\frac{\xi_a}{r_a}\right) t - \frac{\omega r_\phi \cos\theta}{r \sin\theta}\left(\frac{\xi_a}{r_a}\right) t^2 + + \frac{1}{3}\left(\frac{\omega}{r}\right)^2 \left[\frac{\partial}{\partial\theta}(r \cos\theta)^2\right]\left(\frac{\xi_a}{r_a}\right) t^3 + \dots, \tag{3.130}$$

$$W = -\frac{r_\phi}{r \sin\theta}\left(\frac{\xi_a}{r_a}\right) t + \frac{\omega}{r}\left[\frac{\partial}{\partial\theta}(r \cos\theta)\right]\left(\frac{\xi_a}{r_a}\right) t^2 + + \frac{2\omega^2 r_\phi}{r^3 \sin\theta}\left(\frac{\xi_a}{r_a}\right) t^3 + \dots \tag{3.131}$$

in agreement with an expansion of the right-hand sides of Equations (3.123)–(3.125) in ascending powers of t.

Inserting next on the left-hand sides of (3.129)–(3.131) from (1.55) and integrating with respect to the time, the former equations disclose that

$$a = a_0 + \frac{1}{2!}\left(\frac{\sec\beta}{r_a}\right)^2 \left(\frac{\xi_a}{r_a}\right) t^2 - \frac{1}{3!}\omega^2 C_1 t^4 + \dots, \tag{3.132}$$

and similarly for θ and ϕ, demonstrating (cf., e.g., Szebehely, 1967; p.193) that the motion from an arbitrary point specified by the Clairaut coordinates a_0, θ_0, ϕ_0 starts in the direction normal to the equipotential a_0 = constant; and its transverse components become significant only in the course of time. It is also of interest to note that whereas the expansions of the right-hand sides of (3.124) and (3.125) for V and W in ascending powers of t contain all powers even as well as odd, only odd powers are encountered on the right-hand side of Equation (3.123) for U; and, consequently, only even powers will occur on the right-hand side of the foregoing Equation (3.132).

This fact gives rise also to another question: is it possible to satisfy the simultaneous system (3.33)–(3.35) of fundamental equations underlying our analysis by choosing one (or two) of their dependent variables in an arbitrary manner? A trivial solution $U = V = W = 0$ obviously does not satisfy (for $r_a < \infty$) Equations (3.33)–(3.35) because of their non-homogeneous nature; and their right-hand sides could not all vanish even if the zero-velocity surfaces would reduce to spheres. Can, however, these be satisfied by motions in the direction of a (i.e., a purely "radial" motion in Clairaut coordinates) while θ and ϕ are kept constant? In such a case, $V = W = 0$ would represent necessary, but not sufficient conditions for purely radial motion; for in order to satisfy Equations (3.34) and (3.35) we should also require that $r = a$ (implying that $r_a = 1$, $r_\theta = r_\phi = 0$); and $\omega = 0$ in Equation (3.35) corresponding to a two-centre problem. All these conditions are tantamount to an assumption that the equipotential surfaces of our configuration are spheres of radius a; and their departures from spherical form would require a to be small enough for its cubes and higher powers to be ignorable. Under such assumptions, Equation (3.33) reduces then to one of free fall in a radial direction, but in any other case our Equations (3.123)–(3.125) represent the only solution of our problem.

V.4 Bibliographical Notes

The non-orthogonal Clairaut coordinates a, θ, ϕ, in three dimensions—in which the inverse gravitational potential a plays the role of the radial coordinate—have made their appearance in the past ten years in two papers by the present writer (Kopal, 1980b, 1983) in which the various properties of such a frame of reference have first been systematically investigated. The gist of their properties has been summarized in Sections V.1 and V.2 of the present chapter; while in Section V.3 these new coordinates have been utilized to investigate the vibrational stability of the configurations of arbitrary structure distorted by axial rotation (Section V.3A) or by the tides (Section V.3B).

These investigations appeared originally in two recent papers by Kopal (1987a, b); the first of which contains also the discussion of "Clairaut harmonics" associated with the rotational distortion; but for tidal distortion their specification still remains almost virgin ground.

Chapter VI

GENERALIZED ROTATION

In the preceding five chapters of this book we have been concerned with the effects produced by uni-axial rotation on the shape of self-gravitating fluids of arbitrary structure. There was a good reason for that; for all stars observed so far in the sky, whether single or components of multiple systems, are found to be endowed with rotation—no doubt as a consequence of the vorticity in pre-existing gas (or plasma), from which the stars originate by contraction. The angular momentum associated with rotation—conserved over long evolutionary epochs—represents as fundamental a property with which stars embark on their celestial careers as their initial masses and chemical composition.

Moreover, if the stars are not born from pre-exisitng substrate as single individuals but in pairs constituting binary systems, their orbital motion alone is bound to endow them also with axial rotation—usually coincident with the Keplerian angular velocity of orbital motion, or not deviating from it by factors of greater than zero-order. In such systems, the parts constituting the total potential and arising from rotation and tides are generally comparable; the case of tides without rotation (corresponding, dynamically speaking, to a "two-centre" instead of "three-body" problem) may be relevant only to studies of motions which are very fast in comparison with those of orbital revolution.

In addition, the rotational motions of the stars need not be restricted to those about one axis only; and investigations limited to uni-axial rotation underestimate, in general, the number of the degrees of freedom which such configurations may actually possess. A freely-rotating star can, to be sure, do so about an axis which is fixed in space (or with an orientation influenced only by encounters with other nearby stars). Should, however, such a star happen to be the component of a binary or multiple system, a dynamical interaction between the orientation of equatorial and orbital planes becomes an additional problem. Within the framework of the classical Roche model (cf. Section II.3A), their orientations can be arbitrary, and remain constant for an indefinite time. However, should the internal structure of such stars exhibit a finite concentration of the density towards their centre, this need no longer be the case, and the mutual relationship of these quantities remains to be investigated; to do so will be the main aim of the present chapter.

To embark on this task, we shall consider first the general problem of *three-axial rotation*, in which the velocity components u, v, w in the direction of *inertial* coordinates x, y, z will be replaced by the angular velocity components ω_x, ω_y, ω_z of rotation about the respective axes. One particular assumption will, however, accompany us throughout this work to restrict its generality: namely, that *all three angular velocity components* $\omega_{x,y,z}$ *do not depend on the coordinates* x, y, z, *and are functions of the time only.* In doing so we restrict, in effect, our configuration to rotate like a rigid body, but with a velocity which may vary with the time; and shall allow also for a time-dependent deformability of its shape.

In contrast with the problem of rigid solids rotating about three axes in an external field of force—which is classical and goes back to Newton (1687)—a mathematical treatment of three-dimensional rotation of *deformable* bodies was, however, rather slow to progress (cf. Liouville, 1858; Gyldén, 1871; Darwin, 1879; or Poincaré, 1910); and is still far from being properly solved for the precession and nutation of the Earth—let alone for fluid components of close binary systems; and the aim of the present chapter will be to provide a more comprehensive treatment than has been done by any previous investigator of the subject. The principal innovation of our treatment will be to depart, not from the Lagrangian equations of rational mechanics as was done by all our predecessors, but from the hydrodynamical equations of viscous fluids, in which the terms arising from viscosity will not necessarily be regarded as small. The effects of viscous friction will, incidentally, tend to lessen any dependence of $\omega_{x,y,z}$ on x, y, z, and thus provide an *a posteriori* justification for treating the angular velocity components dependent only on the time.

VI.1 Equations of Motion for Deformable Bodies

As in previous chapters of this book, the Eulerian fundamental equations of hydrodynamics governing the motion of compressible viscous fluids can be expressed in rectangular coordinates in the symmetrical form

$$\rho\frac{Du}{Dt} = \rho\frac{\partial\Psi}{\partial x} - \frac{\partial P}{\partial x} + \frac{\partial_{xx}}{\partial x} + \frac{\partial\sigma_{xy}}{\partial y} + \frac{\partial\sigma_{xz}}{\partial z}, \tag{1.1}$$

$$\rho\frac{Dv}{Dt} = \rho\frac{\partial\Psi}{\partial y} - \frac{\partial P}{\partial y} + \frac{\partial\sigma_{yx}}{\partial x} + \frac{\partial\sigma_{yy}}{\partial y} + \frac{\partial\sigma_{yz}}{\partial z}, \tag{1.2}$$

$$\rho\frac{Dw}{Dt} = \rho\frac{\partial\Psi}{\partial z} - \frac{\partial P}{\partial z} + \frac{\partial\sigma_{zx}}{\partial x} + \frac{\partial\sigma_{zy}}{\partial y} + \frac{\partial\sigma_{zz}}{\partial z}, \tag{1.3}$$

where u, v, w denote the velocity components of fluid motion, at the time t, in the direction of increasing coordinates x, y, z;

$$\frac{D}{Dt} \equiv \frac{\partial}{\partial t} + u\frac{\partial}{\partial x} + v\frac{\partial}{\partial y} + w\frac{\partial}{\partial z} \tag{1.4}$$

represents the Lagrangian time-derivative (following the motion); ρ stands for the local density of the fluid; P, for its pressure; Ψ, for the total potential of forces acting upon it; and

$$\sigma_{xx} = \frac{2}{3}\mu\left\{3\frac{\partial u}{\partial x} - \Delta\right\}, \tag{1.5}$$

$$\sigma_{yy} = \frac{2}{3}\mu\left\{3\frac{\partial v}{\partial y} - \Delta\right\} , \tag{1.6}$$

$$\sigma_{zz} = \frac{2}{3}\mu\left\{3\frac{\partial w}{\partial z} - \Delta\right\} , \tag{1.7}$$

$$\sigma_{xy} = \mu\left\{\frac{\partial v}{\partial x} + \frac{\partial u}{\partial y}\right\} = \sigma_{yx} , \tag{1.8}$$

$$\sigma_{yz} = \mu\left\{\frac{\partial w}{\partial y} + \frac{\partial v}{\partial z}\right\} = \sigma_{xy} , \tag{1.9}$$

$$\sigma_{zx} = \mu\left\{\frac{\partial u}{\partial z} + \frac{\partial w}{\partial x}\right\} = \sigma_{xz} , \tag{1.10}$$

are the respective components of the viscous stress-tensor, where μ denotes the coefficient of viscosity; and, lastly,

$$\Delta \equiv \frac{\partial u}{\partial x} + \frac{\partial v}{\partial y} + \frac{\partial w}{\partial z} \tag{1.11}$$

stands for the divergence of the velocity vector of the fluid.

Inasmuch as the principal aim of this chapter will be to investigate the absolute motions in space of the principal axes of inertia of deformable components of close binary systems, the rectangular coordinates x, y, z representing independent variables of the system (1.1)–(1.3) are referred to the *inertial* (unprimed) XYZ system introduced in Section 3A of Chapter II; while the primed system $X'Y'Z'$ of rectangular coordinates with the same origin are displaced with respect to the space axes XYZ by the Eulerian angles θ, ϕ, ψ diagramatically shown on Figure 2.3.

A: Velocities and Accelerations

A transformation of coordinates between these systems is governed by Equations (3.2)–(3.5) of Chapter II, in which primes on the direction cosines a_{ij} will be dropped (in subsections A to C alone) to avoid confusion between ordinary and partial derivatives. Differentiating these coordinates with respect to t, we find that the space velocity components

$$\dot{x} \equiv u, \quad \dot{y} \equiv v, \quad \dot{z} \equiv w \tag{1.12}$$

are related with the primed velocity components

$$\dot{x}' \equiv u', \quad \dot{y}' \equiv v', \quad \dot{z}' \equiv w' \tag{1.13}$$

by the equations

$$\dot{x} = u = \dot{a}_{11}x' + \dot{a}_{12}y' + \dot{a}_{13}z' + a_{11}\dot{x}' + a_{12}\dot{y}' + a_{13}\dot{z}' , \tag{1.14}$$

$$\dot{y} = v = \dot{a}_{21}x' + \dot{a}_{22}y' + \dot{a}_{23}z' + a_{21}\dot{x}' + a_{22}\dot{y}' + a_{23}\dot{z}' , \tag{1.15}$$

$$\dot{z} = w = \dot{a}_{31}x' + \dot{a}_{32}y' + \dot{a}_{33}z' + a_{31}\dot{x}' + a_{32}\dot{y}' + a_{33}\dot{z}' ; \tag{1.16}$$

or

$$\begin{aligned}
\dot{x}' &= u' = \dot{a}_{11}x + \dot{a}_{21}y + \dot{a}_{31}z + a_{11}\dot{x} + a_{21}\dot{y} + a_{31}\dot{z}\ , && (1.17)\\
\dot{y}' &= v' = \dot{a}_{12}x + \dot{a}_{22}y + \dot{a}_{32}z + a_{12}\dot{x} + a_{22}\dot{y} + a_{32}\dot{z}\ , && (1.18)\\
\dot{z}' &= w' = \dot{a}_{13}x + \dot{a}_{23}y + \dot{a}_{33}z + a_{13}\dot{x} + a_{23}\dot{y} + a_{33}\dot{z}\ ; && (1.19)
\end{aligned}$$

in which

$$\begin{aligned}
\dot{a}_{11} &= a_{12}\dot{\psi} - a_{21}\dot{\phi} + a_{31}\dot{\theta}\sin\phi = a_{31}\omega_y - a_{21}\omega_z = a_{12}\omega_{z'} - a_{13}\omega_{y'}, && (1.20)\\
\dot{a}_{12} &= -a_{11}\dot{\psi} - a_{22}\dot{\phi} + a_{32}\dot{\theta}\sin\phi = a_{32}\omega_y - a_{22}\omega_z = a_{13}\omega_{x'} - a_{11}\omega_{z'}, && (1.21)\\
\dot{a}_{13} &= - a_{23}\dot{\phi} + a_{33}\dot{\theta}\sin\phi = a_{33}\omega_y - a_{23}\omega_z = a_{11}\omega_{y'} - a_{12}\omega_{x'}; && (1.22)\\
\dot{a}_{21} &= a_{22}\dot{\psi} + a_{11}\dot{\phi} - a_{31}\dot{\theta}\cos\phi = a_{11}\omega_z - a_{31}\omega_x = a_{22}\omega_{z'} - a_{23}\omega_{y'}, && (1.23)\\
\dot{a}_{22} &= -a_{21}\dot{\psi} + a_{12}\dot{\phi} - a_{32}\dot{\theta}\cos\phi = a_{12}\omega_z - a_{32}\omega_x = a_{23}\omega'_x - a_{21}\omega'_z, && (1.24)\\
\dot{a}_{23} &= + a_{13}\dot{\phi} - a_{33}\dot{\theta}\cos\phi = a_{13}\omega_z - a_{33}\omega_x = a_{21}\omega_{y'} - a_{22}\omega_{x'}; && (1.25)\\
\dot{a}_{31} &= a_{32}\dot{\psi} + \dot{\theta}\sin\psi\cos\theta = a_{21}\omega_x - a_{11}\omega_y = a_{32}\omega_{z'} - a_{33}\omega_{y'}, && (1.26)\\
\dot{a}_{32} &= -a_{31}\dot{\psi} + \dot{\theta}\cos\psi\cos\theta = a_{22}\omega_x - a_{12}\omega_y = a_{33}\omega_{x'} - a_{31}\omega_{z'}, && (1.27)\\
\dot{a}_{33} &= - \dot{\theta}\sin\theta = a_{23}\omega_x - a_{13}\omega_y = a_{31}\omega_{y'} - a_{32}\omega_{x'}. && (1.28)
\end{aligned}$$

If, moreover, we take advantage of the fact that

$$\left.\begin{aligned}
a_{11}\dot{a}_{11} + a_{12}\dot{a}_{12} + a_{13}\dot{a}_{13} &= 0\ ,\\
a_{21}\dot{a}_{21} + a_{22}\dot{a}_{22} + a_{23}\dot{a}_{23} &= 0\ ,\\
a_{31}\dot{a}_{31} + a_{32}\dot{a}_{32} + a_{33}\dot{a}_{33} &= 0
\end{aligned}\right\} \qquad (1.29)$$

and

$$\left.\begin{aligned}
a_{11}\dot{a}_{11} + a_{21}\dot{a}_{21} + a_{31}\dot{a}_{31} &= 0\ ,\\
a_{12}\dot{a}_{12} + a_{22}\dot{a}_{22} + a_{32}\dot{a}_{32} &= 0\ ,\\
a_{13}\dot{a}_{13} + a_{23}\dot{a}_{23} + a_{33}\dot{a}_{33} &= 0\ ,
\end{aligned}\right\} \qquad (1.30)$$

the angular velocity components ω_x, ω_y, ω_z about the space axes X, Y, Z are found to be given by

$$\begin{aligned}
\omega_x &= +(a_{21}\dot{a}_{31} + a_{22}\dot{a}_{32} + a_{23}\dot{a}_{33}) = -(a_{31}\dot{a}_{21} + a_{32}\dot{a}_{22} + a_{33}\dot{a}_{23}), && (1.31)\\
\omega_y &= +(a_{31}\dot{a}_{11} + a_{32}\dot{a}_{12} + a_{33}\dot{a}_{13}) = -(a_{11}\dot{a}_{31} + a_{12}\dot{a}_{32} + a_{13}\dot{a}_{33}), && (1.32)\\
\omega_z &= +(a_{11}\dot{a}_{21} + a_{12}\dot{a}_{22} + a_{13}\dot{a}_{23}) = -(a_{21}\dot{a}_{11} + a_{22}\dot{a}_{12} + a_{22}\dot{a}_{13}); && (1.33)
\end{aligned}$$

or

$$\begin{aligned}
\omega_{x'} &= +(a_{13}\dot{a}_{12} + a_{23}\dot{a}_{22} + a_{33}\dot{a}_{32}) = -(a_{12}\dot{a}_{13} + a_{22}\dot{a}_{23} + a_{32}\dot{a}_{33}), && (1.34)\\
\omega_{y'} &= +(a_{11}\dot{a}_{13} + a_{21}\dot{a}_{23} + a_{31}\dot{a}_{33}) = -(a_{13}\dot{a}_{11} + a_{23}\dot{a}_{21} + a_{33}\dot{a}_{31}), && (1.35)\\
\omega_{z'} &= +(a_{12}\dot{a}_{11} + a_{22}\dot{a}_{21} + a_{32}\dot{a}_{31}) = -(a_{11}\dot{a}_{12} + a_{21}\dot{a}_{22} + a_{31}\dot{a}_{32}), && (1.36)
\end{aligned}$$

about the body axes; the pairs of alternative equations arising from the fact that, by a time-differentiation of the relations $a_{ij}a_{ik} = \delta_{jk}$ it follows that $a_{ij}\dot{a}_{ik} + a_{ik}\dot{a}_{ij} = 0$.

If we insert in Equations (1.31)–(1.36) from (1.20)–(1.28) it follows that, in terms of the Eulerian angles,

$$\omega_x = \dot{\theta}\cos\phi + \dot{\psi}\sin\theta\sin\phi , \tag{1.37}$$
$$\omega_y = \dot{\theta}\sin\phi - \dot{\psi}\sin\theta\cos\phi , \tag{1.38}$$
$$\omega_z = + \dot{\psi}\cos\theta + \dot{\phi} ; \tag{1.39}$$

while

$$\omega_{x'} = \dot{\phi}\sin\theta\ \sin\psi + \dot{\theta}\cos\psi , \tag{1.40}$$
$$\omega_{y'} = \dot{\phi}\sin\theta\ \cos\psi - \dot{\theta}\sin\psi , \tag{1.41}$$
$$\omega_{z'} = \dot{\phi}\cos\theta + \dot{\psi} , \tag{1.42}$$

as could be also directly verified by an application of the inverse of the transformation (3.2) of Chapter II, in accordance with which

$$\left.\begin{aligned} \omega_{x'} &= a_{11}\omega_x + a_{21}\omega_y + a_{31}\omega_z , \\ \omega_{y'} &= a_{12}\omega_x + a_{22}\omega_y + a_{32}\omega_z , \\ \omega_{z'} &= a_{13}\omega_x + a_{23}\omega_y + a_{33}\omega_z . \end{aligned}\right\} \tag{1.43}$$

With the aid of the preceding results the Equations (1.14)–(1.16) or (1.17)–(1.19) for the velocity-components with respect to the space or body axes can be reduced to the forms

$$u = z\omega_y - y\omega_z + u_0' , \tag{1.44}$$
$$v = x\omega_z - z\omega_x + v_0' , \tag{1.45}$$
$$w = y\omega_x - x\omega_y + w_0' , \tag{1.46}$$

or

$$u' = -z'\omega_{y'} + y'\omega_{z'} + u_0 , \tag{1.47}$$
$$v' = -x'\omega_{z'} + z'\omega_{x'} + v_0 , \tag{1.48}$$
$$w' = -y'\omega_{x'} + x'\omega_{y'} + w_0 , \tag{1.49}$$

where

$$\left.\begin{aligned} u_0 &= a_{11}u + a_{21}v + a_{31}w , \\ v_0 &= a_{12}u + a_{22}v + a_{32}w , \\ w_0 &= a_{13}u + a_{23}v + a_{33}w , \end{aligned}\right\} \tag{1.50}$$

are the *space* velocity components in the direction of the *rotating* axes x', y', z'; and

$$\left.\begin{aligned} u_0' &= a_{11}u' + a_{12}v' + a_{13}w' , \\ v_0' &= a_{21}u' + a_{22}v' + a_{23}w' , \\ w_0' &= a_{31}u' + a_{32}v' + a_{33}w' , \end{aligned}\right\} \tag{1.51}$$

are the *body* velocity components in the direction of the *fixed* axes x, y, x.

In order to specify the appropriate forms of the components of *acceleration*, let us differentiate the foregoing expressions (1.44)–(1.49) for the velocity components with respect to the time. Doing so we find that those with respect to the *space* axes assume the forms

$$\dot{u} = w\omega_y + z\dot{\omega}_y - v\omega_z - y\dot{\omega}_z + \dot{u}'_0 , \tag{1.52}$$

$$\dot{v} = u\omega_z + x\dot{\omega}_z - z\omega_x - z\dot{\omega}_x + \dot{v}'_0 , \tag{1.53}$$

$$\dot{w} = v\omega_x + y\dot{\omega}_x - u\omega_y - x\dot{\omega}_y + \dot{w}'_0 , \tag{1.54}$$

where the velocity components u, v, w have already been given by Equations (1.44)–(1.46); and where, by differentiation of (1.51),

$$\dot{u}'_0 = a_{11}\dot{u}' + a_{12}\dot{v}' + a_{13}\dot{w}' + \dot{a}_{11}u' + \dot{a}_{12}v' + \dot{a}_{13}w' , \tag{1.55}$$

$$\dot{v}'_0 = a_{21}\dot{u}' + a_{22}\dot{v}' + a_{23}\dot{w}' + \dot{a}_{21}u' + \dot{a}_{22}v' + \dot{a}_{23}w' , \tag{1.56}$$

$$\dot{w}'_0 = a_{31}\dot{u}' + a_{32}\dot{v}' + a_{33}\dot{w}' + \dot{a}_{31}u' + \dot{a}_{32}v' + \dot{a}_{33}w' . \tag{1.57}$$

The first three terms in each of these expressions represent obviously the body accelerations with respect to the space axes; and we shall abbreviate them as

$$\left.\begin{aligned} a_{11}\dot{u}' + a_{12}\dot{v}' + a_{13}\dot{w}' &= (\dot{u})'_0 , \\ a_{21}\dot{u}' + a_{22}\dot{v}' + a_{23}\dot{w}' &= (\dot{v})'_0 , \\ a_{31}\dot{u}' + a_{32}\dot{v}' + a_{33}\dot{w}' &= (\dot{w})'_0 . \end{aligned}\right\} \tag{1.58}$$

Since, moreover, by insertion from (1.20)–(1.22) and (1.51),

$$\begin{aligned} \dot{a}_{11}u' &+ \dot{a}_{12}v' + \dot{a}_{13}w' = (a_{31}\omega_y - a_{21}\omega_z)u' + \\ &+ (a_{32}\omega_y - a_{22}\omega_z)v' + (a_{33}\omega_y - a_{23}\omega_z)w' = \\ &= \omega_y(a_{31}u' + a_{32}v' + a_{33}w') - \omega_z(a_{21}u' + a_{22}v' + a_{23}w') = \\ &= \omega_y w'_0 - \omega_z v'_0 ; \end{aligned} \tag{1.59}$$

and, similarly,

$$\dot{a}_{21}u' + \dot{a}_{22}v' + \dot{a}_{23}w' = \omega_z u'_0 - \omega_x w'_0 \tag{1.60}$$

while

$$\dot{a}_{31}u' + \dot{a}_{32}v' + \dot{a}_{33}w' = \omega_x v_0 - \omega_y u'_0 , \tag{1.61}$$

Equations (1.52)–(1.54) can be rewritten in a more explicit form

$$\begin{aligned} \dot{u} = & -x(\omega_y^2 + \omega_z^2) + y(\omega_x\omega_y - \dot{\omega}_z) + z(\omega_x\omega_z + \dot{\omega}_y) + \\ & + (\dot{u})'_0 + 2(w'_0\omega_y - v'_0\omega_z) , \end{aligned} \tag{1.62}$$

$$\begin{aligned} \dot{v} = & -y(\omega_z^2 + \omega_x^2) + z(\omega_y\omega_z - \dot{\omega}_x) + x(\omega_x\omega_y + \dot{\omega}_z) + \\ & + (\dot{v})'_0 + 2(u'_0\omega_z - w'_0\omega_x) , \end{aligned} \tag{1.63}$$

and

$$\begin{aligned} \dot{w} = & -z(\omega_x^2 + \omega_y^2) + x(\omega_x\omega_z - \dot{\omega}_y) + y(\omega_y\omega_z + \dot{\omega}_x) + \\ & + (\dot{w})'_0 + 2(v'_0\omega_x - u'\omega_y) . \end{aligned} \tag{1.64}$$

The foregoing equations refer to accelerations with respect to the inertial system of space axes. Those with respect to the (rotating) *body* axes can be obtained by an analogous process from the equations

$$\dot{u}' = -w'\omega_{y'} - z'\dot{\omega}_{y'} + v'\omega_{z'} + y'\dot{\omega}_{z'} + \dot{u}_0 , \tag{1.65}$$

$$\dot{v}' = -u'\omega_{z'} - x'\dot{\omega}_{z'} + w'\omega_{x'} + z'\dot{\omega}_{x'} + \dot{v}_0 \tag{1.66}$$

$$\dot{w}' = -v'\omega_{x'} - y'\dot{\omega}_{x'} + u'\omega_{y'} + x'\dot{\omega}_{y'} + \dot{w}_0 , \tag{1.67}$$

equivalent to (1.52)–(1.54); which on being treated in the same way as the latter can eventually be reduced to the form

$$\begin{aligned} \dot{u}' &= -x'(\omega_{y'}^2 + \omega_{z'}^2) + y'(\omega_{x'}\omega_{y'} + \dot{\omega}_{z'}) + z'(\omega_{x'}\omega_{z'} - \dot{\omega}_{y'}) + \\ &\quad + (\dot{u})_0 - 2(w_0\omega_{y'} - v_0\omega_{z'}) , \end{aligned} \tag{1.68}$$

$$\begin{aligned} \dot{v}' &= -y'(\omega_{z'}^2 + \omega_{x'}^2) + z'(\omega_{y'}\omega_{z'} + \dot{\omega}_{x'}) + x'(\omega_{x'}\omega_{y'} - \dot{\omega}_{x'}) + \\ &\quad + (\dot{v})_0 - 2(u_0\omega_{z'} - w_0\omega_{x'}) , \end{aligned} \tag{1.69}$$

$$\begin{aligned} \dot{w}' &= -z'(\omega_{x'}^2 + \omega_{y'}^2) + x'(\omega_{x'}\omega_{z'} + \dot{\omega}_{y'}) + y'(\omega_{y'}\omega_{z'} - \dot{\omega}_{x'}) + \\ &\quad + (\dot{w})_0 - 2(v_0\omega_{x'} - u_0\omega_{y'}) , \end{aligned} \tag{1.70}$$

where the space velocity components u_0, v_0, w_0 in the direction of increasing x', y', z' continue to be given by Equations (1.50); while the corresponding components of the accelerations are given by

$$\left. \begin{aligned} (\dot{u})_0 &= a_{11}\dot{u} + a_{21}\dot{v} + a_{31}\dot{w} , \\ (\dot{v})_0 &= a_{12}\dot{u} + a_{22}\dot{v} + a_{32}\dot{w} , \\ (\dot{w})_0 &= a_{13}\dot{u} + a_{23}\dot{v} + a_{33}\dot{w} . \end{aligned} \right\} \tag{1.71}$$

B: Eulerian Equations

With the aid of the results obtained so far our next task should be to rewrite the fundamental Equations (1.1)–(1.3) of our problem in terms of the angular velocity components ω_x, ω_y, ω_z about the inertial space axes, in a form suitable for our subsequent work. In order to do so, multiply Equations (1.62)–(1.64) by x, y, z and form their following differences:

$$\begin{aligned} y\dot{w} - z\dot{v} &= (y^2 + z^2)\dot{\omega}_x + (y^2 - z^2)\omega_y\omega_z - \\ &\quad - xy(\dot{\omega}_y - \omega_x\omega_z) - xz(\dot{\omega}_z + \omega_x\omega_y) - yz(\omega_y^2 - \omega_z^2) + \\ &\quad + \{y(\dot{w})_0' - z(\dot{v})_0'\} + 2y\{v_0'\omega_x - u_0'\omega_y\} - 2z\{u_0'\omega_z - w_0'\omega_x\} , \end{aligned} \tag{1.72}$$

$$\begin{aligned} z\dot{u} - x\dot{w} &= (z^2 + x^2)\dot{\omega}_y + (z^2 - x^2)\omega_x\omega_z - \\ &\quad yz(\dot{\omega}_z - \omega_y\omega_x) - yx(\dot{\omega}_x + \omega_y\omega_z) - zx(\omega_z^2 - \omega_x^2) + \\ &\quad + \{z(\dot{u})_0' - x(\dot{w})_0'\} + 2z\{w_0'\omega_y - v_0'\omega_z\} - 2x\{v_0'\omega_x - u_0'\omega_y\} , \end{aligned} \tag{1.73}$$

$$\begin{aligned} x\dot{v} - y\dot{u} &= (x^2 + y^2)\dot{\omega}_z + (x^2 - y^2)\omega_x\omega_y - \\ &\quad - zx(\dot{\omega}_x - \omega_y\omega_z) - zy(\dot{\omega}_y + \omega_x\omega_z) - xy(\omega_x^2 - \omega_y^2) + \\ &\quad + \{x(\dot{v})_0' - y(\dot{u})_0'\} + 2x\{u_0'\omega_z - w_0'\omega_x\} - 2y\{w_0'\omega_y - v_0'\omega_z\} . \end{aligned} \tag{1.74}$$

If so, however, Equations (1.1)–(1.3) can be combined to yield

$$y\dot{w} - z\dot{v} + \frac{1}{\rho}\left\{y\frac{\partial}{\partial z} - z\frac{\partial}{\partial y}\right\}P - \left\{y\frac{\partial}{\partial z} - z\frac{\partial}{\partial y}\right\}\Omega = y\mathcal{H} - z\mathcal{G}\,, \quad (1.75)$$

$$z\dot{u} - x\dot{w} + \frac{1}{\rho}\left\{z\frac{\partial}{\partial x} - x\frac{\partial}{\partial z}\right\}P - \left\{z\frac{\partial}{\partial x} - x\frac{\partial}{\partial z}\right\}\Omega = z\mathcal{F} - x\mathcal{H}\,, \quad (1.76)$$

$$x\dot{v} - y\dot{u} + \frac{1}{\rho}\left\{x\frac{\partial}{\partial y} - y\frac{\partial}{\partial x}\right\}P - \left\{x\frac{\partial}{\partial y} - y\frac{\partial}{\partial x}\right\}\Omega = x\mathcal{G} - y\mathcal{F}\,, \quad (1.77)$$

where

$$\rho\mathcal{F} = \frac{\partial\sigma_{xx}}{\partial x} + \frac{\partial\sigma_{xy}}{\partial y} + \frac{\partial\sigma_{xz}}{\partial z}\,, \quad (1.78)$$

$$\rho\mathcal{G} = \frac{\partial\sigma_{yx}}{\partial x} + \frac{\partial\sigma_{yy}}{\partial y} + \frac{\partial\sigma_{yz}}{\partial z}\,, \quad (1.79)$$

$$\rho\mathcal{H} = \frac{\partial\sigma_{zx}}{\partial x} + \frac{\partial\sigma_{zy}}{\partial y} + \frac{\partial\sigma_{zz}}{\partial z}\,, \quad (1.80)$$

represent the effects of viscosity.

Next, let us rewrite the foregoing expressions in terms of the respective velocity components. Inserting for the components σ_{ij} of the viscous stress tensor from (1.5)–(1.10) we find the expressions on the right-hand sides of Equations (1.78)–(1.80) assume the more explicit forms

$$\frac{\partial\sigma_{xx}}{\partial x} + \frac{\partial\sigma_{xy}}{\partial y} + \frac{\partial\sigma_{xz}}{\partial z} = \mu\nabla^2 u + \frac{\mu}{3}\frac{\partial\Delta}{\partial x} + 2\left\{\frac{\partial u}{\partial x} - \frac{\Delta}{3}\right\}\frac{\partial\mu}{\partial x} + \left\{\frac{\partial v}{\partial x} + \frac{\partial u}{\partial y}\right\}\frac{\partial\mu}{\partial y} + \left\{\frac{\partial u}{\partial z} + \frac{\partial w}{\partial x}\right\}\frac{\partial\mu}{\partial z}\,, \quad (1.81)$$

$$\frac{\partial\sigma_{yx}}{\partial x} + \frac{\partial\sigma_{yy}}{\partial y} + \frac{\partial\sigma_{yz}}{\partial z} = \mu\nabla^2 v + \frac{\mu}{3}\frac{\partial\Delta}{\partial y} + 2\left\{\frac{\partial v}{\partial y} - \frac{\Delta}{3}\right\}\frac{\partial\mu}{\partial y} + \left\{\frac{\partial w}{\partial y} + \frac{\partial v}{\partial z}\right\}\frac{\partial\mu}{\partial z} + \left\{\frac{\partial v}{\partial x} + \frac{\partial u}{\partial y}\right\}\frac{\partial\mu}{\partial x}\,, \quad (1.82)$$

and

$$\frac{\partial\sigma_{zx}}{\partial x} + \frac{\partial\sigma_{zy}}{\partial y} + \frac{\partial\sigma_{zz}}{\partial z} = \mu\nabla^2 w + \frac{\mu}{3}\frac{\partial\Delta}{\partial z} + 2\left\{\frac{\partial w}{\partial z} - \frac{\Delta}{3}\right\}\frac{\partial\mu}{\partial z} + \left\{\frac{\partial u}{\partial z} + \frac{\partial w}{\partial x}\right\}\frac{\partial\mu}{\partial x} + \left\{\frac{\partial w}{\partial y} + \frac{\partial v}{\partial z}\right\}\frac{\partial\mu}{\partial y}\,, \quad (1.83)$$

where Δ denotes, as before, the divergence (1.11) of the velocity vector; and ∇^2 stands for the Laplacian operator.

Next, let us insert for the velocity components u, v, w from (1.44)–(1.46); by doing so we find that

$$\nabla^2 u = 2\nabla^2\omega_y - y\nabla^2\omega_z + \nabla^2 u_0' + 2\left\{\frac{\partial\omega_y}{\partial z} - \frac{\partial\omega_z}{\partial y}\right\}\,, \quad (1.84)$$

$$\nabla^2 v = x\nabla^2\omega_z - z\nabla^2\omega_x + \nabla^2 v_0' + 2\left\{\frac{\partial\omega_z}{\partial x} - \frac{\partial\omega_x}{\partial z}\right\}, \tag{1.85}$$

$$\nabla^2 w = y\nabla^2\omega_x - x\nabla^2\omega_y + \nabla^2 w_0' + 2\left\{\frac{\partial\omega_x}{\partial y} - \frac{\partial\omega_y}{\partial x}\right\}, \tag{1.86}$$

and

$$\begin{aligned}\Delta &= \left\{y\frac{\partial}{\partial z} - z\frac{\partial}{\partial y}\right\}\omega_x + \left\{z\frac{\partial}{\partial x} - x\frac{\partial}{\partial z}\right\}\omega_y + \\ &+ \left\{x\frac{\partial}{\partial y} - y\frac{\partial}{\partial x}\right\}\omega_z + \frac{\partial u_0'}{\partial x} + \frac{\partial v_0'}{\partial y} + \frac{\partial w_0'}{\partial z}.\end{aligned} \tag{1.87}$$

Before proceeding further let us survey what we accomplished so far. By virtue of the assumptions made at the outset, the angular velocity components $\omega_{x,y,z}$ are allowed to be functions of the time only and will, therefore, describe rigid-body rotation of deformable systems; and the velocity components u_0', v_0', w_0' describe the deformations—be these due to free or forced oscillations—which may depend on spatial coordinates as well as on the time. However, inasmuch as the ω's are functions of t alone, Equations (1.84)–(1.86) permit us to assert that

$$\left.\begin{aligned}\nabla^2 u &= \nabla^2 u_0', \\ \nabla^2 v &= \nabla^2 v_0', \\ \nabla^2 w &= \nabla^2 w_0';\end{aligned}\right\} \tag{1.88}$$

and, similarly, the divergence (1.87) of the velocity vector will reduce to

$$\nabla_0' = \frac{\partial u_0'}{\partial x} + \frac{\partial v_0'}{\partial y} + \frac{\partial w_0'}{\partial z}. \tag{1.89}$$

In consequence, the corresponding expressions on the right-hand sides of Equations (1.81)–(1.83) obtain if the velocity components u, v, w present there are replaced by u_0', v_0', w_0' and Δ by Δ_0'; and if so, we are entitled to set

$$\begin{aligned}\rho\{y\mathcal{H} - z\mathcal{G}\} &= \mu\left\{y\nabla^2 w_0' - z\nabla^2 v_0' + \frac{1}{3}D_1\nabla_0'\right\} + \\ &+ \frac{\partial\mu}{\partial x}\left\{D_1 u_0' + \frac{\partial}{\partial x}(yw_0' - zv_0')\right\} + \\ &+ \frac{2}{3}\frac{\partial\mu}{\partial y}\{2D_1 v_0' + D_4 w_0'\} + \frac{2}{3}\frac{\partial\mu}{\partial z}\{2D_1 w_0' - D_4 v_0'\} \\ &- \frac{2}{3}\frac{\partial u_0'}{\partial x}D_1\mu + \frac{1}{3}\xi D_4\mu,\end{aligned} \tag{1.90}$$

$$\begin{aligned}\rho\{z\mathcal{G} - x\mathcal{H}\} &= \mu\left\{z\nabla^2 u_0' - x\nabla^2 w_0' + \frac{1}{3}D_2\Delta_0'\right\} + \\ &+ \frac{2}{3}\frac{\partial\mu}{\partial x}\{2D_2 2u_0' - D_5 w_0'\} + \\ &+ \frac{\partial\mu}{\partial y}\left\{D_2 v_0' + \frac{\partial}{\partial y}(zu_0' - xw_0')\right\} + \frac{2}{3}\frac{\partial\mu}{\partial z}\{2D_2 w_0' + D_5 u_0'\} - \\ &- \frac{2}{3}\frac{\partial v_0'}{\partial y}D_2\mu + \frac{1}{3}\eta D_5\mu,\end{aligned} \tag{1.91}$$

and

$$\begin{aligned}\rho\{x\mathcal{G} - y\mathcal{F}\} &= \mu\left\{x\nabla^2 v_0' - y\nabla^2 u_0' + \frac{1}{3}D_3\Delta_0'\right\} + \frac{2}{3}\frac{\partial\mu}{\partial x}\{2D_3 u_0' + D_6 v_0'\} + \\ &+ \frac{2}{3}\frac{\partial\mu}{\partial y}\{2D_3 3v_0' - D_6 u_0'\} + \frac{\partial\mu}{\partial z}\left\{D_3 e_0' + \frac{\partial}{\partial z}(xv_0' - yu_0')\right\} - \\ &+ \frac{2}{3}\frac{\partial w_0'}{\partial z}D_3\mu + \frac{1}{3}\zeta D_6\mu\ , \end{aligned} \tag{1.92}$$

where the symbols D_j $(j = 1, \dots 6)$ stand for the following operators:

$$D_1 \equiv y\frac{\partial}{\partial z} - z\frac{\partial}{\partial y}\ , \tag{1.93}$$

$$D_2 \equiv z\frac{\partial}{\partial x} - x\frac{\partial}{\partial z}\ , \tag{1.94}$$

$$D_3 \equiv x\frac{\partial}{\partial y} - y\frac{\partial}{\partial x}\ , \tag{1.95}$$

$$D_4 \equiv z\frac{\partial}{\partial z} + y\frac{\partial}{\partial y}\ , \tag{1.96}$$

$$D_5 \equiv x\frac{\partial}{\partial x} + z\frac{\partial}{\partial z}\ , \tag{1.97}$$

$$D_6 \equiv x\frac{\partial}{\partial x} + y\frac{\partial}{\partial y}\ ; \tag{1.98}$$

and where

$$\xi = \frac{\partial w_0'}{\partial y} - \frac{\partial v_0'}{\partial z}\ , \tag{1.99}$$

$$\eta = \frac{\partial u_0'}{\partial z} - \frac{\partial w_0'}{\partial x}\ , \tag{1.100}$$

$$\zeta = \frac{\partial v_0'}{\partial x} - \frac{\partial u_0'}{\partial y}\ , \tag{1.101}$$

denote the components of vorticity of the deformation vector.

As the next step of our analysis, let us integrate both sides of the Equations (1.75)–(1.77) over the entire mass of our configuration with respect to the mass element

$$dm = \rho dV = \rho dx\, dy\, dz\ . \tag{1.102}$$

If, as usual,

$$A = \int (y^2 + z^2) dm\ , \tag{1.103}$$

$$B = \int (x^2 + z^2) dm\ , \tag{1.104}$$

$$C = \int (x^2 + y^2) dm \tag{1.105}$$

denote the *moments of inertia* of our configuration with respect to the axes x, y, z; and

$$D = \int yz\, dm \,, \tag{1.106}$$

$$E = \int xz\, dm \,, \tag{1.107}$$

$$F = \int xy\, dm \tag{1.108}$$

stand for the respective *products of inertia*, the mass integrals of the Equations (1.75)–(1.77) combined with (1.72)–(1.74) will assume the forms

$$\begin{aligned} A\dot{\omega}_x &+ (C-B)\omega_y\omega_z - D(\omega_y^2-\omega_z^2) - E(\dot{\omega}_z+\omega_x\omega_y) - F(\dot{\omega}_y-\omega_x\omega_y) + \\ &+ 2\omega_x \int (yv_0' + zw_0')dm - 2\omega_y \int yu_0' dm - 2\omega_z \int zu_0' dm + \\ &+ \int D_1 P dV - \int D_1 \Psi dm = \int \{z(\dot{v})_0' - y(\dot{w}_0')\} dm + \int \rho\{y\mathcal{H} - z\mathcal{G}\} dV \,, \end{aligned} \tag{1.109}$$

$$\begin{aligned} B\dot{\omega}_y &+ (A-C)\omega_x\omega_z - D(\dot{\omega}_z-\omega_x\omega_y) - E(\omega_z^2-\omega_x^2) - F(\dot{\omega}_x+\omega_y\omega_z) + \\ &+ 2\omega_y \int (xu_0' + zw_0')dm - 2\omega_z \int zv_0' dm - 2\omega_x \int xv_0' dm + \\ &+ \int D_2 P dV - \int D_2 \Psi dm = \int \{x(\dot{w})_0' - z(\dot{u})_0'\} dm + \int \rho\{z\mathcal{F} - x\mathcal{H}\} dV \,, \end{aligned} \tag{1.110}$$

and

$$\begin{aligned} C\dot{\omega}_z &+ (B-A)\omega_x\omega_y - D(\dot{\omega}_y+\omega_x\omega_z) - E(\dot{\omega}_x-\omega_y\omega_z) - F(\omega_x^2-\omega_y^2) + \\ &+ 2\omega_z \int (yv_0' + xu_0')dm - 2\omega_x \int xw_0' dm - 2\omega_y \int yw_0' dm + \\ &+ \int D_3 P dV - \int D_3 \Psi dm = \int \{y(\dot{u}_0') - x(\dot{v}_0')\} dm + \int \rho\{x\mathcal{G} - y\mathcal{F}\} dV \,. \end{aligned} \tag{1.111}$$

The preceding three equations represent the generalized Eulerian equations which govern the precession and nutation of self-gravitating configurations consisting of viscous fluid. They constitute a simultaneous system of three ordinary differential equations for ω_x, ω_y, ω_z with the time t as the sole independent variable; with the velocities u_0', v_0' and w_0' of deformation representing *dynamical tides*. These will, in general, likewise depend on the time (and so will the contributions to the moments and products of inertia arising from them); and the coefficients of our equations should, therefore, be regarded as variable.

Should, however, the tides reduce to those of the *equilibrium* type (such as, for instance, arise in the case of synchronism between rotation and revolution, when both equators become co-planar with the orbit), the velocities u_0', v_0' and w_0' of deformation vanish (and the moments A, B, $\ldots E$, F freeze into constants). The vanishing of the velocity components will annihilate also all effects due to viscosity; and the equations (1.109)–(1.111) will reduce to

$$A\dot{\omega}_x + (C-B)\omega_y\omega_z - D(\omega_y^2-\omega_z^2) - E(\dot{\omega}_z+\omega_x\omega_y) -$$

$$\begin{aligned} - \quad & F(\dot{\omega}_y - \omega_x\omega_z) + \int D_1 P dV - \int D_1 \Psi_0 dm = \int D_1 \Psi_1 dm\,, \quad (1.112) \\ B\dot{\omega}_y \quad + \quad & (A - C)\omega_x\omega_z - D(\dot{\omega}_z - \omega_x\omega_y) - E(\omega_z^2 - \omega_x^2) - \\ - \quad & F(\dot{\omega}_x + \omega_y\omega_z) + \int D_2 P dV - \int D_2 \Psi_0 dm = \int D_2 \Psi_1 dm\,, \quad (1.113) \end{aligned}$$

and

$$\begin{aligned} C\dot{\omega}_z \quad + \quad & (B - A)\omega_x\omega_y - D(\dot{\omega}_y + \omega_x\omega_z) - E(\dot{\omega}_x - \omega_y\omega_z) - \\ - \quad & F(\omega_x^2 - \omega_y^2) + \int D_3 P dV - \int D_3 \Psi_0 dm = \int D_3 \Psi_1 dm\,, \quad (1.114) \end{aligned}$$

where we have decomposed the total gravitational potetial

$$\Psi = \Psi_0 + \Psi_1 \tag{1.115}$$

into its part arising from the mass of the respective body (Ψ_0) and that arising from external disturbing forces (Ψ_1) if any.

Moreover, the maintenance of hydrostatic equilibrium requires that

$$\int D_i\, P\, dV = \int D_i\, \Psi\, dm \tag{1.116}$$

for $i = 1, 2, 3$. If, furthermore, our system of coordinates XYZ were oriented in space so as to coincide in direction with the principal axes of inertia of our configuration, all three products D, E, F of inertia can be made to vanish, Equations (1.112)–(1.114) reduce further to

$$\left. \begin{aligned} A\dot{\omega}_x + (C - B)\omega_y\omega_z &= \textstyle\int D_1 \Psi_1 dm \equiv F_x\,, \\ B\dot{\omega}_y + (A - C)\omega_x\omega_z &= \textstyle\int D_2 \Psi_1 dm \equiv F_y\,, \\ C\dot{\omega}_z + (B - A)\omega_x\omega_y &= \textstyle\int D_3 \Psi_1 dm \equiv F_z\,; \end{aligned} \right\} \tag{1.117}$$

where $F_{x,y,z}$ are the components of the forces acting on our body from outside—equations governing three-dimensional rotation of rigid bodies, well known from celestial mechanics.

C: Moments of Inertia

Let us return now to the fundamental Equations (1.109)–(1.111) of our problem, and set out to specify the contributions to the moments of inertia A, B, C and the products D, E, F arising from the departures of the respective configurations from spherical shape. The definitions of these moments have already been given by Equations (1.103)–(1.108) earlier in this section; and our present task will be to evaluate them in a closed form.

In order to do so, the coordinates should be adopted in which the limits of the requisite integrals assume the simplest possible form—i.e., with axes so oriented that at least one of them becomes identical with that of symmetry of the respective configuration; which for the rotational distortion should be the Z'-axis; and for the tidal distortion, the X''-axis (see Section II.3A) related with the XYZ-system by the transformation (3.6) of Chapter II. Moreover, in order to carry out the actual integration, we find it expedient to resort to the Clairaut coordinates a, θ, ϕ of the preceding Chapter V, for which the mass element

$$\begin{aligned} dm &\equiv \rho\, dV = \\ &= (g_{11}g_{22}g_{33})^{1/2}\left\{1 - \frac{g_{12}^2}{g_{11}g_{22}} - \frac{g_{13}^2}{g_{11}g_{33}} - \right. \\ &\quad \left. - \frac{g_{23}^2}{g_{22}g_{33}} + \frac{2g_{12}g_{13}g_{23}}{g_{11}g_{22}g_{33}}\right\}^{1/2} da\, d\theta\, d\phi \end{aligned} \tag{1.118}$$

on insertion for the metric coefficients g_{ij} from Equations (1.9)–(1.14) of Chapter V reduces to

$$dm = \rho r^2 r_a \sin\theta\, da\, d\theta\, d\phi\,. \tag{1.119}$$

Consider first the contributions invoked by partial *tides* which are symmetrical with respect to the X''-axis. By Equations (1.1) of Chapter V,

$$y^2 + z^2 = r^2(1-\lambda^2) = r^2\{1 - (a''_{11}\lambda'' + a''_{12}\mu'' + a''_{13}\nu'')^2\}\ ;, \tag{1.120}$$

$$x^2 + z^2 = r^2(1-\mu^2) = r^2\{1 - (a''_{21}\lambda'' + a''_{22}\mu'' + a''_{23}\nu'')^2\}\,, \tag{1.121}$$

$$x^2 + y^2 = r^2(1-\nu^2) = r^2\{1 - (a''_{31}\lambda'' + a''_{32}\mu'' + a''_{33}\nu'')^2\} \tag{1.122}$$

by (3.6) of Chapter II; and, similarly,

$$yz = r^2\mu\nu = r^2(a''_{21}\lambda'' + a''_{22}\mu'' + a''_{23}\nu'')(a''_{31}\lambda'' + a''_{32}\mu'' + a''_{33}\nu''), \tag{1.123}$$

$$xz = r^2\lambda\nu = r^2(a''_{11}\lambda'' + a''_{12}\mu'' + a''_{13}\nu'')(a''_{31}\lambda'' + a''_{32}\mu'' + a''_{33}\nu''), \tag{1.124}$$

$$xy = r^2\lambda\mu = r^2(a''_{11}\lambda'' + a''_{12}\mu'' + a''_{13}\mu'')(a''_{21}\lambda'' + a''_{22}\mu'' + a''_{23}\nu''), \tag{1.125}$$

where the direction cosines a''_{ij} continue to be given by Equations (3.7)–(3.9) of Chapter II, and the radius-vector r of a tidally-distorted equipotential can be expressed (cf. Section IV.3) as

$$r = a(1 + \sum f_j) \equiv a(1+f)\,, \tag{1.126}$$

where

$$f_j \equiv (1+2k_j)\frac{m_2}{m_1}\left(\frac{r}{R}\right)^{j+1} P_j(\lambda'') \equiv K_j r^{j+1} P_j(\lambda'') \tag{1.127}$$

for $j = 2, 3, 4$ (to the first order in surficial distortion), where $m_{1,2}$ denote the masses of the distorted (m_1) and disturbing (m_2) component, and

$$R \equiv \frac{A(1-e^2)}{1 + e\cos v} \tag{1.128}$$

denotes the radius-vector of the relative orbit of the two stars, of semi-major axis A and eccentricity e, representing their mutual separation at the true anomaly v;

while the k_j's are constants depending on the internal structure of mass m_1 (cf. Section IV.5).

Inasmuch as the right-hand sides of the integrands (1.120)–(1.125) are expressible in terms of the surface harmonics of second degree ($j = 2$) in λ'', μ'' or ν'', an application of the integral operator

$$\int \ldots dm \equiv \int_0^{a_1}\int_0^{\pi}\int_0^{2\pi} \ldots \rho\, r^2 r_a \sin\theta'' da\, d\theta'' d\phi'' \tag{1.129}$$

will annihilate all terms on the right-hand sides of (1.103)–(1.108) except for those corresponding to $j = 2$; and, as a result, the contributions to the moments of inertia due to the tides will (to the first order in small quantities) be of the form

$$A'' = -\frac{16}{15}K_2 M_2 P_2(a''_{11})\,, \tag{1.130}$$

$$B'' = -\frac{16}{15}K_2 M_2 P_2(a''_{21})\,, \tag{1.131}$$

$$C'' = -\frac{16}{15}K_2 M_2 P_2(a''_{31})\,, \tag{1.132}$$

where we have abbreviated

$$M_2 \equiv 4\pi \int_0^{a_1} \rho(a) a^7\, da\,; \tag{1.133}$$

and, similarly, the contributions to the corresponding products of inertia are found to be

$$D'' = \frac{8}{5}K_2 M_2 a''_{21} a''_{31}\,, \tag{1.134}$$

$$E'' = \frac{8}{5}K_2 M_2 a''_{11} a''_{31}\,, \tag{1.135}$$

$$F'' = \frac{8}{5}K_2 M_2 a''_{11} a''_{21}\,. \tag{1.136}$$

In addition to tidal distortion, however, our configuration of mass m_1 must also suffer a flattening at the poles, due to its rotation about the Z'-axis with an angular velocity ω_z. The (first-order) distortion arising from this cause will be of the form

$$r = a(1+g),\;\; g = -\frac{\omega_z^2 r^3}{3Gm_1} P_2(\mu')\,, \tag{1.137}$$

where

$$\mu' \equiv \cos\theta' = a'_{13}\lambda + a'_{23}\mu + a'_{33}\nu\,; \tag{1.138}$$

and on performing the requisite integration as in the case of the tides we find that the contributions to the moments and products of inertia due to polar flattening will assume the forms

$$A' = \frac{16}{45}\left(\frac{\omega_z^2}{Gm_1}\right) M_2 P_2(a'_{13}) , \tag{1.139}$$

$$B' = \frac{16}{45}\left(\frac{\omega_z^2}{Gm_1}\right) M_2 P_2(a'_{33}) , \tag{1.140}$$

$$C' = \frac{16}{45}\left(\frac{\omega_z^2}{Gm_1}\right) M_2 P_2(a'_{33}) \tag{1.141}$$

and

$$D' = -\frac{16}{15}\left(\frac{\omega_z^2}{Gm_1}\right) M_2 a'_{23} a'_{33} , \tag{1.142}$$

$$E' = -\frac{16}{15}\left(\frac{\omega_z^2}{Gm_1}\right) M_2 a'_{13} a'_{33} , \tag{1.143}$$

$$F' = -\frac{16}{15}\left(\frac{\omega_z^2}{Gm_1}\right) M_2 a'_{13} a'_{23} \tag{1.144}$$

where the singly-primed direction cosines a'_{ij} continue to be given by Equations (3.3)–(3.5) of Chapter II.

In the absence of rotational or tidal distortion,

$$A_0 = B_0 = C_0 = \frac{8}{3}\pi \int_0^{a_1} \rho a^4 da \equiv m_1 h_1^2 , \tag{1.145}$$

where h_1 defines the radius of gyration of the respective configuration, and

$$D_0 = E_0 = F_0 = 0 . \tag{1.146}$$

As, moreover, for configurations of pronounced central condensation, it is possible (cf. Section IV.5) to approximate

$$4\pi \int_0^{a_1} \rho a^{2j+3} da \doteq \frac{2j+1}{j+2} k_j a_1^{2j+1} m_1 , \tag{1.147}$$

where the k_j's are the "apsidal motion" constants as defined by Equation (5.62) of Chapter IV and, therefore,

$$M_2 \doteq \frac{5}{4} k_2 a_1^5 m_1 , \tag{1.148}$$

a combination of Equations (1.130)–(1.136) and (1.139–(1.144) with (1.145) eventually yields that

$$A \equiv A_0 + A' + A'' = m_1 h_1^2 - \frac{4}{3} k_2 a_1^5 \left\{ \frac{m_2}{R^3} P_2(a''_{11}) - \frac{\omega_z^2}{3G} P_2(a'_{13}) \right\} , \tag{1.149}$$

$$B \equiv B_0 + B' + B'' = m_1 h_1^2 - \frac{4}{3} k_2 a_1^5 \left\{ \frac{m_2}{r^3} P_2(a''_{21}) - \frac{\omega_z^2}{3G} P_2(a'_{23}) \right\} , \tag{1.150}$$

$$C \equiv C_0 + C'_0 + C''_0 = m_1 h_1^2 - \frac{4}{3} k_2 a_1^5 \left\{ \frac{m_2}{R^3} P_2(a''_{31}) - \frac{\omega_z^2}{3G} P_2(a'_{33}) \right\} \tag{1.151}$$

and

$$D \equiv D' + D'' = k_2 a_1^5 \left\{ \frac{m_2}{R^3} a''_{21} a''_{31} - \frac{\omega_z^2}{3G} a'_{23} a'_{33} \right\} , \tag{1.152}$$

$$E \equiv E' + E'' = k_2 a_1^5 \left\{ \frac{m_2}{R^3} a''_{11} a''_{31} - \frac{\omega_z^2}{3G} a'_{13} a'_{33} \right\} , \tag{1.153}$$

$$F \equiv F' + F'' = k_2 a_1^5 \left\{ \frac{m_2}{R^3} a''_{11} a''_{21} - \frac{\omega_z^2}{3G} a'_{13} a'_{23} \right\} ; \tag{1.154}$$

correctly to quantities of first order in surficial distortion, but without any restriction on the direction cosines a'_{ij} or a''_{ij}. The reader may also note that while the cosines a'_{i3} depend only on the Eulerian angles θ and ϕ which are constant (or slowly-varying functions of the time), the cosines a''_{i1} involve the true anomaly u in the plane of the orbit, which runs from 0° to 360° in the course of each orbital revolution and is, therefore, a fast-varying function of the time.

D: Gyroscopic Terms

The next task confronting us in the solution of our problem is to evaluate the mass-integrals of terms in the fundamental Equations (1.109)–(1.111) containing the products of the velocity components $\omega_{x,y,z}$ with the body velocity components u'_0, v'_0, w'_0 in the inertial axes XYZ, caused by rotation and tides.

In accordance with Equations (1.51), these velocity components can be expressed by the matrix equation

$$\left\{ \begin{matrix} u'_0 \\ v'_0 \\ w'_0 \end{matrix} \right\} = \left\{ \begin{matrix} a'_{11} & a'_{12} & a'_{13} \\ a'_{21} & a'_{22} & a'_{23} \\ a'_{31} & a'_{32} & a'_{33} \end{matrix} \right\} \left\{ \begin{matrix} u'f + x'\dot{f} \\ v'f + y'\dot{f} \\ w'f + z'\dot{f} \end{matrix} \right\} \tag{1.155}$$

where the direction cosines a'_{ij} continue to be given by Equations (3.3)–(3.5) of Chapter II. Moreover,

$$u' \equiv \dot{x}', \quad v' \equiv \dot{y}', \quad w' \equiv \dot{z}' ; \tag{1.156}$$

and f stands for the tidal distortion of the radius-vector (1.126) (to be replaced by g from (1.137) for the case of rotational distortion).

In order to specify the velocity components u', v', w', let us return to the transformation (3.10) of Chapter II; and remember that phenomena appearing as *dynamical* tides in the inertial (XYZ) or rotating ($X'Y'Z'$) systems of rectangular axes will reduce to the *equilibrium* tides in the revolving $X''Y''Z''$ axes—so that

$$\dot{x}'' = \dot{y}'' = \dot{z}'' = 0 . \tag{1.157}$$

Accordingly, if we differentiate Equations (3.10), Chapter II with respect to the time and repeated use of (3.2) discloses that

$$\left\{\begin{array}{c} u' \\ v' \\ w' \end{array}\right\} = \left\{\begin{array}{ccc} \dot\lambda_1'' & \dot\lambda_2'' & \dot\lambda_3'' \\ \dot\mu_1'' & \dot\mu_2'' & \dot\mu_3'' \\ \dot\nu_1'' & \dot\nu_2'' & \dot\nu_3'' \end{array}\right\}\left\{\begin{array}{c} x'' \\ y'' \\ z'' \end{array}\right\} =$$

$$= \left\{\begin{array}{ccc} \dot\lambda_1'' & \dot\lambda_2'' & \dot\lambda_3'' \\ \dot\mu_1'' & \dot\mu_2'' & \dot\mu_3'' \\ \dot\nu_1'' & \dot\nu_2'' & \dot\nu_3'' \end{array}\right\}\left\{\begin{array}{ccc} \lambda_1'' & \mu_1'' & \nu_1'' \\ \lambda_2'' & \mu_2'' & \nu_2'' \\ \lambda_3'' & \mu_3'' & \nu_3'' \end{array}\right\}\left\{\begin{array}{c} x' \\ y' \\ z' \end{array}\right\}; \tag{1.158}$$

or, more explicitly,

$$\begin{aligned} u' &= (\lambda_1\dot\lambda_1'' + \lambda_2''\dot\lambda_2'' + \lambda_3''\dot\lambda_3'')x' + \\ &\quad + (\mu_1''\dot\lambda_1'' + \mu_2''\dot\lambda_2'' + \mu_3''\dot\lambda_3'')y' + \\ &\quad + (\nu_1''\dot\lambda_1'' + \nu_2''\dot\lambda_2'' + \nu_3''\dot\lambda_3'')z' \,, \end{aligned} \tag{1.159}$$

$$\begin{aligned} v' &= (\lambda_1''\dot\mu_1'' + \lambda_2''\dot\mu_2'' + \lambda_3''\dot\mu_3'')x' + \\ &\quad + (\mu_1\dot\mu_1'' + \mu_2''\dot\mu_2'' + \mu_3''\dot\mu_3'')y' + \\ &\quad + (\nu_1''\dot\mu_1'' + \nu_2''\dot\mu_2'' + \nu_3''\dot\mu_3'')z' \,, \end{aligned} \tag{1.160}$$

$$\begin{aligned} w' &= (\lambda_1''\dot\nu_1'' + \lambda_2''\dot\nu_2'' + \lambda_3''\dot\nu_3'')x' + \\ &\quad + (\mu_2''\dot\nu_1'' + \mu_2''\dot\nu_2'' + \mu_3''\dot\nu_3'')y' + \\ &\quad + (\nu_1''\dot\nu_1'' + \nu_2''\dot\nu_2'' + \nu_3''\dot\nu_3'')z' \,. \end{aligned} \tag{1.161}$$

In order to simplify this latter equation let us observe, first, that

$$\left.\begin{aligned} \lambda_1''\dot\lambda_1'' + \lambda_2''\dot\lambda_2'' + \lambda_3''\dot\lambda_3'' &= 0 \,, \\ \mu_1''\dot\mu_1'' + \mu_2''\dot\mu_2'' + \mu_3''\dot\mu_3'' &= 0 \,, \\ \nu_1''\dot\nu_1'' + \nu_2''\dot\nu_2'' + \nu_3''\dot\nu_3'' &= 0 \,. \end{aligned}\right\} \tag{1.162}$$

Next, in differentiating the direction cosines λ_j'', μ_j'', ν_j'' $(j = 1, 2, 3)$ as given by Equation (3.11) of Chapter II with respect to the time, let us assume that the time-dependence of the direction cosines a_{ij} as given by Equations (3.7)–(3.9) of that chapter arises solely from the presence in them of periodic functions of the true anomaly u of the secondary component in its relative orbit, measured from the node Ω. If we regard, accordingly, the longitude Ω of the node as well as the inclination i of the orbital plane to the invariable plane $Z = 0$ of the system as constant, we readily find that

$$\left.\begin{aligned} \dot a_{i1}'' &= +\dot u a_{i2}'' \,, \\ \dot a_{i2}'' &= -\dot u a_{i1}'' \,, \\ \dot a_{i3}'' &= 0 \,, \end{aligned}\right\} \tag{1.163}$$

$i = 1, 2, 3$; where, for circular orbit, $\dot u \equiv n$ denotes the mean daily motion of the revolving system with respect to the space axes.

Remembering, furthermore, that the time-derivatives of the direction cosines a_{ij} can be expressed in terms of the angular velocity components $\omega_{x,y,z}$ by Equations (1.20)–(1.27), we find that

$$\left.\begin{array}{rcl} \lambda_1''\dot{\mu}_1'' + \lambda_2''\dot{\mu}_2'' + \lambda_3''\dot{\mu}_3'' & = & -\omega_z' + n\nu_3'' , \\ \lambda_1''\dot{\nu}_1'' + \lambda_2''\dot{\nu}_2'' + \lambda_3''\dot{\nu}_3'' & = & +\omega_y' - n\mu_3'' ; \end{array}\right\} \qquad (1.164)$$

$$\left.\begin{array}{rcl} \mu_1''\dot{\lambda}_1'' + \mu_2''\dot{\lambda}_2'' + \mu_3''\dot{\lambda}_3'' & = & +\omega_z' - n\nu_3'' , \\ \mu_1''\dot{\nu}_1'' + \mu_2''\dot{\nu}_2'' + \mu_3''\dot{\nu}_3'' & = & -\omega_x' + n\lambda_3'' ; \end{array}\right\} \qquad (1.165)$$

$$\left.\begin{array}{rcl} \nu_1''\dot{\lambda}_1'' + \nu_2\dot{\lambda}_2'' + \nu_3''\dot{\lambda}_3'' & = & -\omega_y' + n\mu_3'' , \\ \nu_1''\dot{\nu}_1'' + \nu_2''\dot{\mu}_2'' + \nu_3\dot{\mu}_3'' & \doteq & +\omega_x' - n\lambda_3'' ; \end{array}\right\} \qquad (1.166)$$

where

$$\left\{\begin{array}{c} \omega_x' \\ \omega_y' \\ \omega_z' \end{array}\right\} = \left\{\begin{array}{ccc} a_{11}' & a_{21}' & a_{31}' \\ a_{12}' & a_{22}' & a_{32}' \\ a_{13}' & a_{23}' & a_{33}' \end{array}\right\} \left\{\begin{array}{c} \omega_x \\ \omega_y \\ \omega_z \end{array}\right\} \qquad (1.167)$$

are the angular velocity-components about the rotating system of body-axes $X'Y'Z'$.

If we insert now Equations (1.162)–(1.165) in (1.159)–(1.161) we find that the body velocity components u', v', w' in the direction of the rotating axes of coordinates $x'y'z'$ should be given by

$$u' = (\omega_z' - n\nu_3'')y' - (\omega_y' - n\mu_3'')z' , \qquad (1.168)$$

$$v' = -(\omega_z' - n\nu_3'')x' + (\omega_x' - n\lambda_3'')z' , \qquad (1.169)$$

$$w' = (\omega_y' - n\mu_3'')x' - (\omega_x' - n\lambda_3'')y' ; \qquad (1.170)$$

and, therefore,

$$u'\frac{\partial f}{\partial x'} + v'\frac{\partial f}{\partial y'} + w'\frac{\partial f}{\partial z'} = -\{(\omega_z' - n\nu_3'')D_3' + (\omega_y' - n\mu_3'')D_2' + (\omega_x' - n\lambda_3'')D_1'\}f , \qquad (1.171)$$

where the D_j''s stand for the cyclic differential operators

$$D_1' \equiv y'\frac{\partial}{\partial z'} - z'\frac{\partial}{\partial y'} , \qquad (1.172)$$

$$D_2' \equiv z'\frac{\partial}{\partial x'} - x'\frac{\partial}{\partial z'} , \qquad (1.173)$$

$$D_3' \equiv x'\frac{\partial}{\partial y'} - y'\frac{\partial}{\partial z'} . \qquad (1.174)$$

Equation (1.171) can be transformed in the system of inertial coordinates XYZ by remembering (cf. Equations (1.93)–(1.95) that

$$\left\{\begin{array}{c} D_1' \\ D_2' \\ D_3' \end{array}\right\} = \left\{\begin{array}{ccc} a_{11}' & a_{21}' & a_{31}' \\ a_{12}' & a_{22}' & a_{32}' \\ a_{13}' & a_{23}' & a_{33}' \end{array}\right\} \left\{\begin{array}{c} D_1 \\ D_2 \\ D_3 \end{array}\right\} , \qquad (1.175)$$

where the D_j's denote the same operators in the inertial coordinates as the D_j's do in the rotating system. Making use of (1.167), and abbreviating

$$\left\{\begin{array}{ccc} a'_{11} & a'_{12} & a'_{13} \\ a'_{21} & a'_{22} & a'_{23} \\ a'_{31} & a'_{32} & a'_{33} \end{array}\right\}\left\{\begin{array}{c} u' \\ v' \\ w' \end{array}\right\} = \left\{\begin{array}{c} u \\ v \\ w \end{array}\right\}, \tag{1.176}$$

we find that Equation (1.171) can be alternatively rewritten as

$$\begin{aligned} u'f_{x'} + v'f_{y'} + w'f_z &= uf_x + vf_y + wf_z = \\ &= -(p_1D_1 + p_2D_2 + p_3D_3)f , \end{aligned} \tag{1.177}$$

where

$$\left.\begin{aligned} u &= p_3y - p_2z , \\ v &= p_1z - p_3x , \\ w &= p_2x - p_1y ; \end{aligned}\right\} \tag{1.178}$$

and

$$\left.\begin{aligned} p_1 &= \omega_x - na''_{13} , \\ p_2 &= \omega_y - na''_{23} , \\ p_3 &= \omega_z - na''_{33} , \end{aligned}\right\} \tag{1.179}$$

with the a''_{j3}'s as given by Equations (3.7)–(3.9) of Chapter II.

Equation (1.177) enables us to express $\dot{f}$ in the form

$$\dot{f} = f_t - (p_1D_1 + p_2D_2 + p_3D_3)f , \tag{1.180}$$

which is exact for any values of f or p_j; and of which frequent use will be made in the sequel. Moreover, from Equations (1.155) and (1.168)–(1.170) it follows, likewise exactly, that,

$$u'_0 = (p_3y - p_2z)f + x\dot{f} , \tag{1.181}$$

$$v'_0 = (p_1z - p_3x)f + y\dot{f} , \tag{1.182}$$

$$w'_0 = (p_2x - p_1y)f + z\dot{f} ; \tag{1.183}$$

and, in accordance with Equations (1.58),

$$\left\{\begin{array}{c} (\dot{u})'_0 \\ (\dot{v})'_0 \\ (\dot{w})'_0 \end{array}\right\} = \left\{\begin{array}{ccc} a'_{11} & a'_{12} & a'_{13} \\ a'_{21} & a'_{22} & a'_{23} \\ a'_{31} & a'_{32} & a'_{33} \end{array}\right\}\left\{\begin{array}{ccccc} \dot{u}'f & + & 2u'\dot{f} & + & x'\ddot{f} \\ \dot{v}'f & + & 2v'\dot{f} & + & y'\ddot{f} \\ \dot{w}'f & + & 2w'\dot{f} & + & z'\ddot{f} \end{array}\right\}, \tag{1.184}$$

where u', v', w' continue to be given by Equations (1.168)–(1.170).

Differentiating the latter with respect to the time, we find that

$$\begin{aligned} \dot{u}' &= (\omega_{z'} - n\nu''_3)v' - (\omega_{y'} - n\nu''_3)w' + \\ &\quad + (\dot{\omega}_{z'} - \dot{n}\nu''_3 - n\dot{\nu}''_3)y' - (\dot{\omega}_{y'} - \dot{n}\mu''_3 - n\dot{\mu}''_3)z' , \end{aligned} \tag{1.185}$$

$$\begin{aligned} \dot{v}' &= (\omega_{y'} - n\lambda''_3)w' - (\omega_{z'} - n\nu''_3)u' + \\ &\quad + (\dot{\omega}_{x'} - n\lambda''_3 - n\dot{\lambda}''_3)z' - (\dot{\omega}_{z'} - \dot{n}\nu''_3 - n\dot{\nu}''_3)x' , \end{aligned} \tag{1.186}$$

$$\begin{aligned} \dot{w}' &= (\omega_{y'} - n\mu''_3)u' - (\omega_{x'} - n\lambda''_3) + \\ &\quad + (\dot{\omega}_{x'} - \dot{n}\mu''_3 - n\dot{\mu}''_3)x' - (\dot{\omega}_{x'} - \dot{n}\lambda''_3 - n\dot{\lambda}''_3)y' ; \end{aligned} \tag{1.187}$$

so that

$$\begin{aligned} a'_{11}\dot{u}' + a'_{12}\dot{v}' + a'_{13}\dot{w}' &= p_3 y - \dot{p}_2 z + \dot{p}_3 v - p_2 w \,, & (1.188) \\ a'_{21}\dot{u}' + a'_{22}\dot{v}' + a'_{23}\dot{w}' &= p_1 z - \dot{p}_3 x + \dot{p}_1 w - p_3 u \,, & (1.189) \\ a'_{21}\dot{u}' + a'_{32}\dot{v}' + a'_{33}\dot{w}' &= p_2 x - \dot{p}_1 y + \dot{p}_2 u - p_1 v \,. & (1.190) \end{aligned}$$

where u, v, w and $p_{1,2,3}$ continue to be given by Equations (1.178) and (1.179). Accordingly, from Equation (1.184) it follows that

$$\begin{aligned} y(\dot{w})'_0 - z(\dot{v})'_0 &= 2\left[(p_1x + p_2y + p_3z)x - p_1r^2\right]\dot{f} + \\ &+ \left[(\dot{p}_2 + p_1p_3)xy + (\dot{p}_3 - p_1p_2)xz + (p_3^2 - p_2^2)yz - \right. \\ &\left. -(\dot{p}_1 - p_2p_3)y^2 - (\dot{p}_1 + p_2p_3)z^2\right] f \,, & (1.191) \\ z(\dot{u})'_0 - x(\dot{w})'_0 &= 2\left[(p_1x + p_2y + p_3z)y - p_2r^2\right]\dot{f} + \\ &\left[(\dot{p}_3 + p_2p_1)yz + (\dot{p}_1 - p_2p_3)yx + (p_1^2 - p_3^2)xz \right. \\ &\left. -(\dot{p}_2 - p_3p_1)z^2 - (\dot{p}_2 + p_3p_1)x^2\right] f \,, & (1.192) \\ x(\dot{v})'_0 - y(\dot{u})'_0 &= 2\left[(p_1x + p_2y + p_3z)z - p_3r^2\right]\dot{f} + \\ &+ \left[(\dot{p}_1 + p_3p_2)zx + (\dot{p}_2 - p_3p_1)zy + +(p_2^2 - p_1^2)xy - \right. \\ &\left. -(\dot{p}_3 - p_1p_2)x^2 - (\dot{p}_3 + p_1p_2)y^2\right] f \,, & (1.193) \end{aligned}$$

again without any restriction on the extent of the tidal distortion f or of the magnitudes of the coefficients $p_{1,2,3}$.

Having specifiedd the velocity components u'_0, v'_0, w'_0 by Equation (1.155) and their time-derivatives $\dot{u}'_0$, $\dot{v}'_0$, $\dot{w}'_0$ by (1.184), we can proceed now to evaluate the gyroscopic terms in Equations (1.109)–(1.111) by the same method as we did with the moments and products of inertia in the earlier part of this section.

In doing so, let us recall that the operators $D_{1,2,3}$ in (1.180) transform into the doubly-primed coordinates in accordance with the scheme

$$\left\{ \begin{array}{c} D_1 \\ D_2 \\ D_3 \end{array} \right\} = \left\{ \begin{array}{ccc} a''_{11} & a''_{12} & a''_{13} \\ a''_{21} & a''_{22} & a''_{23} \\ a''_{31} & a''_{32} & a_{33} \end{array} \right\} \left\{ \begin{array}{c} D''_1 \\ D''_2 \\ D''_3 \end{array} \right\}, \qquad (1.194)$$

where

$$\begin{aligned} D''_1 &\equiv y''\frac{\partial}{\partial z''} - z''\frac{\partial}{\partial y''} \equiv +\mu''\frac{\partial}{\partial \nu''} \,, & (1.195) \\ D''_2 &\equiv z''\frac{\partial}{\partial x''} - x''\frac{\partial}{\partial z''} \equiv \nu''\frac{\partial}{\partial \lambda''} - \lambda''\frac{\partial}{\partial \nu''} \,, & (1.196) \\ D''_3 &\equiv x''\frac{\partial}{\partial y''} - y''\frac{\partial}{\partial z''} \equiv -\mu''\frac{\partial}{\partial \lambda''} \,; & (1.197) \end{aligned}$$

so that

$$D_1'' f_2 = 0 , \quad (1.198)$$
$$D_2'' f_2 = 3K_2 r^3 \lambda'' \nu'' , \quad (1.199)$$
$$D_3'' f_2 = -3K_2 r^3 \lambda'' \mu'' ; \quad (1.200)$$

$$\int xu_0' dm = \alpha \left\{ \frac{2}{3} H_2 P_2(a_{11}'') + K_2 a_{11}'' (p_2 a_{31}'' - p_3 a_{21}'') \right\} , \quad (1.201)$$
$$\int yu_0' dm = \alpha \left\{ H_2 a_{11}'' a_{21}'' + K_2 \left[\frac{2}{3} p_3 P_2(a_{11}'') - p_1 a_{11}'' a_{31}'' \right] \right\} , \quad (1.202)$$
$$\int zu_0' dm = \alpha \left\{ H_2 a_{11}'' a_{31}'' - K_2 \left[\frac{2}{3} p_2 P_2(a_{11}'') - p_1 a_{11}'' a_{21}'' \right] \right\} ; \quad (1.203)$$
$$\int xv_0' dm = \alpha \left\{ H_2 a_{11}'' a_{21}'' - K_2 \left[\frac{2}{3} p_3 P_2(a_{21}'') - p_2 a_{21}'' a_{31}'' \right] \right\} , \quad (1.204)$$
$$\int yv_0' dm = \alpha \left\{ \frac{2}{3} H_2 P_2(a_{21}'') + K_2 a_{21}'' (p_3 a_{11}'' - p_1 a_{31}'') \right\} , \quad (1.205)$$
$$\int zv_0' dm = \alpha \left\{ H_2 a_{21}'' a_{31}'' + K_2 \left[\frac{2}{3} p_2 P_2(a_{21}'') - p_2 a_{11}'' a_{21}'' \right] \right\} ; \quad (1.206)$$
$$\int xw_0' dm = \alpha \left\{ H_2 a_{11}'' a_{31}'' + K_2 \left[\frac{2}{3} p_2 P_2(a_{31}'') - p_3 a_{21}'' a_{31}'' \right] \right\} , \quad (1.207)$$
$$\int yw_0' dm = \alpha \left\{ H_2 a_{21}'' a_{31}'' - K_2 \left[\frac{2}{3} p_1 P_2(a_{31}'') - p_3 a_{11}'' a_{31}'' \right] \right\} , \quad (1.208)$$
$$\int zw_0' dm = \alpha \left\{ \frac{2}{3} H_2 P_2(a_{31}'') + K_2 a_{31}'' (p_1 a_{21}'' - p_2 a_{11}'') \right\} ; \quad (1.209)$$

where we have abbreviated

$$\alpha = \frac{1}{4} k_2 m_1 a_1^5 \equiv \frac{1}{5} M_2 , \quad (1.210)$$

and where H_2 stands for the coefficient in the equation

$$f_t = H_j \, r^{j+1} \, P_j(\lambda'') . \quad (1.211)$$

The only factor in f_j (as given by Equation 1.127) which renders it explicitly dependent on time is the radius-vector R; and a partial differentiation with respect to t discloses that, for $j = 2$,

$$H_2 = -3K_2 \frac{\dot{R}}{R} = -3 \frac{K_2}{R^3} \left\{ \frac{e \sin v}{1 + e \cos v} \dot{v} \right\} . \quad (1.212)$$

Equations (2.104)–(2.112) make it, moreover, evident that

$$\int (yv_0' + zw_0') dm = \alpha \left\{ K_2 a_{11}'' (p_3 a_{21}'' - p_2 a_{31}'') - \frac{2}{3} H_2 P_2(a_{11}'') \right\} , \quad (1.213)$$
$$\int (xu_0' + zw_0') dm = \alpha \left\{ K_2 a_{21}'' (p_1 a_{31}'' - p_3 a_{11}'') - \frac{2}{3} H_2 P_2(a_{21}'') \right\} , \quad (1.214)$$
$$\int (yv_0' + xu_0') dm = \alpha \left\{ K_2 a_{31}'') p_2 a_{11}'' - p_1 a_{21}'') - \frac{2}{3} H_2 P_2(a_{31}'') \right\} . \quad (1.215)$$

As the next step of our analysis, it remains for us to evaluate similarly the mass integrals of the differences in body accelerations with respect to the space axes, as given by Equations (1.191)–(1.193). Proceeding exactly in the same way as we did in evaluating the coefficients of deformation (1.201)–(1.209) we establish, after some algebra, that

$$\int \{y(\dot{w})'_0 - z(\dot{v})'_0\} dm = 2\alpha \left\{ a''_{11}\omega_{x''} - \frac{1}{3}p_1 \right\} H_2 + \tag{1.216}$$
$$+ \alpha \left\{ a''_{11}(\dot{\omega}_{x''} - n\omega_{y''}) + a''_{12}(\omega_{z''} - n)\omega_{x''} - a''_{13}\omega_{x''}\omega_{y''} - \frac{1}{3}\dot{p}_1 \right\} K_2,$$

$$\int \{z(\dot{u})'_0 - x(\dot{w})'_0\} dm = 2\alpha \left\{ a''_{21}\omega_{x''} - \frac{1}{3}p_2 \right\} H_2 + \tag{1.217}$$
$$+ \alpha \left\{ a''_{21}(\dot{\omega}_{x''} - n\omega_{y''}) + a''_{22}(\omega_{z''} - n)\omega_{x''} - a''_{23}\omega_{x''}\omega_{y''} - \frac{1}{3}\dot{p}_2 \right\} K_2,$$

$$\int \{x(\dot{v})'_0 - y(\dot{u})'_0\} dm = 2\alpha \left\{ a''_{31}\omega_{x''} - \frac{1}{3}p_3 \right\} H_2 + \tag{1.218}$$
$$+ \alpha \left\{ a''_{31}(\dot{\omega}_{x''} - n\omega_{y''}) + a''_{32}(\omega_{z''} - n)\omega_{x''} - a''_{33}\omega_{x''}\omega_{y''} - \frac{1}{3}\dot{p}_3 \right\} K_2,$$

where

$$\begin{Bmatrix} \omega_{x''} \\ \omega_{y''} \\ \omega_{z''} \end{Bmatrix} = \begin{Bmatrix} a''_{11} & a''_{21} & a''_{31} \\ a''_{12} & a''_{22} & a''_{32} \\ a''_{13} & a''_{23} & a''_{33} \end{Bmatrix} \begin{Bmatrix} \omega_x \\ \omega_y \\ \omega_z \end{Bmatrix} = \tag{1.219}$$
$$= \begin{Bmatrix} \lambda''_1 & \mu''_1 & \nu''_1 \\ \lambda''_2 & \mu''_2 & \nu''_2 \\ \lambda''_3 & \mu''_3 & \nu''_3 \end{Bmatrix} \begin{Bmatrix} \omega_{x'} \\ \omega_{y'} \\ \omega_{z'} \end{Bmatrix}$$

stand for the angular velocity components with respect to the revolving (doubly-primed) axes of coordinates.

It may also be noted that, in the coefficients of K_2 on the right-hand sides of Equations (1.213)–(1.215),

$$\left.\begin{aligned} p_3a''_{21} - p_2a''_{31} &= a''_{13}\omega_{y''} - a''_{12}(\omega_{z''} - n)\,, \\ p_1a''_{31} - p_3a''_{11} &= a''_{23}\omega_{y''} - a''_{22}(\omega_{z''} - n)\,, \\ p_2a''_{11} - p_1a''_{21} &= a''_{33}\omega_{z''} - a''_{32}(\omega_{z''} - n)\,. \end{aligned}\right\} \tag{1.220}$$

All transformations used, and results obtained, in the present section continue to be exact, and hold good without any restriction on the fractional height of the dynamical tides f, on the magnitude of the coefficients $p_{1,2,3}$; or on the values of the direction cosines a''_{ij}.

VI.2 Dissipative Forces: Effects of Viscosity

In the preceding section of this chapter we have evaluated in explicit form all coefficients of the dependent variables in the fundamental equations (1.109)–(1.111) of our problem, except for the last terms on the right-hand sides of these equations which arise from the effects of viscosity—representing the only dissipative terms in our equations which may introduce long-period changes in the dependent variables. In order to complete our task, our aim will be to evaluate these terms to the degree of approximation adhered to in the preceding section.

Before we proceed, however, to do so, a few words should be prefaced concerning the order of magnitude which the coefficient μ of viscosity can be expected to attain in stellar interiors. Apart from turbulent viscosity (which constitutes a subject in itself), its principal constituents in stellar interiors are the plasma (μ_H) viscosity, and radiative viscosity (μ_R). As is well known (cf., e.g., Chapman, 1954; or Oster, 1957) the coefficient of viscosity of stellar plasma (consisting essentially of ionized hydrogen) is sensibly equal to

$$\mu_H = \frac{5}{4}\frac{\sqrt{m_H}(kT)^{5/2}}{\sqrt{\pi}\epsilon^4 A_2(\xi)}\ \frac{\text{g}}{\text{cm.sec}}, \tag{2.1}$$

in which T denotes the temperature; k, the Boltzmann constant; m_H, the mass of the proton; and

$$A_2(\xi) = \log(1+\xi^2) - \frac{\xi^2}{1+\xi^2}, \tag{2.2}$$

where the parameter

$$\xi \equiv \frac{4kT}{\epsilon^2}\left(\frac{m_H}{\rho}\right)^{1/3}; \tag{2.3}$$

ϵ denoting the electronic charge.

If we insert for $k = 1.379 \times 10^{-16}$ erg. deg^{-1}, $m_H = 1.672 \times 10^{-24}$g and $\epsilon = 4.802 \times 10^{-10}$ e.s.u., we find that

$$\mu_H = \frac{3.68 \times 10^{-15} T^{5/2}}{\log(1+\xi^2)} - \frac{\xi^2}{1+\xi^2}\ \frac{\text{g}}{\text{cm.sec}} \tag{2.4}$$

where

$$\xi \equiv 2.84 \times 10^{-5} T \rho^{-1/3}. \tag{2.5}$$

On the other hand, the radiative viscosity of photon gas is known (cf. Jeans, 1925, 1926; Thomas, 1930; or Hazlehurst and Sargent, 1958) to be given by

$$\mu_R = \frac{4\sigma T^4}{15 c\kappa\rho}\ \frac{\text{g}}{\text{cm.sec}}, \tag{2.6}$$

where $\sigma = 7.55 \times 10^{-15}$ erg. cm^{-3} deg^{-4} denotes the Stefan constant; $c = 2.998 \times 10^{10}$ cm.sec^{-1}, the velocity of light; and κ, the absorption coefficient of stellar material per unit mass.

For densities ρ and temperatures T prevalent in the interiors of the stars the viscosity may turn out to be remarkably high in comparison with the terrestrial standards—of the order of 10^2–10^{41}—and much greater in degenerate gas (cf. Nishimura and Mori, 1961). The relative magnitude of plasma and radiative contributions to total viscosity depends sensitively on the mass of the respective configuration: whereas in stars of masses comparable with (or smaller than) that of the Sun, the radiative viscosity remains unimportant in comparison with plasma viscosity, in massive stars ($m > 5\odot$) $\nu_R >> \mu_H$ not only in the star's deep interior, but also in the far interior (since the ratio $T^4/\kappa\rho$ diminishes outwards less rapidly than $T^{5/2}$.

A: Tidal Friction

The integrands of the volume integrands on the right-hand sides of the fundamental equations (1.109)–(1.111) of our problems involving viscosity have already been formulated by Equations (1.90)–(1.92) in terms of the velocity components u_0', v_0' and w_0', and in the preceding section have been rewritten in terms of the amplitudes f of dynamical tides.

The first groups of terms on the right-hand sides of (1.90)–(1.92), factored by the viscosity coefficient μ, involve the Laplacians of the velocity components and their divergence (representing the effects of compressibility). These components (referred to the space axes) have already been given by Equations (1.181)–(1.183). An application of the Laplacian ∇^2 in the inertial coordinates then discloses that

$$\begin{aligned} \nabla^2 u_0' &= (p_3 y - p_2 z)\nabla^2 f + 2p_3 f_y - 2p_2 f_z + x\nabla^2 \dot{f} + 2x\dot{f}\,, &(2.7)\\ \nabla^2 v_0' &= (p_1 z - p_3 x)\nabla^2 f + 2p_1 f_z - 2p_3 f_x + y\nabla^2 \dot{f} + 2y\dot{f}\,, &(2.8)\\ \nabla^2 w_0' &= (p_2 x - p_1 y)\nabla^2 f + 2p_2 f_x - 2p_1 f_y + z\nabla^2 \dot{f} + 2z\dot{f}\,; &(2.9)\end{aligned}$$

so that

$$\begin{aligned} y\nabla^2 w_0' - z\nabla^2 v_0' &= 2D_1\dot{f} - 2p_1 D_4 f + 2(p_2 y + p_3 z) f_x - \\ &\quad - [p_1(y^2+z^2) - (p_2 y + p_3 z)x]\nabla^2 f\,, &(2.10)\\ z\nabla^2 u_0' - x\nabla^2 w_0' &= 2D_2\dot{f} - 2p_2 D_5 f + 2(p_1 x + p_3 z) f_y - \\ &\quad - [p_2(z^2+x^2) - (p_1 x + p_3 z)y]\nabla^2 f\,, &(2.11)\\ x\nabla^2 v_0' - y\nabla^2 u_0' &= 2D_3\dot{f} - 2p_3 D_6 f + 2(p_1 x + p_2 y) f_z - \\ &\quad - [p_3(x^2+y^2) - (p_1 x + p_2 y)z]\nabla^2 f\,, &(2.12)\end{aligned}$$

where the operators $D_{1,2,3}$ continue to be given by Equations (1.93)–(1.95);

$$D_4 \equiv z\frac{\partial}{\partial z} + y\frac{\partial}{\partial y} \equiv r\frac{\partial}{\partial r} - x\frac{\partial}{\partial x}\,, \qquad (2.13)$$

[1] Thus bearing out an earlier surmise by Eddington (1926) that "...For hydrodynamical purposes, one must think of the star as a thick oily liquid. This applies even to the regions of low density ...I suppose that even the photosphere will be rather sticky" (op.cit., p.281)

$$D_5 \equiv x\frac{\partial}{\partial x} + z\frac{\partial}{\partial z} \equiv r\frac{\partial}{\partial r} - y\frac{\partial}{\partial y}, \tag{2.14}$$

$$D_6 \equiv x\frac{\partial}{\partial x} + y\frac{\partial}{\partial y} \equiv r\frac{\partial}{\partial r} - z\frac{\partial}{\partial z}; \tag{2.15}$$

and the coefficients $p_{1,2,3}$ continue to be given by (1.179).

The divergence

$$\Delta'_0 \equiv \frac{\partial u'_0}{\partial x} + \frac{\partial v'_0}{\partial y} + \frac{\partial w'_0}{\partial z}, \tag{2.16}$$

arising from compressibility can be expressed in terms of the tidal deformation f with equal ease. Since a differentiation of (2.38)–(2.40) yields

$$\frac{\partial u'_0}{\partial x} = (p_3 y - p_2 z) f_x + \dot{f} + x\dot{f}_x, \tag{2.17}$$

$$\frac{\partial v'_0}{\partial y} = (p_1 z - p_3 x) f_y + \dot{f} + y\dot{f}_y, \tag{2.18}$$

$$\frac{\partial w'_0}{\partial z} = (p_2 x - p_1 y) f_z + \dot{f} + z\dot{f}_z; \tag{2.19}$$

an insertion of the preceding equations in (2.16) discloses that

$$\begin{aligned} \nabla'_0 &= r\dot{f}_r + 3\dot{f} - (p_1 D_1 + p_2 D_2 + p_3 D_3) f = \\ &= (r f_r + 3f)_t - (p_1 D_1 + p_2 D_2 + p_3 D_3)(r f_r + 4f), \end{aligned} \tag{2.20}$$

by virtue of (1.180).

Let us next turn our attention to terms on the right-hand sides of Equations (1.90)–(1.92) which are factored by the derivatives of μ. An application of the operators $D_{1,2,\ldots,6}$ to the expressions (1.181)–(1.183) for u'_0, v'_0, w'_0 discloses that

$$D_1 u'_0 = (-p_2 y - p_3 z) f + (p_3 y - p_2 z) D_1 f + x D_1 \dot{f}, \tag{2.21}$$

$$D_2 u'_0 = (+p_2 x \qquad) f + (p_3 y - p_2 z) D_2 f + x D_2 \dot{f} + z\dot{f}, \tag{2.22}$$

$$D_3 u'_0 = (\qquad +p_3 x) f + (p_3 y - p_2 z) D_3 f + x D_3 \dot{f} - y\dot{f}; \tag{2.23}$$

$$D_4 u'_0 = (-p_2 z - p_3 y) f + (p_3 y - p_2 z) D_4 f + x D_4 \dot{f}, \tag{2.24}$$

$$D_5 u'_0 = (-p_2 z \qquad) f + (p_3 y - p_2 z) D_5 f + x D_5 \dot{f} + x\dot{f}, \tag{2.25}$$

$$D_6 u'_0 = (\qquad +p_3 y) f + (p_3 y - p_2 z) D_6 f + x D_6 \dot{f} + x\dot{f}; \tag{2.26}$$

$$D_1 v'_0 = (+p_1 y \qquad) f + (p_1 z - p_3 x) D_1 f + y D_1 \dot{f} - z\dot{f}, \tag{2.27}$$

$$D_2 v'_0 = (-p_1 x - p_3 z) f + (p_1 z - p_3 x) D_2 f + y D_2 \dot{f}, \tag{2.28}$$

$$D_3 v'_0 = (\qquad +p_3 y) f + (p_1 z - p_3 x) D_3 f + y D_3 \dot{f} + x\dot{f}; \tag{2.29}$$

$$D_4 v'_0 = (+p_1 z - p_3 x) f + (p_1 z - p_3 x) D_4 f + y d_4 \dot{f} + y\dot{f}, \tag{2.30}$$

$$D_5 v'_0 = (+p_1 z - p_3 x) f + (p_1 z - p_3 x) D_5 f + y D_5 \dot{f}, \tag{2.31}$$

$$D_6 v'_0 = (\qquad -p_3 x) f + (p_1 z - p_3 x) D_6 f + y D_6 \dot{f} + y\dot{f}; \tag{2.32}$$

$$
\begin{aligned}
D_1 w_0' &= (+p_1 z \qquad)f + (p_2 x - p_1 y) D_1 f + z D_1 \dot{f} + y \dot{f}\,, && (2.33)\\
D_2 w_0' &= (\qquad +p_2 z)f + (p_2 x - p_1 y) D_2 f + z D_2 \dot{f} - x \dot{f}\,, && (2.34)\\
D_3 w_0' &= (-p_1 x - p_2 y)f + (p_2 x - p_1 y) D_3 f + z D_3 \dot{f}\,, && (2.35)\\
D_4 w_0' &= (-p_1 y \qquad)f + (p_2 x - p_1 y) D_4 f + z D_4 \dot{f} + z \dot{f}\,, && (2.36)\\
D_5 w_0' &= (\qquad +p_2 z)f + (p_2 x - p_1 y) D_5 f + z D_5 \dot{f} + z \dot{f}\,, && (2.37)\\
D_6 w_0' &= (-p_1 x + p_2 y)f + (p_2 x - p_2 y) - D_6 f + z D_6 \dot{f}\,. && (2.38)
\end{aligned}
$$

Accordingly,

$$
\begin{aligned}
\frac{\partial\mu}{\partial x} D_1 u_0' + \frac{\partial\mu}{\partial y} D_1 v_0' + \frac{\partial\mu}{\partial z} D_1 w_0' &= -(p_2 y + p_3 z) f \frac{\partial\mu}{\partial x} + p_1 f D_4 \mu - \\
&- (p_1 D_1 \mu + p_2 D_2 \mu + p_3 D_3 \mu) D_1 f + r \frac{\partial\mu}{\partial r} D_1 \dot{f} + \dot{f} D_1 \mu\,, && (2.39)\\
\frac{\partial\mu}{\partial x} D_2 u_0' + \frac{\partial\mu}{\partial y} D_2 v_0' + \frac{\partial\mu}{\partial z} D_2 w_0' &= -(p_1 x + p_3 z) f \frac{\partial\mu}{\partial y} + p_2 f D_5 \mu - \\
&- (p_1 D_1 \mu + p_2 D_2 \mu + p_3 D_3 \mu) D_d f + r \frac{\partial\mu}{\partial r} D_2 \dot{f} + \dot{f} D_2 \mu\,, && (2.40)\\
\frac{\partial\mu}{\partial x} D_3 u_0' + \frac{\partial\mu}{\partial y} D_3 v_0' + \frac{\partial\mu}{\partial z} D_3 w_0' &= -(p_1 x + p_2 y) f \frac{\partial\mu}{\partial z} + p_1 f D_6 \mu - \\
&- (p_1 D_1 \mu + p_2 D_2 \mu + p_3 D_3 \mu) D_3 f + r \frac{\partial\mu}{\partial r} D_3 \dot{f} + \dot{f} D_3 \mu\,; && (2.41)
\end{aligned}
$$

and

$$
\begin{aligned}
\frac{\partial\mu}{\partial y} D_4 w_0' - \frac{\partial\mu}{\partial z} D_4 v_0' &= -p_1 (f + D_4 f) D_4 \mu + \\
&+ x \left(p_2 \frac{\partial\mu}{\partial y} + p_3 \frac{\partial\mu}{\partial z} \right) D_4 f - (\dot{f} + D_4 \dot{f}) D_1 \mu\,, && (2.42)\\
\frac{\partial\mu}{\partial x} D_5 w_0' - \frac{\partial\mu}{\partial z} D_5 u_0' &= +p_2 (f + D_5 f) D_5 \mu - \\
&- y \left(p_1 \frac{\partial\mu}{\partial x} + p_3 \frac{\partial\mu}{\partial z} \right) D_5 f + (\dot{f} + D_5 \dot{f}) D_2 \mu\,, && (2.43)\\
\frac{\partial\mu}{\partial x} D_6 v_0' - \frac{\partial\mu}{\partial y} D_6 u_0' &= -p_3 (f + D_6 f) D_6 \mu + \\
&+ z \left(p_1 \frac{\partial\mu}{\partial x} + p_2 \frac{\partial\mu}{\partial y} \right) D_6 f - (\dot{f} + D_6 \dot{f})\,. && (2.44)
\end{aligned}
$$

Moreover, the curl-components

$$
\xi = \frac{\partial w_0'}{\partial y} - \frac{\partial v_0'}{\partial z}\,, \quad \eta = \frac{\partial u_0'}{\partial z} - \frac{\partial w_0'}{\partial x}\,, \quad \zeta = \frac{\partial v_0'}{\partial x} - \frac{\partial u_0'}{\partial y}\,, \qquad (2.45)
$$

follow on differentiation of (1.181)–(1.183) as

$$
\begin{aligned}
\xi &= -p_1 [2f + D_4 f] + x (p_2 f_y + p_3 f_z) - D_1 \dot{f}\,, && (2.46)\\
\eta &= -p_2 [2f + D_5 f] + y (p_1 f_x + p_3 f_z) - D_2 \dot{f}\,, && (2.47)\\
\zeta &= -p_3 [2f + D_6 f] + z (p_1 f_x + p_2 f_y) - D_3 \dot{f}\,, && (2.48)
\end{aligned}
$$

and, similarly,

$$\frac{\partial}{\partial x}(yw_0' - zv_0') = (p_2y + p_3z)(f + xf_x) - p_1(y^2 + z^2)f_x , \tag{2.49}$$

$$\frac{\partial}{\partial y}(zu_0' - xw_0') = (p_3z + p_1x)(f + yf_y) - p_2(x^2 + z^2)f_y , \tag{2.50}$$

$$\frac{\partial}{\partial z}(xv_0' - yu_0') = (p_1x + p_2y)(f + zf_z) - p_3(x^2 + y^2)f_z . \tag{2.51}$$

Inserting now the results represented by Equations (2.7)–(2.51) in (1.90)–(1.92), the latter assume the more explicit forms

$$\begin{aligned} \rho\{y\mathcal{F} - z\mathcal{G}\} = \mu\{2[D_1\dot{f} - p_1D_4f + (p_2y + p_3z)f_x]- \\ - [p_1r^2 - x(p_1x + p_2y + p_3z)]\nabla^2 f + \frac{1}{3}D_1\Delta_0'\} - \\ - p_1\{A(y^2 + z^2) + \frac{1}{3}D_1fD_1\mu\} + p_2\{A\,xy - \frac{2}{3}BZ - \frac{1}{3}D_1fD_2\mu\} + \\ + p_3\{A\,xz + \frac{2}{3}By - \frac{1}{3}D_1fD_3\mu\} + \left(r\frac{\partial\mu}{\partial r}\right)D_1\dot{f} - \frac{2}{3}\left(r\frac{\partial\dot{f}}{\partial r}\right)D_1\mu , \end{aligned} \tag{2.52}$$

$$\begin{aligned} \rho\{z\mathcal{F} - x\mathcal{G}\} = \mu\{2[D_2\dot{f} - p_2D_5f + (p_1x + p_3z)f_y]- \\ - [p_2r^2 - y(p_1x + p_2y + p_3z)]\nabla^2 f + \frac{2}{3}D_2\Delta_0'\} + \\ + p_1\left\{A\,xy + \frac{2}{3}Bz - \frac{1}{3}D_2f\,D_1\mu\right\} - p_2\left\{A(x^2 + z^2) + \frac{1}{3}D_2f\,D_2\mu\right\} + \\ + p_3\left\{A\,yz - \frac{2}{3}Bx - \frac{1}{3}D_2f\,D_3\mu\right\} + \left(r\frac{\partial\mu}{\partial r}\right)D_2\dot{f} - \frac{2}{3}\left(r\frac{\partial\dot{f}}{\partial r}\right)D_2\mu , \end{aligned} \tag{2.53}$$

and

$$\begin{aligned} \rho\{x\mathcal{G} - y\mathcal{F}\} = \mu\{2[D_3\dot{f} - p_3D_6f + (p_1x + p_2y)f_z]- \\ - [p_3r^2 - z(p_1x + p_2y + p_3z)]\nabla^2 f + \frac{1}{3}D_3\Delta_0'\} + \\ + p_1\left\{A\,xz - \frac{2}{3}By - \frac{1}{3}D_3f\,D_1\mu\right\} + p_2\left\{A\,yz + \frac{2}{3}Bx - \frac{1}{3}D_3f\,D_2\mu\right\} - \\ - p_3\left\{A(x^2 + y^2) + \frac{1}{3}D_3f\,D_3\mu\right\} + \left(r\frac{\partial\mu}{\partial r}\right)D_3\dot{f} - \frac{2}{3}\left(r\frac{\partial\dot{f}}{\partial r}\right)D_3\mu , \end{aligned} \tag{2.54}$$

where we have abbreviated

$$A \equiv \frac{\partial\mu}{\partial x}\frac{\partial f}{\partial x} + \frac{\partial\mu}{\partial y}\frac{\partial f}{\partial y} + \frac{\partial\mu}{\partial z}\frac{\partial f}{\partial z} \tag{2.55}$$

and

$$B \equiv \frac{\partial\mu}{\partial x}D_1f + \frac{\partial\mu}{\partial y}D_2f + \frac{\partial\mu}{\partial z}D_3f . \tag{2.56}$$

The foregoing expressions (2.52)–(2.56) are exact to quantities of first order in the tidal distortion f, but without any restriction on the magnitude of the coefficients $p_{1,2,3}$ as defined by Equations (1.179), or on the behaviour of the viscosity coefficient $\mu(x,y,z)$ throughout the interior of our configuration.

With the aid of the results obtained so far, we are now in a position to proceed with the evaluation of the volume-integrals on the right-hand sides of Equations (2.52)–(2.54) involved in the terms arising from viscosity. In order to do so, we can avail ourselves of the same technique as led us earlier in this section to evaluate the moments of inertia. Since the viscosity μ is a function of the state variables ρ and T which are constant over the equipotential surfaces of our configurations, we can regard also $\mu \equiv \mu(a)$ only; and, therefore, abbreviating

$$4\pi \int_0^{a_1} \mu a^{2j+1} da \equiv \mathcal{M}_j \tag{2.57}$$

we find that

$$\int \mu x^2 \nabla^2 f \, dV = \frac{4}{5} K_2 \mathcal{M}_2 P_2(a_{11}''), \tag{2.58}$$

$$\int \mu y^2 \nabla^2 f \, dV = \frac{4}{5} K_2 \mathcal{M}_2 P_2(a_{21}''), \tag{2.59}$$

$$\int \mu z^2 \nabla^2 f \, dV = \frac{4}{5} K_2 \mathcal{M}_2 P_2(a_{31}''), \tag{2.60}$$

$$\int \mu xy \, \nabla^2 f \, dV = \frac{6}{5} K_2 \mathcal{M}_2 a_{11}'' a_{21}'', \tag{2.61}$$

$$\int \mu xz \, \nabla^2 f \, dV = \frac{6}{5} K_2 \mathcal{M}_2 a_{21}'' a_{31}'', \tag{2.62}$$

$$\int \mu yz \, \nabla^2 f \, dV = \frac{6}{5} K_2 \mathcal{M}_2 a_{21}'' a_{31}''; \tag{2.63}$$

so that

$$\int \mu (x^2 + y^2) \nabla^2 f \, dV = -\frac{4}{5} K_2 \mathcal{M}_2 P_2(a_{31}''), \tag{2.64}$$

$$\int \mu (x^2 + y^2) \nabla^2 f \, dV = -\frac{4}{5} K_2 \mathcal{M}_2 P_2(a_{21}''), \tag{2.65}$$

$$\int \mu (y^2 + z^2) \nabla^2 f \, dV = -\frac{4}{5} K_2 \mathcal{M}_2 P_2(a_{11}''). \tag{2.66}$$

Moreover, by a similar process we establish that

$$\int \mu D_4 f \, dV = \int \mu x^2 \nabla^2 f \, dV, \tag{2.67}$$

$$\int \mu D_5 f \, dV = \int \mu y^2 \nabla^2 f \, dV, \tag{2.68}$$

$$\int \mu D_6 f \, dV = \int \mu z^2 \nabla^2 f \, dV; \tag{2.69}$$

and, turning to the integrals of terms factored by the radial derivative of μ, we find that

$$\int x^2 \frac{\partial \mu}{\partial r}\frac{\partial f}{\partial r} dV = \frac{2}{5} K_2 P_2(a''_{11}) \int_0^{a_1} \frac{\partial \mu}{\partial a} a^6 da , \tag{2.70}$$

$$\int y^2 \frac{\partial \mu}{\partial r}\frac{\partial f}{\partial r} dV = \frac{2}{5} K_2 P_2(a''_{21}) \int_0^{a_1} \frac{\partial \mu}{\partial a} a^6 da , \tag{2.71}$$

$$\int z^2 \frac{\partial \mu}{\partial r}\frac{\partial f}{\partial r} dV = \frac{2}{5} K_2 P_2(a''_{31}) \int_0^{a^1} \frac{\partial \mu}{\partial a} a^6 da , \tag{2.72}$$

and

$$\int xy \frac{\partial \mu}{\partial r}\frac{\partial f}{\partial r} dV = \frac{3}{5} K_2 a''_{11} a''_{21} \int_0^{a_1} \frac{\partial \mu}{\partial a} a^6 da , \tag{2.73}$$

$$\int xz \frac{\partial \mu}{\partial r}\frac{\partial f}{\partial r} dV = \frac{3}{5} K_2 a''_{11} a''_{31} \int_0^{a_1} \frac{\partial \mu}{\partial a} a^6 da , \tag{2.74}$$

$$\int yz \frac{\partial \mu}{\partial r}\frac{\partial f}{\partial r} dV = \frac{3}{5} K_2 a''_{21} a''_{31} \int_0^{a_1} \frac{\partial \mu}{\partial a} a^6 da . \tag{2.75}$$

These are the only terms which make nonvanishing contributions to the viscous terms of the right-hand sides of Equations (1.109)–(1.111). If, moreover, by partial integration

$$\int_0^{a_1} \frac{\partial \mu}{\partial a} a^6 da = \mu_1 a_1^6 - \frac{3}{2\pi} \mathcal{M}_2 , \tag{2.76}$$

where $\mu_1 \equiv \mu(a_1)$, an insertion of all foregoing results in the right-hand sides of Equations (2.52)–(2.54) we find that

$$\begin{aligned} \int \rho\{y\mathcal{H} - z\mathcal{G}\} dV &= \frac{12}{5} K_2 \mathcal{M}_2 \{a''_{11} p_1 - \omega_{x''}\} a''_{11} + \\ &+ \frac{4\pi}{5} K_2 \mu_1 a_1^6 \{3 a''_{11} \omega_{x''} - p_1\} , \end{aligned} \tag{2.77}$$

$$\begin{aligned} \int \rho\{z\mathcal{F} - x\mathcal{H}\} dV &= \frac{12}{5} K_2 \mathcal{M}_2 \{a''_{21} p_2 - \omega_{x''}\} a''_{21} + \\ &+ \frac{4\pi}{5} K_2 \mu_1 a_1^6 \{3 a''_{21} \omega_{x''} - p_2\} \end{aligned} \tag{2.78}$$

and

$$\begin{aligned} \int \rho\{x\mathcal{G} - y\mathcal{F}\} dV &= \frac{12}{5} K_2 \mathcal{M}_2 \{a''_{31} p_3 - \omega_{x''}\} a''_{31} + \\ &+ \frac{4\pi}{5} K_2 \mu_1 a_1^6 \{3 a''_{31} \omega_{x''} - p_3\} , \end{aligned} \tag{2.79}$$

where $p_{1,2,3}$ continue to be given by Equations (1.179) and

$$\omega_{x''} \equiv a''_{11} p_1 + a''_{21} p_2 + a''_{31} p_3 . \tag{2.80}$$

B: Dissipation of Energy

With the aid of the formulae established in the preceding parts of this section we wish to explore first the extent to which the axial rotation of the components in close binary systems can be slowed down (or possibly accelerated!) by dynamical tides through viscous friction.

In order to consider this problem in the simplest form, let us identify the equator of the rotating configuration and the plane of its orbit with the inertial plane $Z = 0$, so that the Eulerian angle θ and inclination i between the planes $Z = 0$ and $Z'' = 0$ in the direction cosines a'_{ij} and a''_{ij} can be set equal to zero. If so, it follows that the angular velocities $\omega_x = \omega_y = 0$ and the product of inertia $D = E = 0$ as well.

If, moreover, the orbital eccentricity e is sufficiently small for the products of H_2 with $k_2 K_2 a_1^3$ to be ignorable, all terms on both sides of Equations (1.109) and (1.110) vanish identically; while the viscous terms for (1.111) have already been established to be of the form (2.79). Therefore—to the first order in small quantities—the system (1.109)-(1.111) can be shown to reduce to a single equation of the form

$$C\frac{d\omega_z}{dt} + \frac{4}{5}\pi\frac{\mu_1 a_1^6}{R^3}(\omega_z - n) \;=\; 0\,, \tag{2.81}$$

in which the moment C of inertia about the Z-axis can be approximated [cf. Equation (1.145)] by $m_1 h_1^2$.

The foregoing Equation (2.81) is linear and of first order in ω_z, but non-homogeneous; and its solution can be expressed as

$$\omega_z(t) \;=\; n + \{\omega_z(0) - n\}e^{-\kappa t}\,, \tag{2.82}$$

in which

$$\kappa \;=\; \frac{4}{5}\pi\left(\frac{m_2}{m_1}\right)\frac{\mu_1 a_1^6}{m_1 h_1^2 A^3}$$

where h_1 stands for the radius of gyration of the respective configuration; and A, for the semi-major axis of its relative orbit. The foregoing Equation (2.82) makes it evident that, whatever the value of ω_z may have been at $t = 0$, it is bound to approach n asymptotically at $t \to \infty$ —in other words, *no matter what the initial disparity* $\omega_z \gtrless n$ *may have been in magnitude (or sign), viscous forces are bound to equalize them in the course of sufficiently long time.*[2]

Before this state is attained, however, the dynamical tides raised on the star of mass m_1 by its companion will sweep around the equator and—should the viscosity of its material be non-vanishing—tidal friction arising from this source will be bound to bring about a dissipation of kinetic energy of the tides into heat, a process which will tend to diminish the mechanical energy of the system.

[2] The foregoing Equation (2.82) represents, of course, only the simplest form of the solution of our problem; and for fuller details cf., e.g., sec.IV-3 of Kopal (1978). It is, however, sufficient to illustrate its salient features, some consequences of which will be explored in the concluding parts of this chapter.

As is well known (cf., e.g., Lamb, 1932) the viscous dissipation function Φ in compressible flow characterized by the rectangular velocity components u, v, w, is expressible in their terms as

$$\begin{aligned}\Phi &= \mu\{2(u_x^2+v_y^2+w_z^2)+(w_y+v_z)^2+\\ &\quad +(u_z+w_x)^2+(v_x+u_y)^2\}-\frac{2}{3}\mu(\Delta_0')^2\,,\end{aligned} \tag{2.83}$$

where subscripts x, y, z signify partial differentiation of the respective velocity components; and the divergence Δ_0' continues to be given by Equation (2.16).

For mass motions caused by dynamical tides, the rectangular velocity components in (2.83) should be identified with u_0', v_0', w_0' as given by Equations (1.181)–(1.183). Inserting the latter in (2.83) we find that

$$\begin{aligned}\mu^{-1}\Phi &= 6\dot{f}^2+2r(\dot{f}^2)_r+(r\dot{f}_r)^2+(r\dot{f}_n)^2-\\ &\quad -4\dot{f}(p_1D_1f+p_2D_2f+p_3D_3f)-2(r\dot{f}_r)(p_1D_1\dot{f}+p_2D_2\dot{f}+\\ &\quad +p_3D_3\dot{f})+(p_1D_1f+p_2D_2f+p_3D_3f)^2+\\ &\quad +(f_n)^2\{r^2(P_1^2+p_2^2+p_3^2)-(p_1x+p_2y+p_3z)^2\}-\frac{2}{3}(\Delta_0')^2=\\ &= \frac{1}{3}(r\dot{f})^2+\frac{1}{3}(p_1D_1f+p_2D_2f+p_3D_3f)^2+\\ &\quad +(r\dot{f}_n)^2+\frac{4}{3}(r\dot{f}_r)(p_1D_1+p_2D_2+p_eD_3)f-\\ &\quad -2(rf_r)(p_1D_1+p_2D_2+p_3D_3)\dot{f}+\\ &\quad +(f_n)^2\{r^2(p_1^2+p_2^2+p_3^2)-(p_1x+p_2y+p_3z)^2\}\end{aligned} \tag{2.84}$$

on insertion for Δ_0' from (2.16), and where we have abbreviated

$$(f_n)^2 = (f_x)^2+(f_y)^2+(f_z)^2\,. \tag{2.85}$$

Equation (2.84) represents the viscous dissipation functions for dynamical tides of any type. In particular, the terms on its right-hand side which arise from the "tidal breathing" due to orbital eccentricity will be factored by the f_t-component of $\dot{f}$ (cf. Equation 1.180); while those arising from asynchronous rotation (which produces the relative motion of the tides in longitude) will be factored by p_3. In order to determine the relative importance of tidal friction arising from these two causes, let us consider separately the following two cases:

(a) equator lying in the plane of a circular orbit (i.e., $p_1=p_2=0$ and $e=0$), but rotation and revolution asynchronous (i.e., $p_3\neq 0$, though constant);

(b) equator and orbit coplanar (i.e., $p_1=p_2=0$), but $e>0$ and, as a result, $p_3\equiv\omega-n$ becomes a fluctuating function of the time (as $n\equiv\dot{v}$!).

In the first case, it follows at once (for $f_t=0$) that

$$\dot{f} = -p_3D_3f_2 = -p_3D+3''f_2 = 3p_3K_2r^3\lambda''\mu''\,, \tag{2.86}$$

so that

$$r\dot{f}_r = 9p_3K_2r^3\lambda''\mu''\,, \tag{2.87}$$

$$(r\dot{f}_r)^2 \;=\; 81p_3^2K_2^2r^6\lambda''^2\mu''^2\,, \tag{2.88}$$

and

$$\begin{aligned}(r\dot{f}_n)^2 &= (r\dot{f}_r)^2 + (1-\lambda''^2)(\dot{f}_{\lambda''})^2 + (1-\nu''^2)(\dot{f}_{\nu''})^2 - 2\lambda''\nu''\dot{f}_{\lambda''}\dot{f}_{\nu''} \\ &= 45p_3^2K_2^2r^6\lambda''^2\mu''^2 + 9p_3^2k_2^2r^6(1-\nu''^2)\,. \end{aligned} \tag{2.89}$$

Moreover, by (2.88)

$$(r\dot{f}_r)p_3D_3f \;=\; 4p_3^2(D_3f_2)^2 \;=\; -27p_3^2K_2^2r^6\lambda''^2\mu''^2 \tag{2.90}$$

and

$$D_3\dot{f} \;=\; D_3''\dot{f}_2 \;=\; 3p_3K_2r^3(\lambda''^2-\mu''^2)\,; \tag{2.91}$$

which leads to

$$-2(rf_r)p_3D_3\dot{f} \;=\; -18p_3^2K_2^2r^6(\lambda''^2-\mu''^2)p_2(\lambda'') \tag{2.92}$$

and, lastly,

$$(f_n)^2 \;=\; 9K_2^2r^4\{[p_2(\lambda'')]^2 + (1-\lambda''^2)\lambda''^2\}\,; \tag{2.93}$$

so that

$$p_3^2(r^2-z^2)(f_n)^2 \;=\; 9p_3^2K_2^2r^6\{[p_2(\lambda'')]^2 + (1-\lambda''^2)\lambda''^2\}(1-\nu''^2)\,. \tag{2.94}$$

Inserting now from (2.86)–(2.90), (2.92) and (2.94) in (2.84), we find that in case (a),

$$\begin{aligned}\mu^{-1}\Phi &= 75p_3^2K_2^2r^6\lambda''^2\mu''^2 + \\ &\quad + \frac{9}{4}p_3^2K_2^2r^6(1-\lambda''^2)\{5(1-\lambda''^2)(1-\nu''^2) + 4(\lambda''^2-\mu''^2)\}. \end{aligned} \tag{2.95}$$

On integrating this expression over the entire volume of the respective configuration between the limits given by (1.129), we find that the rate of energy dissipation E due to longitudinal viscous tides should be given by

$$\frac{dE}{dt} \;=\; \frac{1}{2}\int \Phi\, dV \;=\; \frac{506}{35}\pi\, p_3^2K_2^2\int_0^{a_1} \mu a^8 da \text{ erg/sec}\,, \tag{2.96}$$

corresponding (on insertion for K_2 and p_3) to an energy loss per orbital cycle equal to

$$E \;=\; \frac{1012}{35n}\left\{\pi(\omega-n)\frac{m_2}{m_1R^3}\right\}^2\int_0^{a_1}\mu a^8\, da \text{ ergs}\,, \tag{2.97}$$

where, for circular orbits, n represents the Keplerian angular velocity $2\pi/P$ of orbital revolution in the period P; and ω stands, as before, for the angular velocity of axial rotation of the tidally-distorted star.

If, however, the orbit of our binary ceases to be circular, terms additional to those included in (2.95) will arise from "tidal breathing", and the angular velocity

n of orbital motion ceases to be constant. The part of viscous dissipation function due to "tidal breathing" will stem from the term f_t in $\dot{f}$, and assume the form

$$\mu^{-1}\Phi_e = \frac{1}{3}(rf_{tr})^2 + (rf_{tn})^2 + \frac{4}{3}(rf_{tr})p_3 D_3 f - 2p_3(rf_r)D_3 f_t \,, \tag{2.98}$$

which on insertion from (2.88) assumes the more explicit form

$$\mu^{-1}\Phi_e = 3H_2^2 r^6\{93\lambda''^2 - 1)^2 + 3\lambda''^2(1-\lambda''^2)\} + 6H_2K_2p_3r^6\lambda''\mu''P_2(\lambda'') \,, \tag{2.99}$$

leading to a rate of dissipation

$$\frac{dE_e}{dt} = \frac{1}{2}\int \Phi_e dV = \frac{36}{5}\pi H_2^2 \int_0^{a_1} \mu a^8 da \text{ erg. sec}^{-1} \,. \tag{2.100}$$

The total dissipation of kinetic energy per cycle should now be given by

$$E = \int_0^P \frac{d}{dt}(E + E_e)dt \,, \tag{2.101}$$

where $\dot{E}$ and $\dot{E}_e$ are given by Equations (2.96) and (2.100), respectively. In evaluating the right-hand side of (2.101) we should, however, keep in mind the fact that the mean daily motion in the relative motion will cese to be constant, but equal to

$$\dot{v} = -2\left(\frac{\dot{R}}{R}\right)n \tag{2.102}$$

and, therefore,

$$nH_2 = \frac{3}{2}K_2\dot{n} \,; \tag{2.103}$$

for the consequences of these facts cf. again sec. IV-3D of Kopal (1978).

VI.3 Precession and Nutation of Deformable Bodies

In the preceding two sub-sections of this chapter we set out to investigate the effects of viscosity of self-gravitating deformable bodies on the basis of a simplifying assumption that the axes Z, Z' and Z'' of our rectangular systems of coordinates are identical with each other—in which case both angular velocity components ω_x and ω_y can be set equal to zero. The aim of this last section of the present chapter will be to remove this assumption, and consider the more general case in which the rotation of a self-gravitating fluid body possesses its full complement of three degrees of freedom

Needless to say, in order to obtain an analytic solution of so generalized a problem, we find it necessary to restrict the range of our dependent variables in such a way that none will appear in our equations (1.109)–(1.111) of motion in higher than first powers. This we shall attain by regarding not only the fractional heights f_j of the respective tides, but also the angles θ and i to be small enough for their squares and higher powers to be ignorable; and the same should be true of the eccentricity e of the binary orbit.

If so, the tides will make the moments A, B, C of inertia differ from their equilibrium values by small quantities of first order, and the same will be true of the product of inertia F''; but D'' and E'' are of second order. Moreover, while the angular velocity ω_z will hereafter be regarded as a quantity of zero order, ω_x and ω_y will be of first order. Lastly, if the rates of change $\dot{\theta}$ and $\dot{\phi}$ of the Eulerian angles θ and ϕ can likewise be regarded as small, $\dot{\omega}_z$ will turn out to be of third order—a fact permitting us thereafter to treat ω_z as a constant.

Within the scheme of such an approximation, the dominant terms of our fundamental equations (1.109) and (1.110) will reduce to

$$\begin{aligned} A\dot{\omega}_x &+ (C-B)\omega_y\omega_z + d\omega_z^2 + F\omega_x\omega_z + \\ &+ 2\alpha K_2\left\{p_3 a''_{11}a''_{31}\omega_x - \frac{2}{3}p_3 P_2(a''_{11})\omega_y + \right. \\ &+ \left.\left[\frac{2}{3}p_2 P_2(a''_{11}) - p_1 a''_{11}a''_{21}\right]\omega_z + \frac{1}{2}p_3 a''_{12}\omega_{x''}\right\} + \\ &+ 8\alpha K_2 a''_{33}a''_{21}\nu''_1\omega_z^2 = \\ &= \frac{12}{5}K_2\mathcal{M}_2\{a''_{11}p_1 - \omega''_x\}a''_{11} + \frac{4}{5}\pi K_2\mu_1 a_1^6\{3a''_{11}\omega''_x - p_1\} \end{aligned} \tag{3.1}$$

and

$$\begin{aligned} B\dot{\omega}_y &+ (A-C)\omega_x\omega_z - E\omega_z^2 - F\omega_y\omega_z - \\ &- 2\alpha K_2\left\{p_3 a''_{11}a''_{21}\omega_y - \frac{2}{3}p_3 P_2(a''_{21})\omega_x + \right. \\ &+ \left.\left[\frac{2}{3}p_1 P_2(a''_{21}) - p_2\_11'' a''_{21}\right]\omega_z - \frac{1}{2}p_3 a''_{22}\omega_{x''}\right\} - \\ &- 8\alpha K_2 a''_{11}a''_{33}\nu''_1\omega_z^2 = \\ &= \frac{21}{5}K_2\mathcal{M}_2\{a''_{21}p_2 - \omega_{x''}\}a_{21} + \frac{4}{5}\pi K_2\mu_1 a_1^6\{3a''_{21}\omega''_x - p_2\}, \end{aligned} \tag{3.2}$$

where the angular velocity components $\omega_{x,y,z}$ continue to be given by Equations (1.37)–(1.39); $p_{1,2,3}$ by (1.179);

$$\omega_{x''} = a''_{11}\omega_x + a''_{21}\omega_y + a''_{31}\omega_z\,; \tag{3.3}$$

and (within the scheme of our approximation),

$$\omega_x = a_{13}\omega_z \quad\text{and}\quad \omega_y = a_{23}\omega_z\,; \tag{3.4}$$

while the corresponding moments and products of inertia are given by Equations (1.139)–(1.144). Inserting them in the left-hand sides of Equations (3.1) and (3.2) we find that

$$(C'-B')\omega_y\omega_z + D'\omega_z^2 + F'\omega_x\omega_z = 0 \tag{3.5}$$

and

$$(A' - C')\omega_x\omega_z - E'\omega_z^2 - F\omega_y\omega_z = 0 . \tag{3.6}$$

Similarly, in the case of tidal distortion (whose axis of symmetry is "locked-in" with the radius vector of direction cosines a''_{11}, a''_{21}, a''_{31}), we have

$$\omega_x = na''_{13}, \ \omega_y = na''_{23}, \ \omega_z = n , \tag{3.7}$$

where n denotes the mean daily motion of the disturbing body. Moreover, the moments and products of inertia of tidally-distorted configurations are given by Equations (1.130)–(1.136) and their insertion in (3.1)–(3.2) discloses that

$$(C'' - B'')\omega_y\omega_z + D''\omega_z^2 + F''\omega_x\omega_z = 0 \tag{3.8}$$

and

$$(A'' - C'')\omega_x\omega_z - E''\omega_z^2 - F''\omega_y\omega_z = 0 \tag{3.9}$$

as well.

Let us multiply Equation (3.1) by cos ϕ, (3.2) by sin ϕ, and add; afterwards multiply (3.1) by sin ϕ, (3.2) by cos ϕ, and subtract. Since, within the scheme of our approximation,

$$a_{13} = \sin\phi \sin\theta, \ a_{23} = -\cos\phi\sin\theta, \ a_{33} = 1 ; \tag{3.10}$$

and

$$\left.\begin{array}{lll} a''_{13} = \sin\Omega\sin i, & a''_{23} = -\cos\Omega\sin i, & a''_{33} = 1 \\ a''_{31} = \sin u \sin i, & a''_{32} = \cos u \sin i ; & \end{array}\right\} \tag{3.11}$$

while

$$\left.\begin{array}{ll} a''_{11} = \cos(u+\omega), & a''_{21} = \sin(u+\omega) , \\ a''_{12} = -\sin(u+\omega), & a''_{22} = \cos(u+\Omega) , \end{array}\right\} \tag{3.12}$$

it follows that, correctly to quantities of second order,

$$A\dot{\omega}_x \cos\phi + B\dot{\omega}_y \sin\phi = a_0\omega_z\dot{\phi}\sin\theta , \tag{3.13}$$

$$A\dot{\omega}_x \sin\phi - B\dot{\omega}_y \cos\phi = A_0\omega_z\dot{\theta}\cos\theta ; \tag{3.14}$$

while

$$2a''_{11}a''_{21}(\omega_x\cos\phi - \omega_y\sin\phi) = \omega_z\sin\theta\sin 2\phi\sin 2(u+\Omega) , \tag{3.15}$$

$$2a''_{11}a''_{21}(\omega_x\sin\phi + \omega_y\cos\phi) = -\omega_z\sin\theta\cos 2\phi\sin 2(u+\Omega) ; \tag{3.16}$$

$$\begin{aligned} \left\{\frac{2}{3}p_2P_2(a''_{11}) - p_1a''_{11}a''_{21}\right\}\cos\phi - \left\{\frac{2}{3}p_1P_2(a''_{21}) - p_2a''_{11}a''_{21}\right\}\sin\phi = \\ = -\frac{2}{3}\sin\theta\, P_2\{\cos(U+\omega-\phi)\} + \\ + n\sin i\{\cos u\cos(u+\Omega-\phi) - \frac{1}{3}\cos(\Omega-\phi)\} , \end{aligned} \tag{3.17}$$

$$\left\{\frac{2}{3}p_2 P_2(a''_{11}) - p_1 a''_{11} a''_{21}\right\} \sin\phi + \left\{\frac{2}{3}p_1 P_2(a''_{21}) - p_2 a''_{11} a''_{21}\right\} \cos\phi =$$

$$= \frac{1}{2}\omega_z \sin\theta \sin 2(u+\Omega-\phi) -$$

$$- n \sin i \cos u \sin(u+\Omega-\phi) + \frac{1}{3} n \sin i \sin(\Omega-\phi) ; \qquad (3.18)$$

$$-\omega_y P_2(a''_{11}) \cos\phi + \omega_x P_2(a''_{21}) \sin\phi =$$

$$= \omega_z \sin\theta \left\{ P_2[\cos(u+\omega-\phi)] - \frac{3}{4}\sin 2\phi \sin 2(u+\Omega)\right\} , \qquad (3.19)$$

$$-\omega_y P_2(a''_{11}) \sin\phi - \omega_x P_2(a''_{21}) \cos\phi =$$

$$= \sin\theta \sin 2\phi \cos 2(u+\Omega) ; \qquad (3.20)$$

and

$$\omega_{x''}(a''_{12}\cos\phi + a''_{22}\sin\phi) =$$

$$= -\omega_z\{\sin u \sin i - \sin\theta \sin(u+\Omega-\phi)\} \sin(u+\Omega-\phi) , \qquad (3.21)$$

$$\omega_{x''}(a''_{12}\sin\phi - a''_{22}\cos\phi) =$$

$$= -\omega_z\{\sin u \sin i - \sin\theta \sin(u+\Omega-\phi)\} \cos(u+\Omega-\phi) , \qquad (3.22)$$

it follows that

$$2\alpha K_2\omega_2 \left\{\frac{2}{3}\omega_z(p_3-\omega_z)\sin\theta\, P_2[\cos(u+\Omega-\phi)] + \right.$$

$$\left. + n \sin i[\cos u \cos(u+\Omega-\phi) - \frac{1}{3}\cos(\Omega-\phi)]\right\} -$$

$$= -\alpha K_2\omega_2 n \left\{\frac{\partial^2 \nu_2''^2}{\partial\theta} + \frac{1}{3}\frac{\partial \nu_3''^2}{\partial\theta}\right\} , \qquad (3.23)$$

and

$$\alpha K_2\omega_z\{(\omega_z - p_3)\sin\theta \sin 2(u+\Omega-\phi) -$$

$$- 2n \sin i \cos u \sin(u+\Omega-\phi) + \frac{2}{3} n \sin i \sin(\Omega-\phi)\} =$$

$$= \alpha K_2\omega_z n \left\{\frac{1}{\sin\theta}\frac{\partial \nu_2^{2''}}{\partial\phi} + \frac{1}{3\sin\theta}\frac{\partial \nu_3^{2''}}{\partial\phi}\right\} , \qquad (3.24)$$

where the direction cosines ν''_j continue to be given exactly by the last one of Equations (3.11) of Chapter II and, within the scheme of our first-order approximation, reduce to

$$\nu''_1 = \sin i \sin u - \sin\theta \sin(u+\Omega-\phi) , \qquad (3.25)$$

$$\nu''_2 = \sin i \cos u - \sin\theta \cos(u+\Omega-\phi) ; \qquad (3.26)$$

so that

$$\sin(u+\Omega-\phi) = -\frac{1}{\sin\theta}\frac{\partial\nu_2''}{\partial\phi} = -\frac{\partial\nu_1''}{\partial\theta}, \tag{3.27}$$

$$\cos(u+\Omega-\phi) = +\frac{1}{\sin\theta}\frac{\partial\nu_1''}{\partial\theta} = -\frac{\partial\nu_2''}{\partial\theta}; \tag{3.28}$$

while from the exact equation (cf. Equation (3.14) of Chapter II)

$$\nu_3'' = \cos\theta\cos i + \sin\theta\sin i\cos(\Omega-\phi) \tag{3.29}$$

it follows that, approximately,

$$\frac{\partial\nu_3''}{\partial\theta} = -\sin\theta + \sin i\cos(\Omega-\phi) \tag{3.30}$$

and

$$\frac{1}{\sin\theta}\frac{\partial\nu_3''}{\partial\theta} = \sin i\sin(\Omega-\phi). \tag{3.31}$$

The foregoing results complete the evaluation of the gyroscopic terms in the equations of motion, factored by different deformation coefficients arising from dynamical tides. The "forcing terms" (cf. Equations (2.131)–(2.133) of Chapter IV, Kopal, 1978) arising from the attraction of the external mass on the equatorial bulge of the rotating configuration will be of the form

$$\begin{aligned}\cos\phi\int D_1\Omega_c dm + \sin\phi\int D_2\Omega_c dm &= \\ &= k_2 a_1^5 m_2 R^{-3}\omega_z^2\nu_1''\sin(u+\Omega-\phi) = \\ &= -\{k_2 a_1^5 m_2 R^{-3}\omega_z^2\}\frac{\partial\nu_1''}{\partial\theta}\end{aligned} \tag{3.32}$$

and

$$\begin{aligned}\sin\phi\int D_1\Omega_c dm - \cos\phi\int D_2\Omega_c dm &= \\ &= k_2 a_1^5 m_2 R^{-3}\omega_z^2\nu_1''\cos(u+\Omega-\phi) = \\ &= \{k_2 a_1^5 m_2 R^{-3}\omega_z^2\}\frac{1}{\sin\theta}\frac{\partial\nu_1''}{\partial\theta}\end{aligned} \tag{3.33}$$

by (3.27) and (3.28).

Lastly, turning our attention to the viscous terms, we shall hereafter assume that $\mu(a_1) = 0$ which makes the integrated part $[\mu a^6 a_1]_0$ on the right-hand side of Equations (3.1) and (3.2) vanish. Since, moreover, to the first order in small quantities

$$\begin{aligned}p_1 a_{11}'' - \omega_{x''} &= \omega_z\{\sin\theta\cos\phi\sin(u+\Omega) - \sin i\sin u\} - \\ &\quad - n\sin i\sin\Omega\cos(u+\Omega)\end{aligned} \tag{3.34}$$

and

$$p_2 a''_{21} - \omega_{x''} = -\omega_z\{\sin\theta \sin\phi \cos(u+\Omega) + \sin i \sin u\} + + n \sin i \cos\Omega \sin(u+\Omega) \; ; \tag{3.35}$$

then abbreviating

$$\frac{12}{5} K_2 (a''_{11} p_1 - \omega_{x''}) a_{11} \mathcal{M}_2 \equiv G \, , \tag{3.36}$$

$$\frac{12}{5} K_2 (a''_{21} p_2 - \omega_{x''}) a''_{21} \mathcal{M}_2 \equiv H \, , \tag{3.37}$$

we find that

$$G \cos\phi + H \sin\phi = \frac{12}{5} K_2 \{-p_3 \sin i \sin u \cos(u+\Omega-\phi) + + \frac{1}{2} \sin 2(u+\Omega)[\omega_z \sin\theta \cos 2\phi - n \sin i \cos(\Omega-\phi)]\} \mathcal{M}_2 \, , \tag{3.38}$$

and

$$G \sin\phi - H \cos\phi = \frac{12}{5} K_2 \{p_3 \sin i \sin u \sin(u+\Omega-\phi) + + \frac{1}{2} \sin 2(u+\Omega)[\omega_z \sin\phi \sin 2\phi + n \sin i \sin(\Omega+\phi)]\} \mathcal{M}_2 \, . \tag{3.39}$$

If we collect all results obtained so far in this section, the linearized Equations (3.1)–(3.2) can ultimately be rewritten as

$$A_0 \omega_z \dot{\phi} \sin\theta + \frac{\alpha K_2 \omega_z}{2} \frac{\partial}{\partial\theta} \left\{ \omega_z \nu_1''^2 - n \nu_2''^2 + \frac{1}{3} n \nu_3''^2 \right\} = = - \left\{ \omega_z k_2 a_1^5 \frac{m_2}{R^3} \right\} \frac{\partial \nu_1''^2}{\partial\theta} + \text{viscous terms} \tag{3.40}$$

and

$$A_0 \omega_z \dot{\theta} \cos\theta - \frac{\alpha K_2 \omega_z}{2} \frac{1}{\sin\theta} \frac{\partial}{\partial\phi} \left\{ \omega_z \nu_1''^2 - n \nu_2''^2 + \frac{1}{3} n \nu_3''^2 \right\} = = \left\{ \omega_z k_2 a_1^5 \frac{m_2}{R^3} \right\} \frac{1}{\sin\theta} \frac{\partial \nu_1''^2}{\partial\phi} + \text{viscous terms} \, , \tag{3.41}$$

where, in accordance with Equation (1.145), $A_0 = m_1 h_1^2$; h_1 denoting the radius of gyration of the respective configuration. The terms in curly brackets on the left-hand sides of the foregoing Equations (3.40) and (3.41) represent the effects of deformation arising from dynamical tides; while the first terms on the right-hand sides arise from the attraction of the companion on the equatorial bulge of the rotating configuration.

If $n = 0$ (corresponding to a "two-centre" problem, in which the rotating configuration is attracted by a fixed mass-point), Equations (3.40) and (3.41) will reduce to

$$A_0\dot{\phi} \sin\theta + \frac{\alpha K_2 \omega_z}{2}\frac{\partial \nu_1''^2}{\partial\theta} = = -\left\{\omega_z k_2 a_1^5 \frac{m_2}{R^3}\right\}\frac{\partial \nu_1''^2}{\partial\theta} + \text{viscous terms} \tag{3.42}$$

and

$$A_0\dot{\theta} - \frac{\alpha K_2 \omega_z}{2}\frac{1}{\sin\theta}\frac{\partial \nu_1''^2}{\partial\phi} = = \left\{\omega_z k_2 a_1^5 \frac{m_2}{R^3}\right\}\frac{1}{\sin\theta}\frac{\partial \nu_1''^2}{\partial\phi} + \text{viscous terms}\,. \tag{3.43}$$

If, moreover, the configuration were rigid, the deformation terms on the left-hand side of Equations (3.42)–(3.43) as well as all viscous terms would be identically zero; and their system would reduce to

$$A_0\dot{\phi}\sin\theta = -\left\{\omega_z k_2 a_1^5 \frac{m_2}{R^3}\right\}\frac{\partial \nu_1''^2}{\partial\theta}\,, \tag{3.44}$$

$$A_0\dot{\theta}\sin\theta\cos\theta = \left\{\omega_z k_2 a_1^5 \frac{m_2}{R^3}\right\}\frac{\partial \nu_1''^2}{\partial\phi}\,, \tag{3.45}$$

in which form the problem was treated previously by Brouwer (1946).

If, on the other hand, $\omega_z = 0$ but $n \neq 0$, Equations (3.40) and (3.41) would reduce to

$$A_0\dot{\phi}\sin\theta - \frac{\alpha K_2 n}{2}\frac{\partial}{\partial\theta}\left\{\nu_2''^2 - \frac{1}{3}\nu_3''^2\right\} = \text{viscous terms}\,, \tag{3.46}$$

$$A_0\dot{\theta}\cos\theta + \frac{\alpha K_2 n}{2}\frac{1}{\sin k\theta}\frac{\partial}{\partial\phi}\left\{\nu_2''^2 - \frac{1}{3}\nu_3''^2\right\} = \text{viscous terms}\,, \tag{3.47}$$

the solution of which represents the "stabilizing" effects of dynamical tides on a non-rotating configuration susceptible of deformation. If, lastly, both ω_z and n are set equal to zero, the case corresponds to a neutral equilibrium of a stationary body, for which the Eulerian angles θ, ϕ, ψ may assume arbitrary constant values.

A: Secular and Long-Period Motion

Among the assumptions committed to linearize our fundamental equations (1.108) and (1.109) was the expectation that the time-derivatives $\dot{\theta}$ and $\dot{\phi}$ of the Eulerian angles are so small that their ratios to $\dot{\psi} \equiv \omega_z$ can be regarded as small quantities of first order. If so, however, all terms in (3.40) and (3.41) containing the true anomaly u of the disturbing components can obviously be *averaged* with respect to u over the orbital cycle, and values of their different trigonometric functions replaced by their averages taken over the interval $(0, 2\pi)$. As, in accordance with the exact Equations (3.12) and (3.13) of Chapter II,

$$\nu_1'' = \Re \sin u + \mathcal{L} \cos u \,, \tag{3.48}$$

$$\nu_2'' = \Re \cos u - \mathcal{L} \sin u \,, \tag{3.49}$$

where

$$\Re = \sin i \cos \theta - \cos i \sin \theta \cos(\Omega - \phi) \tag{3.50}$$

$$\mathcal{L} = - \sin \theta \sin(\Omega - \phi) \,, \tag{3.51}$$

in agreement with Equations (3.12) — (3.16) of Chapter II, the average values over a cycle (denoted hereafter by the square brackets [...] of $\nu_{1,2}''^2$ are obviously given by

$$[\nu_1''^2] = \frac{1}{2}(\Re^2 + \mathcal{L}^2) = [\nu_2''^2] \,. \tag{3.52}$$

Since, moreover,

$$\nu_1''^2 + \nu_2''^2 + \nu_3''^2 = 1 \tag{3.53}$$

and, by (3.29),

$$[\nu_3''] = \nu_3'' \text{ and } [\nu_3''^2] = \nu_3''^2 \,; \tag{3.54}$$

so that

$$[\nu_1''^2] = [\nu_2''^2] = \frac{1}{2}(1 - \nu_3''^2) \,, \tag{3.55}$$

it follows that

$$[\omega_z \nu_1''^2 - n\nu_2''^2 + \frac{1}{3}\nu_3''^2] = \frac{p_3}{2}(1 - \nu_3''^2) + \frac{n}{3}\nu_3''^2 \,; \tag{3.56}$$

and, likewise, from (3.38) and (3.39) that

$$[G \cos \phi + H \sin \phi] = \frac{6}{5} K_2 p_3 \sin i \sin(\Omega - \phi) \mathcal{M}_2 \tag{3.57}$$

and

$$[G \sin \phi - H \cos \phi] = \frac{6}{5} K_2 p_3 \sin i \cos(\Omega - \phi) \mathcal{M}_2 \,, \tag{3.58}$$

Accordingly, on insertion for $p_3 = \omega_z - n$, $\alpha = \frac{1}{4}k_2 m_1 a_1^5$, $A_0 = m_1 h_1^2$ and $K_2 = (m_2/m_1)R^3$, the secular part of the linearized equations (3.40) and (3.41) can be expressed in a more concise form as

$$\dot{\phi} \sin \theta = -\frac{\kappa}{2} \frac{\partial \nu_3''^2}{\partial \theta} + \lambda \sin i \sin(\Omega - \phi) \,, \tag{3.59}$$

$$\dot{\theta} \sin \theta = +\frac{\kappa}{2} \frac{\partial \nu_3''^2}{\partial \phi} + \lambda \sin i \cos(\Omega - \phi) \sin \theta \,, \tag{3.60}$$

where we have abbreviated

$$\kappa = \frac{27\omega - 5n}{24} k_2 \left(\frac{m_2}{m_1}\right) \left(\frac{a_1}{R}\right)^3 \left(\frac{a_1}{h_1}\right)^2 \tag{3.61}$$

and

$$\lambda = \frac{6}{5m_1h_1^2R^3}\left(\frac{m_2}{m_1}\right)\frac{\omega_z - n}{\omega_z}\mathcal{M}_2 \,. \tag{3.62}$$

In order to proceed further, let us set

$$p = \sin\theta\sin\phi, \quad q = \sin\theta\cos\phi\,; \tag{3.63}$$

$$r = \sin i\sin\Omega, \quad s = \sin i\cos\Omega\,; \tag{3.64}$$

and adopt these direction cosines, rather than the constituent angles $\theta\,\phi$ or i, Ω, as the dependent variables of our problem—such that

$$\nu_3'' = pr + qs + \sqrt{(1-p^2-q^2)(1-r^2-s^2)}\,. \tag{3.65}$$

By a differentiation of (3.63) it readily follows that

$$\dot{\theta}\sin\theta\cos\theta = p\dot{p} + q\dot{q}\,, \tag{3.66}$$

$$\dot{\phi}\sin^2\theta = q\dot{p} - p\dot{q}\,, \tag{3.67}$$

and, moreover, the operators

$$\sin\theta\frac{\partial}{\partial\theta} = p\frac{\partial}{\partial p} + q\frac{\partial}{\partial q}\,, \tag{3.68}$$

$$\frac{\partial}{\partial\phi} = q\frac{\partial}{\partial p} - p\frac{\partial}{\partial q}\,. \tag{3.69}$$

Accordingly, the system of Equations (3.59)–(3.60) can be alternatively rewritten in the more symmetrical form

$$\dot{p} = +\frac{\kappa}{2}\frac{\partial\nu_2''^2}{\partial q} + \lambda r\,, \tag{3.70}$$

$$\dot{q} = -\frac{\kappa}{2}\frac{\partial\nu_3''^2}{\partial p} + \lambda s\,, \tag{3.71}$$

By a differentiation of Equation (3.65) we find that

$$\frac{\partial\nu_3''^2}{\partial p} = 2\nu_3''\left\{r - \frac{p(1-r^2-s^2)}{\nu_3''-pr-qs}\right\}\,, \tag{3.72}$$

$$\frac{\partial\nu_3''^2}{\partial q} = 2\nu_3''\left\{s - \frac{q(1-r^2-s^2)}{\nu_3''-pr-qs}\right\}\,; \tag{3.73}$$

but if we remember that the direction cosines p, q, r, s constitute small quantities of first order whose squares and higher powers can be ignored, the foregoing expressions reduce to

$$\frac{\partial\nu_3''}{\partial p} = r - p\,, \tag{3.74}$$

$$\frac{\partial\nu_3''}{\partial q} = s - q\,. \tag{3.75}$$

On insertion of (3.74) and (3.75) in (3.70)–(3.71) the latter pair of equations assume the neat forms

$$\dot{p} = \kappa(s - q) + \lambda r \,, \tag{3.76}$$
$$\dot{q} = \kappa(p - r) + \lambda s \,, \tag{3.77}$$

where κ and λ continue to be given by Equations (3.61) and (3.62).

The foregoing Equations (3.76)–(3.77) do not define yet the variables p and q uniquely; for they involve also the direction cosines r and s. Therefore, in order to render the problem determinate, an additional two equations must be adjoined to specify r and s. These are the variational equations for the angular variables Ω and i, well known from celestial mechanics (cf., e.g., section V-1 of Kopal, 1978); and their linearized form can be reduced to

$$\dot{r} = \gamma(q - s) \,, \tag{3.78}$$

and

$$\dot{s} = \gamma(r - p) \,, \tag{3.79}$$

where we have abbreviated

$$\gamma = \frac{k_2 a_1^5 \omega_z^2 n}{G m_1 R^2} \,. \tag{3.80}$$

The simultaneous differential equations for p, q, r, s of the linearized system governing the precession and nutation of a rotating deformable configuration attracted by an external mass are, therefore, of the form,

$$\left.\begin{aligned} \dot{p} &= \kappa(s - q) + \lambda r \,, \\ \dot{q} &= \kappa(p - r) + \lambda s \,, \\ \dot{r} &= \gamma(q - s) \,, \\ \dot{s} &= \gamma(r - p) \,, \end{aligned}\right\} \tag{3.81}$$

where the dependent variables p, q, r and s have already been defined by Equations (3.76)–(3.79); while κ, λ and γ are constants given by Equations (3.61), (3.62) and (3.80).

In order to construct a solution of the foregoing simultaneous system (2.113), let us assume that

$$p = A_1 e^{\alpha_1 t}, \quad q = A_2 e^{\alpha_2 t}, \quad r = A_3^{\alpha_3 t}, \quad s = A_4 e^{\alpha_4 t} \tag{3.82}$$

where the A_j's and α_j's are constants satisfying the system of linear algebraic equations

$$\left.\begin{aligned} A_1\alpha + A_2\kappa - A_3\lambda - A_4\kappa &= 0 \,, \\ -A_1\kappa + A_2\alpha + A_3\kappa - A_4\lambda &= 0 \,, \\ -A_2\gamma + A_3\alpha + A_4\gamma &= 0 \,, \\ A_1\gamma \qquad - A_3\gamma + A_4\alpha &= 0 \,. \end{aligned}\right\} \tag{3.83}$$

This system can obviously possess a nontrivial solution only if the determinant

$$\begin{vmatrix} \alpha & \kappa & -\lambda & -\kappa \\ -\kappa & \alpha & \kappa & -\lambda \\ 0 & -\gamma & \alpha & \gamma \\ \gamma & 0 & -\gamma & \alpha \end{vmatrix} = 0 ; \tag{3.84}$$

this latter equation being equivalent to

$$\alpha^4 + (\sigma\alpha + \rho)^2 = 0 , \tag{3.85}$$

where we have abbreviated

$$\sigma = \kappa + \gamma \tag{3.86}$$

and

$$\rho = -\gamma\lambda . \tag{3.87}$$

If we take the square-root of Equation (3.85), the latter reduces to the quadratic equation

$$\alpha^2 \pm i(\sigma\alpha + \rho) = 0 \tag{3.88}$$

with real and imaginary coefficients and, therefore, complex roots of the form

$$\alpha = x + iy . \tag{3.89}$$

Inserting (3.89) in (3.88) and splitting up the latter in its real and imaginary parts we find that the quantities x and y on the right-hand side of (3.89) should satisfy the simultaneous equations

$$\left.\begin{aligned} x^2 - y^2 \mp \sigma y &= 0 , \\ 2xy \pm (\sigma x + \rho) &= 0 , \end{aligned}\right\} \tag{3.90}$$

or, on separation of variables,

$$y = \mp\frac{\sigma x + \rho}{2x} \tag{3.91}$$

where x is a root of the equation

$$\frac{x^2}{2} = \left(\frac{\sigma}{4}\right)^2 \left\{\sqrt{1 + (4\rho/\sigma^2)^2} - 1\right\} . \tag{3.92}$$

If $\rho = 0$ (i.e., no viscous dissipation) the real part x of the complex root (3.89) disappears, but in such a way that

$$\lim(\rho/x) = 0 , \quad \rho \to 0 . \tag{3.93}$$

In consequence, the imaginary part of the root (3.89) will become equal (from 3.91) to $\mp\sigma$ and, therefore, $\alpha = \pm\iota\sigma$. In such a case, the solutions (3.82) of the system (3.81) will consist of terms which are purely periodic in time.

If, however, $\rho^2 > 0$ [3], the real part x of the complex root (3.89) follows from (3.92) in the form

$$x = \pm\frac{\sigma}{2\sqrt{2}}\left\{\sqrt{1+\left(\frac{4\rho}{\sigma^2}\right)^2}-1\right\}^{1/2} = \pm\frac{|\rho|}{\sigma}+\ldots, \tag{3.94}$$

and, from (3.91),

$$\begin{aligned} y &= \mp\left\{\frac{\sigma}{2}\mp\frac{\rho}{\frac{\sigma}{\sqrt{2}}\left[\sqrt{1+\left(\frac{4\rho}{\sigma^2}\right)^2}-1\right]^{1/2}}\right\} \\ &= \mp\sigma\left\{1+\frac{2\rho^2}{\sigma^4}+\ldots\right\}; \end{aligned} \tag{3.95}$$

but the upper algebraic signs in (3.90) and (3.95) are ruled out by physical considerations[4] rendering x and y single-valued.

Moreover, for values of α which satisfy the determinantal equation (3.84) the ratios of the coefficients $A_{2,3,4}/A_1$ can be determined from the system (3.83) in the form

$$\frac{A_2}{A_1} = \frac{\alpha\kappa-\rho}{\alpha(\alpha\mp\iota\gamma)}, \tag{3.96}$$

$$\frac{A_3}{A_1} = \mp\frac{\iota\gamma}{\alpha\mp\iota\gamma}, \tag{3.97}$$

$$\frac{A_4}{A_1} = -\frac{\gamma}{\alpha\mp\iota\gamma}; \tag{3.98}$$

which on insertion from (3.89) yield

$$\begin{aligned} \frac{A_2}{A_1} &= \frac{\kappa(x^2+y^2)-\rho(x^2-y^2+\gamma y)}{(x^2+y^2)[x^2+(y-\gamma)^2]} - \\ &\quad -\iota\left\{\frac{\kappa(y-\gamma)(x^2+y^2)-\rho(y^2-\gamma)}{(x^2+y^2)[x^2+(y-\gamma)^2]}\right\}, \end{aligned} \tag{3.99}$$

$$\frac{A_3}{A_1} = -\frac{\gamma(y-\gamma)}{x^2+(y-\gamma)^2}-\frac{\iota\gamma x}{x^2+(y-\gamma)^2}, \tag{3.100}$$

and

$$\frac{A_4}{A_1} = -\frac{\gamma x}{x^2+(y-\gamma)^2}+\frac{\iota\gamma(y-\gamma)}{x^2+(y-\gamma)^2}. \tag{3.101}$$

---

[3] The value of ρ itself (i.e., those of $\lambda_{1,2}$) can be positive or negative, depending on whether $\omega_z \gtrless n$.

[4] I.e., the need to render the XY-plane of our inertial coordinates the invariable plane of the binary system.

In the inviscid case, when

$$\rho = x = 0 \tag{3.102}$$

and

$$\alpha = \iota\sigma = \iota(\kappa + \gamma) , \tag{3.103}$$

Equations (3.99)–(3.101) reduce to

$$\frac{A_2}{A_1} = -\iota, \quad \frac{A_3}{A_1} = -\frac{\gamma}{\kappa}, \quad \frac{A_4}{A_1} = \iota\frac{\gamma}{\kappa} ; \tag{3.104}$$

and, accordingly,

$$\left.\begin{aligned} p &= A_1(\cos \sigma t + \iota \sin \sigma t) , \\ q &= -\iota A_1(\cos \sigma t + \iota \sin \sigma t) , \end{aligned}\right\} \tag{3.105}$$

$$\left.\begin{aligned} r &= -(\gamma/\kappa) A_1(\cos \sigma t + \iota \sin \sigma t) , \\ s &= \iota(\gamma/\kappa) A_1(\cos \sigma t + \iota \sin \sigma t) . \end{aligned}\right\} \tag{3.106}$$

When $\rho^2 > 0$, we note that

$$p^2 + q^2 = \sin^2 \theta = \tilde{A}^2 e^{-2xt} \tag{3.107}$$

and

$$r^2 + s^2 = \sin^2 i = (\gamma/\kappa)^2 \tilde{A}^2 e^{-2xt} , \tag{3.108}$$

so that, at all times,

$$\sin i = (\gamma/\kappa) \sin \theta . \tag{3.109}$$

Moreover, by use of the definitions (3.64)–(3.65) and of Equations (3.104)–(3.105) we find that

$$\sin \phi = -\sin \Omega \quad \text{and} \quad \cos \phi = -\cos \Omega , \tag{3.110}$$

disclosing that

$$\Omega - \phi = \pm\pi \tag{3.111}$$

and that, moreover,

$$\phi(t) = \phi(0) - yt , \tag{3.112}$$

$$\Omega(t) = \Omega(0) - yt . \tag{3.113}$$

Therefore, the nodal lines of the equatorial plane of the rotating configuration, and of the orbital plane of its companion, recede secularly at the same uniform rate; and complete their regression in a period U which bears to the orbital period the ratio

$$\frac{P}{U} = \frac{y}{\omega_K} , \tag{3.114}$$

where ω_K denotes the Keplerian angular velocity (mean daily motion) of the relative orbit. Moreover—unlike in the inviscid case, the angle θ of inclination of the axis of rotation to the invariable plane of the system diminishes secularly in accordance with the equation

$$\sin\theta = \tilde{A}\, e^{-xt} ; \tag{3.115}$$

while

$$\sin i = (\gamma/\kappa)\, \tilde{A}\, e^{xt} . \tag{3.116}$$

As, moreover, the ratio γ/n is of the order of the square of the fractional radius of gyration $(h_1/a_1)^2$, it follows that, in general,

$$\theta > i ; \tag{3.117}$$

the foregoing inequality being the stronger, the higher the degree of central condensation of the respective configuration.

All results obtained so far are relevant to the simple case of a binary system in which the secondary (disturbing) component can be regarded as a mass-point, without finite dimensions or rotational momentum of its own. Should, however, the secondary be a finite celestial body in its own right, our entire analysis as given so far in this chapter continues to hold good, provided only that the indices of different quantities referring to respective components be appropriately interchanged.If, moreover, both components of a close pair are regarded as configurations of finite size and angular momentum, their precession and nutation become mutually interlocked; and an analysis of their combined motions can be performed as follows.

Let the subscripts $j = 1$ and 2 refer hereafter to the values of θ_j, ϕ_j as well as to the constants κ_j and λ_j appropriate for each individual component. If so, the system (3.81) of the fundamental equations of our problem should be generalized to a simultaneous system

$$\left.\begin{aligned} \dot{p}_1 &= \kappa_1(s - q_1) + \lambda_1 r , \\ \dot{p}_2 &= \kappa_2(s - q_2) + \lambda_2 r , \\ \dot{q}_1 &= \kappa_1(p_1 - r) + \lambda_1 s , \\ \dot{q}_2 &= \kappa_2(p_2 - r) + \lambda_2 s , \\ \dot{r} &= \gamma_1(q_1 - s) + \gamma_2(q_2 - s) , \\ \dot{s} &= \gamma_2(r - p_1) + \gamma_2(r - p_2) , \end{aligned}\right\} \tag{3.118}$$

of sixth order, which (because of its linearity) can again be expected to admit of exponential solutions of the form

$$\left.\begin{aligned} p_1 &= A_1 e^{\alpha t} , & p_2 &= A_2 e^{\alpha t} , \\ q_1 &= A_3 e^{\alpha t} , & q_2 &= A_4 e^{\alpha t} , \\ r &= A_5 e^{\alpha t} , & s &= A_6 e^{\alpha t} \end{aligned}\right\} \tag{3.119}$$

where the constants $A_{1,2,\ldots,6}$ are constrained to obey the linear system of homogeneous equations

$$\left.\begin{array}{llllllcl}
+\alpha A_1 & & +\kappa_1 A_3 & & -\lambda_1 A_5 & -\kappa_1 A_6 & = & 0\,, \\
& +\alpha A_2 & & +\kappa_2 A_4 & -\lambda_2 A_5 & -\kappa_2 A_6 & = & 0\,, \\
-\kappa_1 A_1 & & +\alpha A_3 & & -\kappa_1 A_5 & -\lambda_1 A_6 & = & 0\,, \\
& -\kappa_2 A_2 & & +\alpha A_4 & +\kappa_2 A_5 & -\lambda_2 A_6 & = & 0\,, \\
& & -\gamma_1 A_3 & -\gamma_2 A_4 & +\alpha A_5 & +(\gamma_1+\gamma_2) A_6 & = & 0\,, \\
+\gamma_1 A_1 & +\gamma_2 A_2 & & & -(\gamma_1+\gamma_2) A_5 & +\alpha A_6 & = & 0\,.
\end{array}\right\} \tag{3.120}$$

Its non-trivial solution requires that the determinant

$$\begin{vmatrix}
\alpha & 0 & \kappa_1 & 0 & -\lambda_1 & -\kappa_1 \\
0 & \alpha & 0 & \kappa_2 & -\lambda_2 & -\kappa_2 \\
-\kappa_1 & 0 & \alpha & 0 & \kappa_1 & -\lambda_1 \\
0 & -\kappa_2 & 0 & \alpha & \kappa_2 & -\lambda_2 \\
0 & 0 & -\gamma_1 & -\gamma_2 & \alpha & \gamma_1+\gamma_2 \\
\gamma_1 & \gamma_2 & 0 & 0 & -\gamma_1-\gamma_2 & \alpha
\end{vmatrix} = 0\,; \tag{3.121}$$

or, which is equivalent, that

$$\begin{aligned}\{\alpha^3 - (\sigma_1\sigma_2 \;-\; \gamma_1\gamma_2)\alpha + (\gamma_1\kappa_2\lambda_1 + \gamma_2\kappa_1\lambda_2)\}^2 &+ \\ + \alpha^2\{(\sigma_1+\sigma_2)\alpha - (\gamma_1\lambda_1+\gamma_2\lambda_2)\}^2 &= 0,\end{aligned} \tag{3.122}$$

where $\sigma_i \equiv \kappa_i + \gamma_i$.

Taking a square-root of the foregoing equation we find that the exponents α in Equations (3.119) must be roots of the cubic equation

$$\begin{aligned}\alpha^3 \mp \iota(\sigma_1+\sigma_2)\alpha^2 \;-\; [(\sigma_1\sigma_2 - \gamma_1\gamma_2) \mp \iota(\gamma_1\lambda_1 + \gamma_2\lambda_2)]\alpha &+ \\ + (\gamma_1\kappa_2\lambda_1 + \gamma_2\kappa_1\lambda_2) &= 0\,,\end{aligned} \tag{3.123}$$

where again the upper sign of the pair $\mp$ alone is admissible for the same reasons as before. The complex nature of some of the coefficients of this equation discloses that its roots must again be expected to be complex and of the form (3.89) which inserted in (3.123) splits up the latter into a pair of simultaneous equations

$$x^3 - (3y^2 + 2Ay + B)x + D \;=\; Cy\,, \tag{3.124}$$

$$y^3 - (3x^2 \qquad - B)y - Ay^2 \;=\; Cx - Ax^2\,, \tag{3.125}$$

where we have abbreviated

$$\left.\begin{aligned}
A &= \sigma_1 + \sigma_2\,, \\
B &= \sigma_1\sigma_2 - \gamma_1\gamma_2\,, \\
C &= \gamma_1\lambda_1 + \gamma_2\lambda_2\,, \\
D &= \gamma_1\lambda_1\kappa_2 + \gamma_2\lambda_2\kappa_1\,.
\end{aligned}\right\} \tag{3.126}$$

In the inviscid case—if $\lambda_1 = \lambda_2 = 0$—both $C = D = 0$; and so is the real part of the complex root (3.89). In such a case, Equation (3.124) is satisfied identically; while (3.125) reduces to

$$y^2 - Ay + B = 0 \tag{3.127}$$

which will possess a pair of real and distinct roots $y_{1,2}$, since

$$(y_1 - y_2)^2 = (\sigma_1 - \sigma_2)^2 + 4\gamma_1\gamma_2 . \tag{3.128}$$

In such a case, by setting

$$p_j = \tilde{A}_j \sin y_1 t + \tilde{B}_j \sin y_2 t , \tag{3.129}$$

$$q_j = \tilde{A}_j \cos y_1 t + \tilde{B}_j \cos y_2 t , \tag{3.130}$$

$j = 1, 2$; and

$$\begin{aligned} r &= -a_1 p_1 - a_2 p_2 \\ &= -\{a_1 \tilde{A}_1 + a_2 \tilde{A}_2\} \sin y_1 t - \{a_1 \tilde{B}_1 + a_2 \tilde{b}_2\} \sin y_2 t , \end{aligned} \tag{3.131}$$

$$\begin{aligned} s &= -a_1 q_1 - a_2 q_2 \\ &= -\{a_1 \tilde{A}_1 + a_2 \tilde{A}_2\} \cos y_1 t - \{a_1 \tilde{B}_1 + a_2 \tilde{B}_2\} \cos y_2 t ; \end{aligned} \tag{3.132}$$

where we have abbreviated

$$a_j \equiv \frac{\gamma_j}{\kappa_j} ; \tag{3.133}$$

and where the integration constants $\tilde{A}_j$, $\tilde{B}_j$ bear to each other the ratios

$$\frac{\tilde{A}_2}{\tilde{A}_1} = \frac{\sigma_2 - \sigma_1 + y_2 - y_1}{2a_2\kappa_1} = \frac{2a_1\kappa_2}{\sigma_1 - \sigma_2 + y_2 - y_1} \tag{3.134}$$

and

$$\frac{\tilde{B}_2}{\tilde{B}_1} = \frac{\sigma_2 - \sigma_1 + y_1 - y_2}{2a_2\kappa_1} = \frac{2a_1\kappa_2}{\sigma_1 - \sigma_2 + y_1 - y_2} . \tag{3.135}$$

In consequence, by virtue of Equations (3.64) and (3.65) we find that

$$\sin^2\theta_j = \tilde{A}_j^2 + \tilde{B}_j^2 + 2\tilde{A}_j\tilde{B}_j \cos(y_1 - y_2)t \tag{3.136}$$

and

$$\begin{aligned} \sin^2 i = & (a_1 \tilde{A}_1 + a_2 \tilde{A}_2)^2 + (a_1 \tilde{B}_1 + a_2 \tilde{B}_2)^2 + \\ & + 2(a_1 \tilde{A}_1 + a_2 \tilde{A}_2)(a_1 \tilde{B}_1 + a_2 \tilde{B}_2) \cos(y_1 - y_2)t . \end{aligned} \tag{3.137}$$

The physical meaning of the constants $\widetilde{A}_j$ and $\widetilde{B}_j$ becomes clear when we consider their role on the right-hand sides of Equations (3.136) and (3.137). Equation (3.128) makes it evident that the difference $y_1 - y_2 > 0$. As, moreover, $\cos(y_1 - y_2)t$ oscillates between ± 1, $\sin \theta_j$ is bound to oscillate between $\widetilde{A}_j \pm \widetilde{B}_j$. Therefore, the constants $\widetilde{A}_{1,2}$ are seen to specify the *mean* values of the *inclination* of the equatorial planes of the two components to the invariable plane of the binary system; while the $\widetilde{B}_{1,2}$'s denote the amplitudes of their *nutations*, in the period U' given by the equation

$$\frac{P}{U'} = \frac{|y_1 - y_2|}{\omega_z}. \tag{3.138}$$

Furthermore, the angle i of inclination between the orbital and invariable planes of the system will (in accordance with Equation 3.137) oscillate about its mean position specified by the value of $a_1 \widetilde{A}_1 + a_2 \widetilde{A}_2$ in the period U' as given by (3.138) and with an amplitude $a_1 \widetilde{B}_1 + a_2 \widetilde{B}_2$ bearing a fixed ratio to that of nutation, but (since $a_{1,2} \ll 1$) generally much smaller.

The nodal line of the system continues to recede—though no longer uniformly—in a period U given now by the equation

$$\frac{2}{U} = \frac{1}{U_1} + \frac{1}{U_2} + \frac{1}{U'}, \tag{3.139}$$

representing the harmonic mean of the periods $U_{1,2}$ of nodal regression due to the action of each component separately (as given by our previous Equation (3.114), and agumented by that of nutation U'. In consequence, in every system consisting of two components of finite size and angular momentum,

$$U' > U \tag{3.140}$$

i.e., the period of nutation will always be longer than that of nodal revolution or precession. Lastly, the difference $\phi(t) - \Omega(t)$ will no longer be equal to π, but will oscillate around this value in the period U' and its submultiples.

All this is strictly true only in the absence of dissipative forces. Should, however, either (or both) constants $\lambda_{1,2}$ be different from zero, the coefficients C and D in Equations (3.124) and (3.125) will cease to vanish and, as a result, the variables x and y in these equations can no longer be conveniently separated. Since, moreover, each equation is one of third degree, their simultaneous solution can proceed only numerically, for specific values of the constants A, B, C, and D. The non-zero values of x will again give rise to real exponential terms in the solutions of the form (3.115) and (3.116), causing a secular decrease in the values of θ_i and i; and will likewise affect (through second-order terms) the periods of precession and nutation. An investigation of the magnitude of these effects requires, however, numerical solutions of the system (3.124)–(3.125) for specific values of its coefficients—a task which, at this stage, still must be left as an exercise for the reader.

VI.4 Bibliographical Notes

The general subject of this chapter follows selected parts of Chapters IV and V of Kopal's treatise on the *Dynamics of Close Binary Systems* (1976) treated in Clairaut's coordinates.

Our treatment of the equations of motion of deformable bodies in Section VI.1A follows largely Kopal (1968a). This appears to be the first instance in which the underlying problem has been treated in terms of Eulerian equations of viscous flow; all earlier work (Liouville, 1858; Darwin, 1879; or Poincaré, 1910) has been based on the Lagrangian equations of motion); its principal physical limitation has been to regard the angular velocities $\omega_{x,y,z}$ as functions of the time only. The Eulerian approach is, in principle, free from such a limitation; if we continue to adhere to a time-dependent but rigid-body rotation, it is because of the fact that, for double-star systems, any finite viscosity of stellar plasma is bound (cf. Section VI.2) to synchronize axial rotation with orbital revolution—so that the resultant state of rotation will approximate that of a rigid body (see Section VI.2A)—especially if the principal source of viscosity is turbulence (or, even more, plasma degeneracy).

Throughout this chapter our subject will be treated in terms of scalar equations; and this accounts to some extent for its length. In subsequent years, Tokis (1974a,b,c; 1975) re-formulated the whole problem in elegant vector form; but as the splitting of the respective vectors into their scalar components becomes a necessary prerequisite for any kind of application, its scalar formulation has been retained throughout this book. The attenuation vector formulation of the present Section VI.1 can, however, be found in Tokis (1974a); while a treatment of tidal friction in vector form can be found in Tokis (1974b,c; or 1975); for dissipation of energy through tidal friction (Section VI.2B) the reader should compare with Tokis (1974c). For subsequent investigations of our problem in vector formalism cf. also Geroyannis and Tokis (1977), or Geroyannis (1980).

Chapter VII

OBSERVABLE EFFECTS OF DISTORTION IN CLOSE BINARY SYSTEMS

In the preceding chapters of this book our aim has been to outline the ways in which an application of the Roche model can facilitate an interpretation of the proximity effects exhibited by close binary systems that may become observable by photometric or spectroscopic means. Because of their distance, such celestial objects can, at all times, be observed as single picture points; and the aim of the investigator can be to develop the information contained in the variable properties of such points which, to the mathematicians, would represent "stationary time series". For the observer, such a task can be best approached by a comparison of the observations with suitable models which should represent such observations; and the reasons have already been put forward before why the Roche model should represent a tool eminently suitable to this end.

In order to provide a basis adequate for this purpose, it is necessary to investigate first the way in which close binary systems would actually appear to us if we could convert the one-channel observations depending on the time into a stationary but two-dimensional projection of the object onto the celestial sphere; and the crucial element in such a comparison is to investigate the distribution of brightness over the apparent disc of a distorted star, the geometry of which we investigated already in Chapters II and IV of this volume. In order to do so, the problem of radiative transfer in semi-transparent fringes of the stars (which in regions of low gravity may become quite deep) will first have to be investigated in all its implications. This we shall do in the first section of this chapter for subsequent use; and to this end we shall find once more the Clairaut coordinates introduced in Chapter V a very valuable tool. Once this has been accomplished, an application of the results to observational manifestations—both photometric and spectroscopic—of close binary systems will conclude the contents of this monograph; a continuation of further tasks arising from its contents will have to be left as a task for the future.

VII.1 Radiative Transfer in Clairaut Coordinates

That the star nearest to us—our Sun—does not exhibit to the naked eye a uniformly bright apparent disc, but one which is progressively darkened towards the limb, seems to have first been noted by Luca Valerio (1552–1618) of the Academy dei Lincei in Rome in 1612[1]; and the phenomenon was given a prominent description in Scheiner's *Rosa Ursina* (1626); though the first actual measurements of this effect were not performed till by Bouguer in 1729.

The cause of the solar limb-darkening invoked much speculation during the 18–19th centuries; but it was not until the beginning of this century that the work of Schuster (1905) and Karl Schwarzschild (1906) placed the theory of radiative transfer of energy in stellar atmospheres on solid physical foundations which will also constitute the basis of this section.

For stars comparable in physical properties with our Sun, the equivalent height of their semi-transparent outer fringes amounts to so small a fraction of the radius of curvature of atmospheric layers that their ratio can be ignored to a very high degree of approximation; and curved surfaces replaced by plane-parallel layers. A theory of stellar limb-darkening caused by such plane-parallel atmospheres has been extensively studied in the past half a century, and the results summarized in several major works (cf. Milne, 1930; Hopf, 1934; Unsöld, 1938; Chandrasekhar, 1950; Kourganoff and Busbridge, 1952) to which the reader can be referred for fuller details; while the post-1950 papers published in the past forty years are too many to be individually quoted.

The sheer magnitude of existing literature should, however, not be allowed to conceal the fact that virtually all of it deals with cases not applicable to close binary systems: this is not so much because it deals with radiative transfer in plane-parallel atmospheres—an approximation sufficient for most "detached" components well interior in size to their Roche limits—but because their "extended atmospheres" in regions of low gravity have consistently been treated as spherical (cf., e.g., Kozyrev, 1934; or Chandrasekhar, 1934); which, in reality, this is far from being the case.

The aim of the present section will, therefore, be to construct the solution of the radiative transfer in Clairaut coordinates (valid regardless of whether the respective atmosphere can be treated as plane-parallel or extended); and then to apply the results to an evaluation of the light- and velocity-changes of close binary systems approximable by the Roche model.

In order to approach our problem in its simplest form, consider the mechanism of radiative transfer to be restricted to absorption-reemission (i.e., ignore the scattering), and let $I(x, y, z)$ denote the intensity of radiation at any point of stellar atmosphere specified by the coordinates x, y, z (with origin at the centre of the respective star); ρ, its density; and k, the absorption coefficient at that point. If $k\rho > 0$, and j stands for the emissivity of the respective material, the passage of a pencil of this radiation through an (arbitrarily oriented) elementary

---

[1] Curiously enough, the reality of solar limb-darkening was vigorously contested by Galileo Galilei in a letter to Prince Cesi dated 25 January 1613 (cf. *Opera di Galileo Galilei*, Edizione Nazionale (ed. A. Favaro, Firenze), vol. 6, p.198); though later in his *Discourses on the Two New Sciences* (Leiden, 1638) he called Valerio "the greatest geometer, and new Archimedes of our time."

cylinder of length ds will alter its intensity by an amount dI specified by the differential equation

$$\frac{dI}{ds} = j\rho - k\rho I\,, \tag{1.1}$$

the right-hand side of which stands for the difference between the light emitted and absorbed within our elementary cylinder.

If the material within it is incapable of emission, $j = 0$; and in this case the foregoing Equation (1.1) can be regarded as a first-order linear differential equation for I, and integrated to yield

$$I(s) = I(0)e^{-\tau(s,0)}\,, \tag{1.2}$$

where $\tau(s, s')$ represents the "optical depth" of the material between s and s', defined by the equation

$$d\tau = -k\rho\, ds\,;, \tag{1.3}$$

such that

$$\tau(s, s') = \int_{s'}^{s} k\rho\, da\,. \tag{1.4}$$

Equations (1.2) and (1.4) represent the case of pure absorption, of limited interest for astrophysical applications, for the characteristic feature of stellar atmospheres is the fact that the absorption of the flux of radiation generated in the interior of the star, passing through the atmosphere, will heat the atmospheric layers to a finite temperature $T \gg 0$; and this fact will cause the return of a part of the absorbed radiation by emission (so that $j > 0$). If, moreover, the material in question can be regarded as being in local thermodynamic equilibrium (i.e., if any temperature change along a distance $(k\rho)^{-1}$ equal to the mean free path of the photons of light can be disregarded), then according to Kirchhoff's law (for unit index or refraction)

$$j = kB(T) \tag{1.5}$$

where, for total radiation,

$$B(T) = \frac{\sigma}{\pi}T^4\,, \tag{1.6}$$

in which Stefan's constant $\sigma = 5.66 \times 10^{-5}$ erg cm^{-2} s^{-1} deg^{-4}.

As has been shown in the pioneer work of Bialobrzewski (1913) and Eddington (1916), in the interiors of the stars local thermodynamic equilibrium obtains a very high degree of approximation (better than can be obtained in any laboratory experiments) because of very high opacity of the material at temperatures of the order of 10^6 degrees (the temperature drop per each mean free path $(k\rho)^{-1}$ of photons existing there is of the order of 0°001); but even in the outer layers of the stars an assumption of the existence of local thermodynamic equilibrium has proved to be a useful approximation; and we shall adhere to it throughout this section.

In such an equilibrium, Equation (1.1) of radiative transfer will, accordingly, be combined with (1.5) to yield

$$\frac{dI}{ds} = k\rho(B - I)\,; \tag{1.7}$$

and the operator d/ds on the left-hand side can, in rectangular coordinates x, y, z, be written (cf. Chandrasekhar, 1950; p.9) as

$$\frac{d}{ds} \equiv \ell_0 \frac{\partial}{\partial x} + m_0 \frac{\partial}{\partial y} + n_0 \frac{\partial}{\partial z}\,, \tag{1.8}$$

where ℓ_0, m_0, n_0 denote the direction cosines of the line of sight of the observer. In spherical polar coordinates r, θ, ϕ this operator can be rewritten, more simply, as

$$\frac{d}{ds} \equiv \cos\gamma \frac{\partial}{\partial r} - \frac{\sin\gamma}{r} \frac{\partial}{\partial \gamma}\,, \tag{1.9}$$

where

$$\cos\gamma = \ell\ell_0 + mm_0 + nn_0 \tag{1.10}$$

stands for the cosine of the angle γ of foreshortening between the surface normal (of direction cosines ℓ, m, n) and the line of sight.

Equation (1.7) as it stands has two dependent variables on its right-hand side: namely the intensity I and the emissivity B; therefore, an additional relation between them must be sought to render our problem determinate. If no light is incident from outside (Section VII.3) and the atmosphere is heated from below by radiation produced in the interior only, the maintenance of radiative equilibrium requires that

$$4\pi B = \oint I\, d\sigma_R\,, \tag{1.11}$$

where

$$d\sigma_R = r^2 \sin\theta\, d\theta\, d\phi \tag{1.12}$$

stands for the surface element perpendicular to the radius-vector; and the limits of integration are to be extended over the entire surface.

The simultaneous system (1.7) and (1.11) of transfer equations of two dependent variables contains three spatial coordinates as independent variables in the presence of distortion through the product $k\rho$ or the temperature T. However, in the state of hydrostatic equilibrium, all these can be made to depend on the potential only—and, therefore, on the *single* Clairaut coordinate a—regardless of the extent to which its form may deviate from spherical shape. Moreover, the advantages stemming from this source continue to transfer to the plane-parallel case as well, obtaining if the semi-transparent fringe of a star constitutes so small a fraction of the star's size that its curvature can be neglected. In such a case, we are clearly entitled to let $r \to \infty$ on the right-hand side of Equation (1.9) for the operator d/ds in spherical polar coordinates; and, as a result, Equation (1.7) can be replaced by

$$\cos\gamma\frac{\partial I}{\partial r} = k\rho(B - I)\,, \tag{1.13}$$

which together with the condition (1.12) of radiative equilibrium then defines the respective transfer problem.

In order to change over from r to a as the radial independent variable,

$$\frac{\partial}{\partial r} \equiv \frac{\partial a}{\partial r}\frac{\partial}{\partial a} + \frac{\partial\theta}{\partial r}\frac{\partial}{\partial\theta} + \frac{\partial\phi}{\partial r}\frac{\partial}{\partial\phi}\,, \tag{1.14}$$

which (since $\theta_r = \phi_r = 0$) discloses that

$$a_r \;=\; \lambda a_x + \mu a_y + \nu a_z \;=\; \frac{1}{r_a} \tag{1.15}$$

in accordance with Equation (1.33) of Chapter V. Accordingly,

$$\frac{\partial}{\partial r} \;=\; \frac{1}{r_a}\frac{\partial}{\partial a}\,, \tag{1.16}$$

where

$$r_a \;=\; 1 + \sum_{j=0}^{\infty}(\eta_j + 1)f_j(a)Y_j^i(\theta,\phi) \tag{1.17}$$

in which $\eta_j(a)$ stands for the logarithmic derivative of $f_j(a)$.

As the next step of our analysis, let us change over from the Clairaut variable a to the optical depth τ as the independent variable, defined (in accordance with Equation 1.3) by the ordinary differential equation

$$d\tau \;=\; -k\rho\, da\,; \tag{1.18}$$

and a combination of (1.16) and (1.18) with (1.3) enables us to rewrite (1.13) as

$$\cos\gamma\frac{\partial I}{\partial\tau} \;=\; (I - B)r_a\,, \tag{1.19}$$

which represents the desired generalization of the Equation (1.13) of radiative transfer, in plane-parallel atmospheres, governing the angular distribution of brightness over the apparent discs of distorted stars for different values of $\cos\gamma$ over their surface.

The foregoing Equation (1.19) differs from its form appropriate for spherical stars in two respects: first, by the presence of the factor r_a on the right-hand side (which, for spherical stars, would reduce to unity); and, secondly, through the direction cosines ℓ, m, n on the r.h.s. of Equation (1.19) which defines the angle γ of foreshortening. As was established already by Equation (1.36) of Chapter V, these are identical with ℓ_1, m_1, n_1 which specify the direction cosines of a normal to the equipotential surfaces a = constant; and as such they are defined by the matrix equation

$$\left\{ \begin{array}{c} \ell_1 \\ m_1 \\ n_1 \end{array} \right\} = \cos\beta\{L^{-1}\} \left\{ \begin{array}{c} 1 \\ r_\theta/r \\ r_\phi/r\sin\theta \end{array} \right\}, \tag{1.20}$$

where $\{L^{-1}\}$ stands for a transpose of the square matrix given by Equation (3.18), Chapter V; the angular elements θ, ϕ of which refer to the "revolving" (doubly-primed) coordinates θ'', ϕ''; and the angle β between the radius-vector and surface normal is given by Equation (1.27) of Chapter V or, alternatively,

$$\cos\beta = \ell_1\lambda'' + m_1\mu'' + n_1\nu'' . \tag{1.21}$$

This angle (zero for spherical stars) will be a quantity of first order in surficial distortion—and, therefore, $\cos\beta$ will deviate from unity by quantities of second order in r_θ or r_ϕ—though the direction cosines ℓ, m, n will do so by terms of the first order.

Moreover, the gravitational acceleration

$$g(r, \theta'', \phi'') \equiv \operatorname{grad}\Psi \tag{1.22}$$

over distorted equipotential surfaces a = constant will vary—in accordance with Equation (1.15) of Chapter V—as

$$g(r, \theta'', \phi'') = r_a\, g(a), \tag{1.23}$$

where r_a continues to be given by Equation (1.17), and $g(a) \equiv g_0$ stands for the mean gravity over the respective equipotential, so that (to quantities of first order in surficial distortion)

$$\frac{g - g_0}{g_0} = \sum_{j=2}^{4}(\eta_j + 1) f_j(a) Y_j^i(\theta'', \phi'') . \tag{1.24}$$

Lastly, the condition (1.11) of radiative equilibrium assumes in Clairaut's coordinates the form

$$4\pi B(\tau) = \oint I\, d\sigma_a , \tag{1.25}$$

where the surface element

$$d\sigma_a = \sqrt{g_{22}g_{33} - g_{23}^2}\, d\theta\, d\phi = r^2 \sec\beta \sin\theta\, d\theta\, d\phi , \tag{1.26}$$

replacing (1.12), in which the metric coefficients g_{ij} continue to be given by Equations (1.3)–(1.8); and $\cos\beta$, by (1.21).

A: Solution of the Equations

In order to construct a solution of the plane-parallel transfer problem posed by Equation (1.19) under the constraint represented by Equation (1.25), let us solve first (1.19) by regarding it as a first-order ordinary linear differential equation for I, rendered nonhomogeneous by the presence of the source-function $B(\tau)$ on its right-hand side which, by (1.25), is a function of τ only. If, moreover, the semi-transparent layers surrounding our distorted star can be regarded as locally stratified in plane-parallel layers—i.e., if the scale-height of the respective atmosphere becomes negligible in comparison with the local radius of curvature of our configuration as a whole—the optical depth τ can increase from 0 to ∞ within a range of a which is negligible within the scheme of plane-parallel approximation; and this, in turn, entitles us to integrate Equation (1.19) *while regarding the function* r_a *on its right-hand side as a constant independent of* τ.

If so, the solution of Equation (1.19) can be written down by standard methods as its particular integral of the form

$$I_+(\tau,\gamma) = r_a \sec\gamma\, e^{r_a\tau\sec\gamma} \int_\tau^\infty B(\tau)e^{-r_a\tau\sec\gamma}d\tau \tag{1.27}$$

for $0 < \gamma < 90°$, and

$$I_-(\tau,\gamma) = -r_a \sec\gamma\, e^{r_a\tau\sec\gamma} \int_0^\tau B(\tau)e^{r_a\tau\sec\gamma}d\tau \tag{1.28}$$

for $0 > \gamma > -90°$, in terms of an arbitrary source-function $B(\tau)$; and, at the boundary of our configuration ($\tau = 0$) visible to a distant observer, the intensity of light $I_=(0,\gamma)$ emerging from the atmosphere can be expressed as

$$I(0,\gamma) \;=\; r_a \sec\gamma \int_0^\infty B(\tau)\, e^{r_a\tau\sec\gamma}d\tau \tag{1.29}$$

—i.e., as a Laplace transform of the source-function $B(\tau)$.

The foregoing Equation (1.29) represents the distribution of brightness over the apparent disc of a distorted star, surrounded by an atmosphere whose scale-height is negligible in comparison with the radius of curvature of the star's surface. In order to evaluate the integral on the right-hand side of (1.29), the source function $B(\tau)$ must now be specified. As, in local thermodynamic equilibrium, $B(\tau)$ is connected with the temperature T of the respective atmospheric layers by Equation (1.6), and this temperature increases inwards with increasing value of τ, we set out to expand the source function $B(\tau)$ in a Taylor series in ascending powers of τ. For $\tau > 0$ this should indeed be legitimate on physical grounds, but not necessarily at the origin $\tau = 0$ where $B(\tau)$ ceases to be analytic (for a discussion of this point cf. Milne, 1930; p.132); or, more precisely, only piecewise discontinuous in τ.

In order to escape difficulties which may arise from this point, suppose that we set out to expand $B(\tau)$ about a point τ_0, such that

$$0 \;<\; \tau_0 \;<\; 1\,. \tag{1.30}$$

At this point, the source function $B(\tau)$ becomes expansible in a Taylor series of the form

$$B(\tau) = \sum_{n=0}^{\infty} \left(\frac{\partial^n B}{\partial \tau^n}\right)_{\tau_0} \frac{(\tau - \tau_0)^n}{n!}, \tag{1.31}$$

where all derivatives on the right-hand side now exist. Such a series can permit us to represent the function $B(\tau)$ for any value of $\tau > \tau_0$; while the surface brightness over the visible hemisphere of the distorted star will, in accordance with Equation (1.27), be given by

$$I(\tau_0, \gamma) = r_a \sec\gamma\, e^{r_a \tau_0 \sec\gamma} \int_{\tau_0}^{\infty} B(\tau)\, e^{r_a \tau \sec\gamma} d\tau\,. \tag{1.32}$$

In physical terms, the adoption of the validity of Equations (1.31)–(1.32) implies that we disregard the effect, on the distribution of brightness $I(\tau_0, \gamma)$, of the outermost atmospheric layers comprised within the range of optical depth given by Equation (1.30); and as long as τ_0 is small, the function $I(\tau_0, \gamma)$ remains sensibly identical with the distribution of light over apparent discs of the stars exposed to a distant observer (cf., e.g., Unsöld, 1938; p.95).

If we insert now in (1.32) for $B(\tau)$ from (1.31), and take advantage of the fact that

$$\int_{\tau_0}^{\infty} e^{-r_a \tau \sec\gamma} \tau^n d\tau = n!(r_a \sec\gamma)^{-n-1} e^{r_a \tau_0 \sec\gamma} \sum_{j=0}^{n} \frac{1}{j!} \left(\frac{r_a \tau_0}{\gamma}\right)^j, \tag{1.33}$$

Equation (1.32) can be rewritten in the form

$$I(\tau_0, \gamma) = \sum_{n=0}^{\infty} \left(\frac{\partial^n B}{\partial \tau^n}\right)_{\tau_0} (r_a \sec\gamma)^{-n}, \tag{1.34}$$

representing a distribution of apparent brightness of the distorted star.

In order to convert $I(\tau_0, \gamma)$ as given by the preceding equation into the ordinary "law of darkening", we still face the task of relating the successive derivatives $(\partial^n B / \partial \tau^n)_{\tau_0}$ with the flux of radiation transmitted through the distorted surface. As is well known, in the case of radiative equilibrium this (vector) flux $\mathbf{F}$ must satisfy the equation

$$\pi \mathbf{F} = -\frac{c}{k\rho} \mathrm{grad}\, p_R\,, \tag{1.35}$$

where c stands for the velocity of light; and p_R, for the radiation pressure which (in local thermodynamic equilibrium) is related with the temperature T by

$$p_R = \frac{1}{3} a T^4\,, \tag{1.36}$$

where a stands for the Stefan-Boltzmann constant ($= 7.55 \times 10^{-15}$ erg cm^{-3} deg^{-4}); so that Equation (1.35) can be alternatively rewritten as

$$\pi \mathbf{F} = -\kappa_R\, \mathrm{grad}\, T\,, \tag{1.37}$$

where (cf. Hazlehurst and Sargent, 1959)

$$\kappa_R = \frac{4}{3}\left(\frac{ac}{k\rho}\right) T^3 \tag{1.38}$$

stands for the coefficient of heat conduction of photon gas at a temperature T.

If we insert from (1.38) in (1.37) and remember that (in local thermodynamic equilibrium) T is related with the emissivity B by Equation (1.5), Equation (1.38) can evidently be rewritten as

$$\pi \mathbf{F} = -\frac{ac}{3k\rho} \operatorname{grad} T^4 = -\frac{4\pi}{3k\rho} \operatorname{grad} B \tag{1.39}$$

if we remember that $ac = 4\sigma$. Moreover, since for configurations in hydrostatic equilibrium k, ρ and T are constant over equipotential surfaces characterized by a given value of a, an appeal to Equations (1.16) and (1.18) permits us to rewrite Equation (1.39) as

$$\frac{\partial B}{\partial \tau} = \frac{3}{4} r_a F_a\,, \tag{1.40}$$

where F_a stands for the component of the flux of radiation passing normally through the level surface at a distance a from its centre.

The foregoing Equation (1.40) holds good for any value of $\tau > 0$; and proves that even at the boundary $\tau = 0$ the source-function possesses a finite first derivative (becoming discontinuous only in derivatives of higher orders). And more: Equation (1.40) discloses that the derivative $\partial B/\partial \tau$ being constant over any equipotential surface, so must be the product $r_a F_a$ on its right-hand side; and since, by Equation (1.23)

$$r_a = g_0/g\,, \tag{1.41}$$

disclosing that the flux must vary in proportion to the gravitational acceleration g—a fact which is responsible for the "gravity-darkening" of distorted stars in radiative equilibrium.

On the other hand, we also know that the flux F_a is related with the source function B by an integral relation, disclosing that

$$\pi F_a(\tau) = \oint I(\tau,\gamma) \cos\gamma \, d\sigma_a\,. \tag{1.42}$$

If, moreover—following Eddington (1926, p.322)—we replace the foreshortening factor $\cos\gamma$ (as defined by (1.10)) in the integral on the r.h.s. of (1.42) for F_a by its mean value of one-half averaged within $0 < \gamma < 90°$, Equation (1.42) discloses that, within this scheme of approximation,

$$\pi F_a(\tau) \simeq \frac{1}{2}\oint I(\tau,\gamma) d\sigma_a = 2\pi B(\tau) \tag{1.43}$$

by Equation (1.11) of radiative equilibrium.

Now the "law of darkening" $I(\tau_0, \gamma)$, as given by Equation (1.34) deduced earlier in this section, can be written as

$$I(\tau_0, \gamma) = B(\tau_0) + \left(\frac{\partial B}{\partial \tau}\right)_{\tau_0} \frac{\cos \gamma}{r_a} + \dots ; \tag{1.44}$$

and if we ignore the higher powers of cos γ on its right-hand side, we can set $\tau_0 = 0$. In such a case, by insertion for $B(0)$ and its first derivative from Equations (1.40) to (1.49) in (1.44) we find that

$$I(0, \gamma) = \frac{3}{4} F_a \left\{\cos \gamma + \frac{2}{3}\right\} = \frac{3}{4} \left(\frac{\partial B}{\partial \tau}\right)_0 \frac{g}{g_0} \left\{\cos \gamma + \frac{2}{3}\right\} , \tag{1.45}$$

corresponding to the "coefficient of limb-darkening" $u = 0.6$; so that the product

$$I(0, \gamma) d\sigma_a = \frac{5}{4} \left(\frac{\partial B}{\partial \tau}\right)_0 \frac{g}{g_0} (1 - u + u \cos \gamma) r^2 \sin \theta \, d\theta \, d\phi. \tag{1.46}$$

The limb-darkening part of Equations (1.45) or (1.46) is identical with that deduced originally by Schwarzschild (1906); and as subsequently demonstrated by Milne (1930) or Unsöld (1938), each term varying as $(\tau - \tau_0)^n$ on the right-hand side for the expansion of $B(\tau)$ corresponds to a term varying as $\cos^n \gamma$ in the "law of limb-darkening" in the form of $I(\tau_0, \gamma)$, observationally attested for the Sun; while photometric effects of this darkening on the light changes of eclipsing binary systems within minima have been under discussion since the commencement of this century (cf. Pannekoek, 1902; or Rödiger, 1902).

Limb-darkening due to semi-transparency of atmospheric layers of the stars represents only one cause of unequal distribution of brightness over their apparent discs. Should the star be distorted, additional photometric phenomena arise which cause apparent surface brightness to vary proportionally to local gravity—a phenomenon now commonly referred to as the "gravity-darkening"—the existence of which was first pointed out by H. von Zeipel (1924); for a modern version of his proof that, in the case of radiative equilibrium, the flux of total light emerging normally from the distorted surface should vary in proportion to the local gravity, cf. pp.191–196 of Kopal (1978).

The fact that von Zeipel's proof of the existence of such gravity-darkening followed as a corollary to a physically improbable assertion that, under the same conditions,the rate ϵ of energy generation in stellar interiors should depend also on the velocity ω of their axial rotation, astronomers were at first disinclined to take seriously not only the theorem relating ϵ with ω, but also its corollary making the flux F dependent on the gravity; but the situation was by no means clear. Thus Eddington (1926a), in a discussion of the physical implications of von Zeipel's theorem in his book on *The Internal Constitution of the Stars*, admitted that ..."the approximation for it" (i.e., gravity-darkening) "is used in a more specialized way than in a discussion of the total radiation of the star, and it would seem necessary to examine how closely the result is bound up with the accuracy of the approximation before we can be sure that it will apply to actual stars. I daresay," and here Eddington revealed again his penetrating physical insight, "it will be found that the approximation still justifies itself, but it is not at all obvious that it is legitimate. The distribution of surface brightness over a tidally disturbed star is of considerable practical importance in the interpretation of the light-curves of eclipsing variables" (op.cit., p.288).

It is regrettable indeed that Eddington's words of this last sentence were not heeded more promptly. To be sure, von Zeipel's theorem itself has long been relegated to oblivion, as it was based on an inconsistent set of equations—von Zeipel assumed the transfer of energy to be accomplished by radiation, but missed the fact established at about the same time (cf. Jeans, 1925) that the same radiation transports also momentum. Earlier in this section we have demonstrated the ease with which the existence of gravity-darkening can be proved by use of Clairaut's coordinates. Moreover, as was pointed out by Eddington, the phenomenon itself is susceptible also to an observational verdict.

The first investigator to work out the effects which should be exerted by gravity-darkening on the light curves of close eclipsing systems was Takeda (1934, 1937), and his work was subsequently extended by Sterne (1941a) and Kopal (1941a) to stars of arbitrary structure; this latter investigator extended also Takeda's work to the case of monochromatic gravity-darkening of the stars radiating like black bodies. A comparison of theory with observations (Kopal, 1941a; or, more recently, 1968b) readily discloses that gravity-darkening must indeed be operative to account for the amplitudes of the light-changes exhibited by close photometric binaries between eclipses—in fact, for very close systems, photometric phenomena arising from this effect appear to be present to an extent exceeding that predicted by the original theory (cf. Kopal, 1979b).

The explanation of this curious phenomenon is not too difficult to trace. As is well known, for spherical stars the flux F of radiation emerging normally from the surface varies as the inverse square of the distance from the centre; and in plane-parallel atmospheres it reduces to a constant. For F = constant, the brightness I of any element of the surface of the star should then depend only on the cosine of the angle γ of inclination of the respective surface element to the line of sight; and the variation of $I(\tau_0, \gamma)$ between the centre and the limb of the apparent (circular) disc then represents the "law of limb-darkening".

When, however, the star becomes distorted, then even in the plane-parallel approximation the flux F of emergent radiation *no longer remains a constant of integration of the respective transfer problem, but varies over the star's apparent disc in proportion to the local gravity*. Now Takeda (1934) and his successors tacitly assumed that, in such a case, the resulting distribution of brightness over apparent discs of distorted stars can be expressed as a *product* of the effects produced by the limb- and gravity-darkening; each part obtained independently of the other.

The validity of such an assumption is, however, far from evident. The effective distribution of surface brightness I as influenced by both limb- *and* gravity-darkening must represent the solution of a transfer problem incorporating *both* phenomena. Such a problem has not been formulated—let alone solved—by any previous investigator; and as long as this is the case, it is impossible to foresee *a priori* whether or not its solution is actually *factorizable* for the effects of limb- and gravity-darkening—as has been tacitly assumed so far.

However, it is also evident from the process by which we arrived at the above Equation (1.44) that the assumptions underlying its proof are the only ones leading to expressions in which the effects of "limb", and "gravity" darkening can be factorized, and the expression for $I(\tau_0, \gamma)$ written out as a product of two functions, each dependent on only one of these effects. These assumptions are:

1. The linearity of the source functions $B(\tau)$ in optical depth τ on the r.h.s. of Equation (1.31); should this not be the case, the terms responsible for limb- and gravity-darkening could no longer be separated by factorization.

2. The existence of radiative as well as hydrostatic equilibrium, which permits us to relate the intensity $I(\tau, \gamma)$ with emissivity $B(\tau)$ by Equations (1.26) and (1.36); and to express all variables of state as functions of a single variable a.

3. The legitimacy of Eddington's approximation underlying Equation (1.43).

If any one of these assumptions ceases to provide valid approximations, the resulting expressions for $I(\tau, \gamma)$ become very much more complicated; and their derivation must be postponed for subsequent investigations.

Apart from these limitations, additional considerations should be kept in mind to serve for further extension of our work. The first concerns the processes by which radiation is being transferred through the atmosphere. Equations (1.19) and (1.125) at the basis of our entire work in this paper entail a tacit assumption that the actual process of its transmission is *absorption–reemission*; and, moreover, Equation (1.25) in its present form requires that all radiation absorbed be re-emitted (i.e., the atmospheric "heat-albedo" to be unity). However, a part of the radiation transfer can also be caused by *scattering* of light in the atmosphere, described by a specific "scattering indicatrix" $p(\cos \Theta)$, which is a function of the angle Θ between incident and scattered rays: for example, for scattering of light on free electrons, $p(\cos \Theta) = \frac{3}{4}(1 + \cos^2 \Theta)$. In such a case, Equation (1.25) should be replaced by

$$4\pi B(\tau) = \oint I(\tau, \gamma) p(\cos \Theta) d\sigma_a \; ; \tag{1.47}$$

and a suitable arithmetic mean (normalized to unity) taken of (1.25) and (1.47) if both absorption–reemission and scattering participate in the total radiation transfer.

The second comment we wish to make in this place concerns the legitimacy of our assumption, adopted throughout this section, that transmitting atmosphere can be regarded as stratified in plane-parallel layers. A sufficiently small ratio of the atmospheric scale-height to the radius of the star surrounded by it would vindicate this assumption regardless of whether the star in question is spherical or moderately distorted. However, if the distortion becomes really large—such as may be the case for components of "contact" systems—low gravity prevalent over the hemisphere facing the companion may extend the atmosphere in this direction, and render the factor r_a in Equation (1.19) no longer independent of the optical depth; and if so, this factor may not be taken out in front of the integral sign in (1.27) or (1.29). This should, moreover, be true not only of the tidal distortion caused by a close proximity of the companion, but also of equatorial regions of rapidly-rotating configurations (Be stars); for these too the solution (1.44) justified in this section may not describe adequately the actual situation.

A star distorted by axial rotation—slow or fast—can exist alone in space. Appreciable tidal distortion requires, however, the presence of a companion of comparable mass in close proximity of the distorted configuration; and this companion will not only raise tides on the neighbouring mass, but also *irradiate* its hemisphere exposed to it. Such an illumination will provide an additional source of energy, incident from outside, which will be progressively absorbed by the illuminated atmosphere, or scattered by it. In such a case, Equation (1.25) of radiative equilibrium should be augmented by the external flux πS incident in a direction at an angle α to surface normal (cf. Section VII.3) and rewritten as

$$4\pi B(\tau) = \oint I(\tau,\gamma)d\sigma_a + \pi S e^{-\tau \sec \alpha} \ . \tag{1.48}$$

Unless the tide-raising object is a "black hole" (or similar highly-degenerate configuration emitting but very little light) it would be inconsistent to consider only the geometry of tidal distortion to affect the radiation transfer, and ignore the contributing effects of energy input produced by the illuminating star and partly intercepted by the atmosphere of its mate.

Last, but not least, throughout this section we have been concerned only with atmospheric transfer of *total* radiation emitted (or intercepted) by a star distorted by axial rotation or tidal action—i.e., radiation integrated over all frequencies. An investigation of spectral distribution of this radiation—i.e., radiative transfer of monochromatic light—can, in principle, be carried out (on the assumption of local thermodynamic equilibrium) in much the same way as we followed in this section; but fuller details must be sought elsewhere. And the same is, moreover, true of the much more difficult problem of radiation transfer in *moving* gas—set (and maintained) in motion by (say) convection currents produced in close binary systems by outside irradiation (cf. Kirbiyik and Smith, 1976; or Kopal, 1980c, 1988). The kinetic energy of gas set in motion by radiative effects must, of course, be subtracted from the energy transmitted by radiation (cf. Hopf, 1934; p.8). Equation (1.19) of transfer is one of great generality; and does not require that radiation is the only mode of energy transfer: it just refers to that part of energy appearing in the form of radiation. But an investigation of the mutual interplay between energy transfer by photons and mass particles still represents an important (though difficult) task for the future.

VII.2 Light Variations in Close Binary Systems

In the preceding section of this chapter, an outline has been given of a theory which governs the distribution of brightnėss over the surfaces of stellar configurations which can be approximated by the Roche model. Provided that the proximity of both components is sufficient to render their distortion appreciable, such light changes are bound to operate throughout the entire orbital cycle regardless of whether our binary happens also to be an eclipsing variable. Should, moreover, the inclination of the orbital plane to the line of sight be such as to cause the components to eclipse each other at the time of conjunction (cf. Section II.2C), the changes of light then exhibited must likewise be affected by distortion of the shadow cylinder which the two stars cast on each other, as well as by the effects of distortion on the distribution of brightness over the eclipsed part. In the first part of this section we shall give an analysis of such effects on the basis of a theory of the Roche model developed in the preceding chapters of this book; while its second part will be concerned with an extension of this theory to the eclipse phenomena.

A: Light Changes of Distorted Stars Outside Eclipses

In order to investigate, in Clairaut coordinates, the light changes produced by the rotation of the Roche model in the course of its orbital cycle, the light $\mathcal{L}$ of its component as seen by the observer at a great distance can be generally expressed as

$$\mathcal{L} = \int_S J \cos\gamma \, d\sigma_a , \tag{2.1}$$

where

$$J \equiv \frac{I(0,\gamma)}{I(0,0)} \tag{2.2}$$

denotes the intensity of light at any point of its surface; $\cos\gamma$ continues to be given by Equation (1.10); the surface element $d\sigma_a$, by (1.26); and the limits of integration S are to be extended over the entire visible hemisphere.

In conformity with the results of the preceding section, the surface brightness J can be generally approximated by

$$J = H(1 - u_1 - u_2 - \ldots - u_n + u_1 \cos\gamma + u_2 \cos^2\gamma + \ldots u_n \cos^n\gamma) , \tag{2.3}$$

where $u_1, u_2, \ldots u_n$ stand for the respective coefficients of limb-darkening; and H denotes the intensity of radiation emerging normally to the surface which, for distorted stars in radiative equilibrium, should (cf. Equation (1.146)) vary in accordance with the equation

$$\frac{H - H_0}{H_0} = \frac{g - g_0}{g_0} , \tag{2.4}$$

the right-hand side of which continues to be given by Equation (1.24).

If, moreover, our star radiates like a black body (the total emission of which is proportional to the fourth power of the absolute temperature T), it follows from (2.4) that

$$\frac{H}{H_0} = \left(\frac{T}{T_0}\right)^4 ; \tag{2.5}$$

while the flux H_λ at any particular wavelength λ should (in accordance with Planck's law) vary as

$$\frac{H_\lambda}{H_0} = \frac{\exp(c_2/\lambda T_0) - 1}{\exp(c_2/\lambda T\ ) - 1}, \tag{2.6}$$

where $c_2 = 1.438$ cm.deg. If we substitute now in (2.6) for T from (2.4)–(2.5) and expand (2.6) in a Taylor series in ascending powers of $(g - g_0)/g_0$ in the neighbourhood of $T = T_0$, then to the first order in surficial distortion

$$\frac{H_\lambda}{H_0} = 1 - \tau\left(1 - \frac{g}{g_0}\right) + \ldots, \tag{2.7}$$

where

$$\tau \equiv \left\{\frac{d \log H_\lambda}{d \log T^4}\right\}_{T_0} = \frac{c_2/\lambda T_0}{4[1 - \exp(-c_2/\lambda T_0)]}. \tag{2.8}$$

Let, moreover, the radius-vector $r(a, \theta, \phi)$ of an equipotential surface $A_1 =$ constant (such that $\rho(a_1) = 0$) be expressible as

$$r(a_1, \theta'', \phi'') = \sum_{i,j} k_{i,j} Y_j^i(\theta'', \phi''), \tag{2.9}$$

in which the $k_{i,j}$'s are constants depending on the nature of distortion, and θ'', ϕ'' are doubly-primed angular coordinates, the surface harmonics $Y_j^i(\theta'', \phi'')$ of which—as homogeneous functions of the direction cosines λ'', μ'', ν'' of j-th degree—satisfy (by Euler's theorem on homogeneous functions) the relation

$$\lambda'' \frac{\partial Y_j^i}{\partial \lambda''} + \mu'' \frac{\partial Y_j^i}{\partial \mu''} + \nu'' \frac{\partial Y_j^i}{\partial \nu''} = j\, Y_j^i . \tag{2.10}$$

If so, we find that the integrand on the right-hand side of Equation (2.1) will consist of a series of terms of the form

$$\begin{aligned} r^2 H \cos^2 \gamma \;=\; & a^2 H_0 N^h \Bigg\{ 1 - \sum_{i,j} \Omega_j^{(h)} k_{i,j} Y_j^i - \\ & - \frac{h}{N} \sum_{i,j} k_{i,j} \left[\ell_0 \frac{\partial Y_j^i}{\partial \lambda''} + m_0 \frac{\partial Y_j^i}{\partial \mu''} + n_0 \frac{\partial Y_j^i}{\partial \nu''} \right] \Bigg\} , \end{aligned} \tag{2.11}$$

where (for Roche model) $h = 1, 2, 3, \ldots$ such that $h - 1 \equiv n$ is the degree of limb-darkening represented by Equation (2.3), and

$$\Omega_j^{(h)} \equiv (j + 2)\tau - hj - 2 , \tag{2.12}$$

$$N \equiv \lambda'' \ell_0 + \mu'' m_0 + \nu'' n_0 \,, \tag{2.13}$$

in which

$$\left. \begin{aligned} \ell_0 &= \cos\psi \sin j \,, \\ m_0 &= \sin\psi \sin j \,, \\ n_0 &= \cos j \end{aligned} \right\} \tag{2.14}$$

are the direction cosines of the line of sight—such that ψ denotes the true anomaly (i.e., phase angle) of the system measurement from the moment of conjunction; and j, the inclination of the orbital plane to the celestial sphere.

If the effects of distortion were ignored and our star considered as spherical (with a distribution of brightness J given by (2.3) for constant $H \equiv H_0$), its luminosity $\mathcal{L}_1$ should be constant and given by

$$\mathcal{L}_1 = \pi \int_0^{a_1} J \sin^{-1}(r/a_1) = \pi a_1^2 H_0 \left\{ 1 - \sum_{\ell=0}^{n} \frac{\ell u_\ell}{\ell+2} \right\} . \tag{2.15}$$

In the presence of distortion, let us write

$$\mathcal{L} = L_1(1+\Delta\mathcal{L}), \quad \Delta\mathcal{L} = \sum_{h=1}^{n+1} \frac{C^{(n)}}{\nu} \Delta\mathcal{L}_1^{(h)} \,, \tag{2.16}$$

where (as before) $\nu \equiv 2/(n+2)$,

$$C^{(h)} = \frac{1-\sum_{\ell=0}^{n} u_\ell}{1-\sum_{\ell=0}^{n} \frac{\ell u_\ell}{\ell+2}} \quad \text{for } h=1 \tag{2.17}$$

$$= \frac{u_{h-1}}{1-\sum_{\ell+2}^{n} \frac{\ell u_\ell}{\ell=0}} \quad \text{for } h>2 \,, \tag{2.18}$$

and

$$\Delta\mathcal{L}_1^{(h)} = -\sum_{i,j} k_{i,j} \int_0^{\pi} \int_{\epsilon-\pi/2}^{\epsilon+\pi/2} N^h \left\{ \Omega_j^{(h)} Y_j^i(\theta'',\phi'') + \right.$$
$$\left. + \frac{h}{N} \left[\ell_0 \frac{\partial Y_j^i}{\partial \lambda''} + m_0 \frac{\partial Y_j^i}{\partial \mu''} + n_0 \frac{\partial Y_j^i}{\partial \nu''} \right] \right\} \sec\beta \sin\theta'' d\theta'' d\phi'', \tag{2.19}$$

in which the angle β continues to be given by (1.22); and ϵ is the angle between the radius-vector R joining the centres of mass of the two components and the line of sight, defined by

$$\cos\epsilon \equiv \ell_0 = \cos\psi \sin j \,. \tag{2.20}$$

In order to perform an integration of all terms on the right-hand side of Equation (1.19), it is necessary to specify the form of the surface harmonics Y_j^i in terms of their angular variables; and this depends on the nature of the forces responsible for the distortion. In the case that these are (non-lagging) *tides*,

$$Y_j^i(\theta'',\phi'') \equiv P_j(\lambda'') \tag{2.21}$$

and

$$k_{0,j} = \frac{m_2}{m_1}\left(\frac{a_1}{R}\right)^{j+1} \equiv w_1^{(j)} \tag{2.22}$$

for $j = 2, 3, 4$ (but not beyond!). On the other hand, for axial *rotation* about an axis perpendicular to the orbital plane, the sole term arising from centrifugal force will be of the form

$$Y_j^i(\theta'', \phi'') \equiv P_2(\nu''), \tag{2.23}$$

factored by

$$k_{0,j} = \frac{\omega_1^2 a_1^3}{3Gm_1} \equiv -\frac{1}{3}v_1^{(2)}, \tag{2.24}$$

where G denotes the constant of gravitation. If, moreover, the angular velocity ω_1 of rotation about the Z''-axis happens to coincide with the Keplerian angular velocity

$$\omega_K^2 = \frac{G(m_1 + m_2)}{R^3} \tag{2.25}$$

of orbital revolution, Equation (2.24) may be rewritten as

$$k_{0,2} = -\frac{1}{3}\left(1 + \frac{m_2}{m_1}\right)\left(\frac{a_1}{R}\right)^3. \tag{2.26}$$

It may be noted that the coefficients $v_1^{(2)}$ and $w_1^{(j)}$ introduced in Equations (2.22) and (2.24) possess simple geometrical meanings: namely, $v^{(2)}$ defines the polar flattening of the respective configuration; while the $w^{(j)}$'s represent contributions to its equatorial ellipticity produced by the j-th partial tide.

For Y_j^i's reducing to the zonal harmonics $P_j(\lambda'')$ or $P_2(\nu'')$ it is of further advantage to note that

$$\lambda P_j'(\lambda) = j\,P_j(\lambda) + P_{j-1}'(\lambda) \tag{2.27}$$

valid for an arbitrary argument (λ or μ); with primes denoting derivatives with respect to the argument in question. In doing so, we may rewrite the r.h.s. of Equation (2.19) in the form

$$\begin{aligned}\Delta\mathcal{L}_1^{(h)} = {} & \frac{1}{3}v_1^{(2)}\int_0^{\pi}\int_{\epsilon-\pi/2}^{\epsilon+\pi/2}\{\Omega_2^{(h)}P_2(\nu'') + h(n_0/N)P_2'(\nu'') - \\ & -h\}\,N^h \sin\theta'' d\theta'' d\phi'' - \\ & -\sum_{j=2}^{4} w_1^{(j)}\int_0^{\pi}\int_{\epsilon-\pi/2}^{\epsilon+\pi/2}\{\Omega_j^{(h)}P_j(\lambda'') + \\ & +h(\ell_0/N)P_j'(\lambda'') - h\,P_{j-1}'(\lambda'')\}\,N^h \sin\theta'' d\theta'' d\phi''.\end{aligned} \tag{2.28}$$

In integrating the r.h.s. of the above equation we may note that *each disturbing term varying as $Y_j^i(\theta, \phi)$ in the expansion* (2.9) *for the radius-vector will give rise to a photometric effect varying as the same harmonic $Y_j^i(\ell_0, n_0)$ in the* *light curve*, factored by coefficients which depend on the extent of limb- and gravity-darkening (i.e., the values of h and $\Omega_j^{(h)}$). The outcome discloses that

$$\begin{aligned}\Delta\mathcal{L}_1^{(h)} &= \frac{h(4+\beta_2)}{3(h+3)} v_1^{(2)} P_2(n_0) - \frac{h(4+\beta_2)}{h+3} w_1^{(2)} P_2(\ell_0) - \\ &- \frac{(h-1)(h+1)}{(h+2)(h+4)}(10+\beta_3) w_1^{(3)} P_3(\ell_0) - \\ &- \frac{h(h-2)}{(h+3)(h+5)}(18+\beta_4) w_1^{(4)} P_4(\ell_0) - \dots, \end{aligned} \tag{2.29}$$

where we have abbreviated

$$\beta_j = \{1+\eta_j(a_1)\}\tau \doteq (j+2)\tau . \tag{2.30}$$

Under these circumstances, an insertion of Equations (2.17)–(2.18) and (2.16) leads to the expression

$$\begin{aligned}\Delta\mathcal{L} &= X_2^{(n)}\left\{1+\frac{1}{4}\beta_2\right\}\left\{\frac{1}{3}v_1^{(2)}P_2(n_0) - w_1^{(2)}P_2(\ell_0)\right\} - \\ &- X_3^{(n)}\left\{1+\frac{1}{10}\beta_3\right\} w_1^{(3)} P_3(\ell_0) - \\ &\quad X_4^{(n)}\left\{1+\frac{1}{18}\beta_4\right\} w_1^{(4)} P_4(\ell_0) - \dots, \end{aligned} \tag{2.31}$$

where

$$X_j^{(n)} = 2(j+2)(j-1)\sum_{h=1}^{n+1} \frac{(j-j+2)(h-j+4)\dots(h+j)}{(h+1)(h+2)\dots(h+j+1)} C^{(h)} , \tag{2.32}$$

and the $C^{(h)}$'s continue to be given by Equations (2.17)–(2.18); with n denoting the degree of the adopted law of limb-darkening of the form (2.3) in powers of $\cos\gamma$. In particular, for uniformly bright discs ($n=0$),

$$X_2^{(1)} = 1 , \; X_3^{(1)} = 0 , \; X_4^{(1)} = -\frac{3}{4} ; \tag{2.33}$$

for a linear law of limb-darkening ($n=1$)

$$X_2^{(2)} = \frac{15+u_1}{5(3-u_1)} , \; X_3^{(2)} = \frac{5u_1}{2(3-u_1)} , \; X_4^{(2)} = \frac{9(1-u_1)}{4(3-u_1)} ; \tag{2.34}$$

for quadratic limb-darkening ($n=2$),

$$\begin{aligned} X_2^{(3)} &= \frac{2(15+u_1)}{5(6-2u_1-3u_2)} , \\ X_3^{(3)} &= \frac{35u_1+48u_2}{7(6-2u_1-3u_2)} \\ X_3^{(3)} &= \frac{9(4u_1+7u_2-4)}{8(6-2u_1-3u_2)} ; \end{aligned} \tag{2.35}$$

etc.

B: Discussion of the Results

A few additional remarks are required to bring this particular section of this chapter to a close. First, we may note that if the density concentration in the components constituting our close binary system is not infinite (and their internal density distribution $\rho(a)$ continuous), Equation (2.12) should be augmented to the form (cf. Kopal, 1942)

$$\Omega_j^{(h)} = \left\{ \frac{2j+1}{\Delta_j} + 1 - j \right\} \tau - jh - 2\,, \tag{2.36}$$

where

$$\Delta_j = \frac{2j+1}{j + \eta_j(a_1)} \tag{2.37}$$

in which $\eta_j(a_1)$ stands for the surface value of the function defined by Equation (5.4) of Chapter IV, subject to the initial condition $\eta_j(0) = j - 2$. The right-hand sides of Equations (2.22) and (2.24) should then likewise be multiplied by Δ_j as well. For the homogeneous model $\eta_j(a) = j - 2$ throughout the entire configuration; rendering

$$\Delta_j = \frac{2j+1}{2j-2}\,. \tag{2.38}$$

However, for the Roche model (cf. Section II-5), Equation (5.7) discloses that $\eta_j(a) = j + 1$, $\Delta_j = 1$; and if so, Equation (2.36) becomes identical with (2.12).

The second comment concerns the orientation of the axis of rotation of our Roche configuration in space. The validity of Equations (2.23) and (2.24) of this section implies this axis to be perpendicular to the orbital plane. If, however, this is not the case, and the Z'-axis is inclined to Z'' by the (Eulerian) angle θ, Equation (2.23) should be replaced by

$$Y_j^i(\theta'', \phi'') \equiv P_2(\cos\Theta)\,, \tag{2.39}$$

where $P_2(\cos\Theta)$ is given by Equation (3.25) of Chapter II; and, as a result, the first term on the right-hand side of Equation (2.29) should be agumented to

$$\begin{aligned} \frac{h(4+\beta_2)}{3(h+3)} v_1^{(h)} \{ P_2(n_0) \; &+ \; 3P_1(\ell_0)P_1(n_0) \sin\theta \sin(\phi - \Omega - u) - \\ & -3P_1(m_0)P_1(n_0) \sin\theta \sin(\phi - \Omega - u) + \ldots \}\,, \end{aligned} \tag{2.40}$$

where for the definition of the Eulerian angles θ and ϕ cf. Figure 2.3. It is noteworthy that—unlike all other terms on the r.h.s. of (2.29)—the above term (2.40) contains (through m_0) the term varying as $\sin\phi$, which should make the light curve *asymmetric* with respect to the conjunctions.

Apart from the terms factored by $\sin\theta$, Equation (2.29) makes it, however, evident that *the polar flattening due to axial rotation will not cause any variation of light between eclipses*; all terms which will do so are due to the *tides*. Since, however, all terms of tidal origin depend on the phase only through the direction cosine $\ell_0 = \cos\psi \sin j$, the light changes produced by them will be *symmetrical* with respect to the moment of conjunctions (when $\psi = 0°$ or $180°$).

The foregoing statements are, to be sure, strictly true only under certain dynamical restrictions. Thus if the orbit and equatorial plane of one (or both) components are not co-planar, the angle j of inclination of the orbital plane to the celestial sphere (and, therefore, the direction cosine n_0) will be subject to short-term as well as long-term periodic perturbations, due to precession and nutation (cf. Section VI.3A) which will affect also the observed variations of light. Similarly, if the two components revolve around the common centre of gravity in eccentric orbits so oriented that the longitude ω of the apsidal line is not 0° or 180°, the phase angle ψ in the direction cosine ℓ_0 becomes identical with the *true* anomaly of the eclipsing component in its eccentric relative orbit; and, as a result, a symmetry of the light changes with respect to the conjunctions is no longer preserved. In general, however, both these complications are likely to be small; and in most cases, negligible; so that symmetry of the light changes should generally be fulfilled within the limits of observational errors—unless, of course, additional complications of a physical nature (not incorporated in our present model) happen to be operative.

Let us consider next the nature of the effects, upon light changes, due to limb- and gravity-darkening of rotating distorted stars. In the case of a linear approximation ($n = 1$) to the actual law of limb-darkening, Equations (2.34) make it evident that no light variation arises from third-harmonic tidal distortion unless there is some limb-darkening; nor any from the fourth unless the darkening is incomplete. If gravity-darkening were absent (i.e., $\tau = 0$) Equation (2.29) would, for $u_1 = 0$, reduce to

$$\Delta\mathcal{L}^{(1)} = \frac{1}{3}v_1^{(2)}P_2(n_0) - w_1^{(2)}P_2(\ell_0) + \frac{3}{4}w_1^{(4)}P_4(\ell_0) + \ldots ; \tag{2.41}$$

while for $u_1 = 1$

$$\Delta\mathcal{L}^{(2)} = \frac{8}{15}v_1^{(2)}P_2(n_0) - \frac{8}{5}w_1^{(2)}P_2(\ell_0) - \frac{5}{4}w_1^{(3)}P_3(\ell_0) + \ldots . \tag{2.42}$$

Equation (2.41) would express the changes of light due to the distorted geometry alone. A comparison of (2.42) with (2.41) discloses, moreover, that limb-darkening tends to exaggerate the variation aising from second-harmonic distortion; gives rise to a significant term varying as third harmonic; and reduces that of the fourth harmonic. On the other hand, a full gravity-darkening ($\tau = 1$) of a centrally-condensed star would, for $u_1 = 0$, lead to

$$\Delta\mathcal{L}^{(1)} = \frac{2}{3}v_1^{(2)}P_2(n_0) - 2w_1^{(2)}P_2(\ell_0) + w_1^{(4)}P_4(\ell_0) + \ldots , \tag{2.43}$$

while, for $u_1 = 1$,

$$\Delta\mathcal{L}^{(2)} = \frac{16}{15}v_1^{(2)}P_2(n_0) - \frac{16}{5}w_1^{(2)}P_2(\ell_0) - \frac{15}{8}w_1^{(3)}P_3(\ell_0) + \ldots . \tag{2.44}$$

The reader may note that, in either case, the effects of full gravity-darkening upon changes of light invoked by the second, third, and fourth tidal-harmonic distortion of a centrally-condensed star is to multiply variations due to distorted geometry alone by the factors of 2, $\frac{3}{2}$ and $\frac{4}{3}$, respectively.

The effects of limb-darkening of higher orders on the variation of light between minima are altogether small: in total light, all terms on the r.h.s. of Equation (2.31) of degree higher than the first (corresponding to $h > 2$) are found to magnify the photometric effects of the second and third tidal harmonics by 1% and 4%, respectively; and to diminish the coefficients of the fourth harmonic by 2% (cf. Kopal, 1949). Therefore, an interpretation of even the most precise light curves between minima should not call for retention of more than the quadratic terms in cos γ on the r.h.s. of the adopted law of limb-darkening; while failure to consider the quadratic term would vitiate the theoretical coefficients of harmonic light variation by more than a few per cent. This is, to be sure, true of the light changes exhibited between minima; while within eclipses (see the next section) a very different situation will be encountered.

Throughout all the foregoing developments we have also confined our attention to the light changes due to distortion of the primary component of mass m_1 and radius a_1, which we placed at the origin of our coordinate system. The total light of a close binary between eclipses will, however, consist of a sum of the luminosities of both components; and the same is true of their light changes. Those of the secondary component are obviously governed by an equation of the same type as (2.31), in which indices referring to the primary and secondary components have been interchanged, and the phase angle ϕ in ℓ_0 shifted by 180°. This difference in phase will change the algebraic sign of the only odd harmonic (corresponding to $j = 2$ and 4). As a result, no odd harmonics can appear in the combined light of a binary system consisting of identical components (for the effects of odd harmonics of one star would be neutralized by those due to its mate). On the other hand, the terms factored by harmonics of even orders will always tend to reinforce each other.

In forming the sum

$$\mathcal{L} = \mathcal{L}_1 + \mathcal{L}_2 \tag{2.45}$$

and factoring out equal powers of cos ψ arising from both components, we find that their combined light outside eclipses should vary as

$$\begin{aligned}
\mathcal{L} &= 1 + \frac{1}{3}\sum_{i=1}^{2} L_i(X_2^{(k)})_i(1+\tau_i)v_i^{(2)}P_2(n_0) - \\
&\quad - \sum_{i=1}^{2}\sum_{j=2}^{4}(-1)^{(i+1)j}L_i(X_j^{(k)})\left\{1+\frac{\tau_1}{j-1}\right\}w_i^{(j)}P_j(\ell_0) = \\
&= 1 + \frac{1}{3}\sum_{i=1}^{2} L_i(X_2^{(k)})_i(1+\tau_i))v_i^{(2)}P_2(n_0) +
\end{aligned}$$

$$+\frac{1}{2}\sum_{i=1}^{2} L_i(X_2^{(k)})_i(1+\tau_i)w_i^{(2)} - \frac{3}{8}\sum_{i=1}^{2} L_i(X_4^{(k)})_i\left(1+\frac{1}{3}\tau_i\right)w_i^{(4)} +$$

$$+\frac{2}{3}\ell_0\sum_{i=1}^{2}(-1)^{i+1}L_i(X_3^{(k)})_i(1+\frac{1}{2}\tau_i)w_i^{(3)} -$$

$$-\frac{3}{2}\ell_0^2\sum_{i=1}^{2} L_i\left\{(X_2^{(k)})_i(1+\tau_i)w_i^{(2)} - \frac{5}{2}(X_4^{(k)})_i(1+\frac{1}{3}\tau_i)w_i^{(4)}\right\} -$$

$$-\frac{5}{2}\ell_0^3\sum_{i=1}^{2}(-1)^{i+1}L_i(X_4^{(k)})_i(1+\frac{1}{2}\tau_i)w_i^{(3)} -$$

$$-\frac{32}{8}\ell_0^4\sum_{i=1}^{2} L_i(X_4^{(k)})_i(1+\frac{1}{3}\tau_i)w_i^{(4)} + \ldots;, \tag{2.46}$$

where, in accordance with Equations (2.24)–(2.26),

$$v_i^{(2)} = \frac{\omega_i^2 a_i^3}{Gm_i} = \left(1+\frac{m_{3-i}}{m_i}\right)\left(\frac{a_i}{R}\right)^3 \tag{2.47}$$

in case of synchronous rotation, and

$$w_i^{(j)} = \frac{m_{3-i}}{m_i}\left(\frac{a_i}{R}\right)^{j+1} \tag{2.48}$$

The ratios a_i/R in these latter equations denote the fractional radii of the two components, expressed in terms of their separation R taken as the unit of length. Lastly, the sum of the luminosities L_1 and L_2 of both components in their undistorted state has been adopted as our unit of light—so that $L_2 = 1 - L_1$.

VII.3 Light Changes of Distorted Systems within Eclipses

In the preceding section we gave an account of the light changes of close binary systems which arise from the distortion of their components caused by rotation and tides. Provided only that the proximity of both components is sufficient to render distortion appreciable, such light changes are bound to arise irrespective of whether or not our binary happens also to be an eclipsing variable. Should, however, the inclination of the orbital plane to the line of sight be such as to cause the components to eclipse each other at the time of conjunctions, the changes of light then exhibited must likewise be affected by distortion of both components of such a couple: a distortion of the secondary (eclipsing) component must cause a corresponding deformation of its shadow cylinder cast in the direction of the line of sight; while a distortion of the component undergoing eclipse will affect not only a proportion of its apparent discs eclipsed by its mate, but also the distribution of brightness over the eclipsed part. The aim of the present section will be to describe such effects to the same order of accuracy to which we dealt with the light changes between minima in the preceding section.

In order to do so, let us introduce for practical use one additional "astrocentric" system of triply-primed rectangular coordinates $X'''Y'''Z'''$, in which the Z'''-axis coincides with the line of sight, of direction cosines ℓ_0, m_0, n_0 in the doubly-primed system of coordinates; and the X'''-axis is made to coincide with the projection of the radius-vector (i.e., the X''-axis) on the plane $Z''' = 0$ tangent to the celestial sphere. Let, moreover, the transformation matrix between the doubly- and triply-primed coordinates be defined as

$$\left\{ \begin{array}{c} x'' \\ y'' \\ z'' \end{array} \right\} = \left\{ \begin{array}{ccc} \ell_2 & \ell_1 & \ell_0 \\ m_2 & m_1 & m_0 \\ n_2 & n_1 & n_0 \end{array} \right\} \left\{ \begin{array}{c} z''' \\ y''' \\ z''' \end{array} \right\} . \tag{3.1}$$

As the X'''-axis has been defined as a projection of the X''-axis on the $Z''' = 0$ plane, it follows that the direction cosines of the X'''-axis in the doubly-primed system of coordinates on the r.h.s. of Equation (3.1) are

$$\ell_0,\ \ell_1 \ = \ 0,\ \ell_2 \ = \ \sqrt{1-\ell_0^2}\ ; \tag{3.2}$$

while the remaining direction cosines m_j and n_j ($j = 1,2$) follow in terms of those for $j = 0$ from the orthogonality conditions

$$\left. \begin{array}{rcl} \ell_0\ell_1 + m_0m_1 + n_0n_1 & = & 0\,, \\ \ell_0\ell_2 + m_0m_2 + n_0n_2 & = & 0\,, \\ \ell_1\ell_2 + m_1m_2 + n_1n_2 & = & 0\,; \end{array} \right\} \tag{3.3}$$

which combined with (3.2) yield

$$m_1 \ = \ -n_0/\ell_2,\ \ n_1 \ = \ m_0/\ell_2, \tag{3.4}$$

and

$$m_2 \ = \ -\ell_0m_0/\ell_2,\ n_2 \ = \ -\ell_0n_0/\ell_2. \tag{3.5}$$

A transformation of the inertial into triply-primed coordinates can be performed with the aid of the matrix equation

$$\left\{ \begin{array}{c} x \\ y \\ z \end{array} \right\} = \left\{ \begin{array}{ccc} a'''_{11} & a'''_{12} & a'''_{13} \\ a'''_{21} & a'''_{22} & a'''_{23} \\ a'''_{31} & a'''_{32} & a'''_{33} \end{array} \right\} \left\{ \begin{array}{c} x''' \\ y''' \\ z''' \end{array} \right\} \tag{3.6}$$

where

$$\left\{ \begin{array}{c} a'''_{1j} \\ a'''_{2j} \\ a'''_{3j} \end{array} \right\} = \left\{ \begin{array}{ccc} a''_{11} & a''_{12} & a''_{13} \\ a''_{21} & a''_{22} & a''_{23} \\ a''_{31} & a''_{32} & a''_{33} \end{array} \right\} \left\{ \begin{array}{c} \ell_{3-j} \\ m_{3-j} \\ n_{3-j} \end{array} \right\} \tag{3.7}$$

$j = 1, 2, 3.$

In Section 3 of Chapter II we introduced three different systems of rectangular coordinates, to which we have now added the fourth. Which ones of the 12 axes specifying these systems are inertial (i.e., possess directions with are invariant in time)? Only four: namely, X, Y, Z and Z'''. The singly- and doubly-primed axes rotate and revolve in space—and so do the X'''- and Y'''-axes of the triply-primed system—because the Eulerian angles ϕ, θ and ψ contained in the direction cosines a'_{ij} as well as the Keplerian elements Ω, i and u contained in the a''_{ij}'s are, in general, functions of the time. An exception is the angle of inclination of the invariable ($Z = 0$) plane of the binary system to the plane $Z''' = 0$ tangent to the celestial sphere, which will hereafter be denoted by I. Since both Z- and Z'''-axes are inertial, the angle I should be independent of the time (or its value may change but slowly as our terrestrial observing station and the respective binary may change their relative positions due to their peculiar motions in the Galaxy).

As was already stated before, the inertial XYZ system of coordinates has been fixed so that the XY-plane coincides with the invariable plane of the system; but the directions of the X- and Y-axes in this plane have not been specified so far. In order to remove this arbitrariness, *let us constrain the X-axis to lie in the ZZ'''-plane*. If so, however, the direction cosines of the Z'''-axis in the inertial frame of reference will be given by

$$a'''_{13} = \sin I, \quad a'''_{23} = 0, \quad a'''_{33} = \cos I \; ; \tag{3.8}$$

and the cosines ℓ_0, m_0, n_0 between the revolving X'', Y'' and Z''-axes and the line of sight Z''' can be expressed as

$$\left.\begin{aligned} \ell_0 &= a''_{11} \sin I + a''_{31} \cos I \, , \\ -m_0 &= a''_{12} \sin I + a''_{32} \cos I \, , \\ n_0 &= a''_{13} \sin I + a''_{33} \cos I \, , \end{aligned}\right\} \tag{3.9}$$

which on insertion for a''_{1j} and a''_{3j} from Equations (2.109)–(2.111) of Chapter II yield

$$\ell_0 = (\cos u \cos \Omega - \sin u \sin \Omega \cos i) \sin I + \sin u \sin i \cos I \, , \tag{3.10}$$

$$m_0 = (\sin u \cos \Omega + \cos u \sin \Omega \cos i) \sin I - \cos u \sin i \cos I \, , \tag{3.11}$$

$$n_0 = \sin \Omega \sin i \sin I + \cos i \cos I \, . \tag{3.12}$$

If, moreover, we define two new angles ψ and j in terms of ℓ_0, m_0 and n_0 by Equations (2.14) of the preceding section, the angle j stands evidently for an *instantaneous* inclination of the orbital plane to the celestial sphere; and ψ, for the true anomaly in their orbit. As we have demonstrated in Section 3A of the preceding chapter, the angles Ω and i in Equations (3.10)–(3.12) will, in general, be functions of the time—Ω secularly regressing and i oscillating on account of nutation—so that, in accordance with Equation (3.12), the actual inclination j of the orbital plane to the celestial sphere should oscillate between $I \pm i$ as Ω

runs from 0 to 2π; and will remain secularly constant only if $i = 0$ (which can, in turn, be true only if the equators of both components of a close binary system are coplanar with the orbit).

A: Geometry of the Eclipses: Special Functions

After these preliminaries, let in what follows $a_{1,2}/R \equiv r_{1,2}$ denote the fractional radii of the respective components expressed in terms of the radius-vector R of their relative orbit, the fractional projection of which on the celestial sphere is given by

$$\delta^2 = \sin^2\psi \sin^2 j + \cos^2 j \equiv \ell_2^2 \tag{3.13}$$

in accordance with (2.14) and (3.2). The equation of the shadow cylinder cast by the secondary component in the direction of the line of sight in the primed coordinates then becomes

$$(\delta - X''')^2 + Y'''^2 = r_2^2(1 + \Delta\overline{r}_2)^2 , \tag{3.14}$$

where $\Delta\overline{r}_2$ represents the deformation of the secondary's shadow cylinder in the $Z''' = 0$ plane. If, furthermore, Δr_1 stands for the corresponding distortion of the primary component undergoing eclipse, the surface of the latter admits of the parametric representation of the form

$$\left.\begin{aligned} X''' &= r_1(1 + \Delta r_1)L , \\ Y''' &= r_1(1 + \Delta r_1)M , \\ Z''' &= r_1(1 + \Delta r_1)N , \end{aligned}\right\} \tag{3.15}$$

where L, M, N denote the direction cosines of an arbitrary radius vector r in the triply-primed system. These direction cosines satisfy the relation

$$L^2 + M^2 + N^2 = 1 , \tag{3.16}$$

and can be regarded as rectangular coordinates, in the primed system, on the unit sphere.

If, however, we retain the radius-vector R of the relative orbit of the two stars as our unit of length, then

$$L = \frac{x}{r_1} , \quad M = \frac{y}{r_1} , \quad N = \frac{z}{r_1} . \tag{3.17}$$

Let us hereafter adopt x, y, z as our new coordinates in terms of which to express the direction cosines λ'', μ'', ν''. The two sets are related by equations of the form (3.1): i.e.,

$$\left.\begin{aligned} r_1\lambda'' &= \ell_0 z && && +\ell_2 x , \\ r_1\mu'' &= m_0 z && +m_1 y && +m_2 x , \\ r_1\nu'' &= n_0 z && +n_1 y && +n_2 x . \end{aligned}\right\} \tag{3.18}$$

By virtue of these relations, the zonal harmonics $P_j(\lambda'')$ or $P_2(\nu'')$ encountered on the right-hand side of Equation (2.28) may now be rewritten as polynomials of the j-th degree in integral powers of x, y and z, with coefficients depending on the amount of distortion and the position of the components in their relative orbit. Since, moreover, the surface element

$$r_1 \sin \theta''' d\theta''' d\phi''' = \frac{dx\, dy}{z}, \tag{3.19}$$

the whole integrand in (2.28) can be rewritten as an algebraic function of x, y and z; the general terms arising from the rotational and tidal distortion being of the form $x^m y z^n$ and $x^m z^n$, respectively, where $m \geq 0$ and $n \geq -1$.

In order to perform the integration on the right-hand side of Equation (2.28) over the fraction of the primary's disc visible at a particular phase of the eclipse, we must first specify the appropriate limits. Turning to the parametric Equations (3.16) we note that N, the cosine of the angle of foreshortening, would vanish at the limb of the primary star if the latter were spherical; and, for distorted bodies, becomes a small quantity whose squares we agreed to ignore. If so, however, then to this order of accuracy the intersection of the primary's surface with the xy-plane reduces to the circle

$$x^2 + y^2 = r_1^2, \tag{3.20}$$

a part of which represents the (dashed) arc QRQ' on Figure VII.1 limiting the eclipsed fraction of the primary's apparent disc.

In order to ascertain the form of the other arc $QR'Q'$ constituting the opposite limb of the eclipsed area, let us solve Equation (3.15) of the shadow cylinder with Equation (3.16) of the surface of the primary star. The equation of the arc $QR'Q'$ in the xy-plane is—to the first order in small quantities—then given by

$$(\delta - x)^2 + y^2 = r_2^2(1 + 2\Delta\bar{r}_2) - 2\{r_2^2 - \delta(\delta - x)\}\Delta r_1 + \dots, \tag{3.21}$$

and the deviation Δy of the arc $QR'Q'$ from the circle

$$(\delta - x)^2 + y^2 = r_2^2 \tag{3.22}$$

assumes the form

$$\Delta y = \frac{r_2^2\, \Delta\bar{r}_2}{\sqrt{r_2^2 - (\delta - x)^2}} - \frac{r_2^2 - \delta(\delta - x)}{\sqrt{r_2^2 - (\delta - x)^2}} \Delta r_1 + \dots. \tag{3.23}$$

The first term on the right-hand side of this latter equation arises from the first-order distortion of the eclipsing component; and the second from that of the eclipsed star.

The fraction of the primary's light lost at any moment during eclipse can be decomposed in two parts: the "circular" one—obtained by extending the limits of integration over the area bounded by the circles (3.20) and (3.22)—and the "boundary corrections" arising from the distortion of the arcs QRQ' and $QR'Q'$. If, moreover,

$$s = \frac{r_1^2 - r_2^2 + \delta^2}{2\delta} \tag{3.24}$$

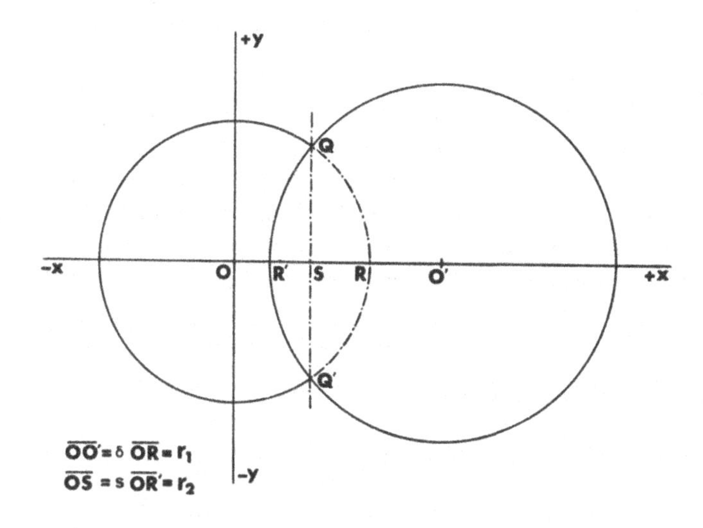

Figure VII.1

denotes the x-coordinate of the common chord of the circles (3.20) and (3.22) at any phase of the eclipse (cf. again Figure VII.1) the "circular" part of the resulting fractional loss of light can be expressed in terms of the "associated α-functions" of the form

$$\pi r_1^{m+n+2}\alpha_n^m = \left\{\int_s^{r_1}\int_{-\sqrt{r_1^2-x^2}}^{+\sqrt{r_1^2-x^2}} + \int_{\delta-r_2}^{s}\int_{-\sqrt{r_2^2-(\delta-x)^2}}^{+\sqrt{r_2^2-(\delta-x)^2}}\right\} x^m z^n \, dx\, dy \qquad (3.25)$$

if the eclipse is partial, and

$$\pi r_1^{m+n+2}\alpha_n^m = \int_{\delta-r_2}^{\delta+r_2}\int_{-\sqrt{r_2^2-(\delta-x)^2}}^{+\sqrt{r_2^2-(\delta-x)^2}} x^m z^n \, dx\, dy \qquad (3.26)$$

if it is annular. The loss of light due to the distortion of the arcs QRQ' and $QR'Q'$ which limits the eclipsed area can, moreover, be obtained by a single integration of $2\Delta y$ with respect to x, within limits which range from $\delta - r_2$ to s as long as the eclipse remains partial, and from $\delta - r_2$ to $\delta + r_2$ when it becomes annular.

In order to be able to perform these latter integrations we must express the quantities Δr_1 as well as $\Delta \overline{r}_2$ on the right-hand side of Equation (3.23) as functions of x. Consistent with Equation (2.9) and its sequel,

$$\Delta r_{1,2} = \sum_{j=2}^{4} w_{1,2}^{(j)} P_j(\lambda) - \frac{1}{3} v_{1,2}^{(2)} P_2(\nu) , \tag{3.27}$$

where the coefficients $v_{1,2}^{(2)}$ and $w_{1,2}^{(j)}$ continue to be given by Equations (2.47) and (2.48).

The arguments λ and ν of the zonal harmonics on the right-hand side of (3.27) are given by Equations (3.18) in which appropriate values for y and z were inserted. Along the arc $QR'Q'$ at which the primary's surface is intersected by the shadow cylinder cast by the secondary component, from (3.16) and (3.17) it follows that

$$z = \sqrt{r_1^2 - x^2 - y^2} = \sqrt{2\delta(s-x)} \tag{3.28}$$

if we insert for y from (3.22); so that

$$\Delta r_1 = \sum_{j=2}^{4} w_1^{(j)} P_j(\lambda_1) - \frac{1}{3} v_1^{(2)} P_2(\nu_1) , \tag{3.29}$$

where

$$r_1 \lambda_1 = \ell_0 \sqrt{2\delta(s-x)} + \ell_2 x , \tag{3.30}$$

$$r_1 \nu_1 = n_0 \sqrt{2\delta(s-x)} + n_1 \sqrt{r_2^2 - (\delta - x)^2} + n_2 x ; \tag{3.31}$$

while along the curve at which the shadow cylinder is tangent to the secondary component (and, therefore, $z = 0$),

$$\Delta \overline{r}_2 = \sum_{j=2}^{4} w_2^{(j)} P_j(\overline{\lambda}_2) - \frac{1}{3} v_2^{(2)} P_2(\overline{\nu}_2) , \tag{3.32}$$

where

$$r_2 \overline{\lambda}_2 = +\ell_2(\delta - x) , \tag{3.33}$$

$$r_2 \overline{\nu}_2 = n_1 \sqrt{r_2^2 - (\delta - x)^2} + n_2(\delta - x) . \tag{3.34}$$

The "boundary corrections" due to the deformation of the arc $QR'Q'$ which arises from the distortion of the eclipsed component should, therefore, be expressible in terms of the functions $\Im_{\beta,\gamma}^m$ as defined by the equation

$$\pi r_1^{\beta+\gamma+3} \Im_{\beta,\gamma}^m = (r_2^2 - \delta^2) I_{0,\beta,\gamma}^m + \delta I_{0,\beta,\gamma}^{m+1} , \tag{3.35}$$

where β, γ and m are integers such that

$$\beta \geq -1, \quad \gamma \geq -1, \quad m \geq 0 ; \tag{3.36}$$

and

$$I^m_{0,\beta,\gamma} = \int_{\delta-r_2}^{c_2} [r_2^2 - (\delta - x)^2]^{\beta/2}[2\delta(s - x)]^{\gamma/2} x^m dx , \tag{3.37}$$

where $c_2 = s$ or $\delta + r_2$ depending on whether the eclipse is partial or annular.

Moreover, the remaining boundary corrections arising from $\Delta\bar{r}_2$ are all expressible in terms of an additional family of integrals of the form

$$\pi r_2^{\beta+\gamma+m+1} I^m_{\beta,\gamma} = \int_{\delta-r_2}^{c_2} [r_2^2 - (\delta - x)^2]^{\beta/2}[2\delta(s - x)]^{\gamma/2}(\delta - x)^m dx . \tag{3.38}$$

Note, in this connection, that while both three-index functions $\Im^m_{\beta,\gamma}$ and $I^m_{\beta,\gamma}$ as defined by the foregoing Equations (3.35) and (3.38) are nondimensional quantities, the $I^m_{0,\beta,\gamma}$'s are not.

In order to formulate explicit expressions for the loss of light during eclipses of distorted components of close binary systems, let us write—in conformity with (2.16)—that

$$\Delta\mathcal{L} = \sum_{h=1}^{n+1} C^{(h)}\{\alpha^0_{h-1} + \Delta\mathcal{L}^{(h)}\} , \tag{3.39}$$

where the coefficients $C^{(h)}$ due to limb-darkening continue to be given by Equations (2.17) and (2.18), and the α^M_n's are associated α-functions as defined by Equations (3.25) or (3.26). The first part of the right-hand side of (2.18)—consisting of $C^{(h)}\alpha^0_{h-1}$—represents the loss of light during eclipses of spherical stars arbitrarily darkened at the limb; while the effects $\Delta\mathcal{L}^{(h)}$ of distortion can be represented by

$$\Delta\mathcal{L}^{(h)} = f^{(h)}_* + f^{(h)}_1 + f^{(h)}_2 , \tag{3.40}$$

where $f^{(h)}_*$ will constitute the contributions of distortion and gravity-darkening over the circular portion of the eclipsed disc and is expressible in terms of the associated α-functions; while $f^{(h)}_{1,2}$ represent the photometric contributions of the "boundary corrections" along $QR'Q'$ arising from the distortion of the primary and secondary component.

In order to establish the explicit form of $f^{(h)}_*$ for limb-darkening of any degree, we rewrite the integrand on the right-hand side of Equation (2.28) in powers of x, y, z by means of Equations (3.18) and remember that, in the rotational terms, odd powers of y vanish on account of symmetry, while even powers can be expressed in terms of those of x and z by means of the relation $y^2 = r_1^2 - x^2 - z^2$ following from (3.18). A term-by-term integration over circular limits then yields

$$\begin{aligned} f^{(h)}_* = & \frac{1}{3}\Big\{\frac{1}{2}\Omega^{(h)}_2[3(n_0^2 - n_1^2)\alpha^0_{h+1} + 3(n_2^2 - n_1^2)\alpha^2_{h-1} + 6n_0n_2\alpha^1_h + \\ & +2P_2(n_1)\alpha^0_{h-1}] + h[2P_2(n_0)\alpha^0_{h-1} + 3n_0n_2\alpha^1_{h-2}]\Big\} v_1^{(2)} - \\ & - \Big\{\frac{1}{2}\Omega^{(h)}_2[3\ell_0^2\alpha^0_{h+1} + 6\ell_0\ell_2\alpha^1_h + 3\ell_2^2\alpha^2_{h-1} - \alpha^0_{h-1}] + \end{aligned}$$

$$
\begin{aligned}
&+h[2P_2(\ell_0)\alpha^0_{h-1} + 3\ell_0\ell_2\alpha^1_{h-2}]\Big\} w_1^{(2)} - \Big\{\frac{1}{2}\Omega_3^{(h)}[5\ell_0^3\alpha^0_{h+2} + \\
&+ 15\ell_0^2\ell_2\alpha^1_{h+1} + 15\ell_0\ell_2^2\alpha^2_h + 5\ell_2^3\alpha^3_{h-1} - 3\ell_0\alpha^0_h - 3\ell_2\alpha^1_{h-1}] + \\
&+h[(\ell_0\alpha^0_h + 2\ell_2\alpha^1_{h-1})P_3'(\ell_0) - \frac{3}{2}\ell_0(\alpha^0_h - 5\ell_2^2\alpha^2_{h-2} + \alpha^0_{h-2})]\Big\} w_1^{(3)} - \\
&- \Big\{\frac{1}{8}\Omega_4^{(h)}[35\ell_0^4\alpha^0_{h+3} + 140\ell_0^3\ell_2\alpha^1_{h+2} + 210\ell_0^2\ell_2^2\alpha^2_{h+1} + 140\ell_0\ell_2^3\alpha^3_h \\
&+ 35\ell_2^4\alpha^4_{h-1} - 30\ell_0^2\alpha^0_{h+1} - 60\ell_0\ell_2\alpha^1_h - 30\ell_2^2\alpha^2_{h-1} + 3\alpha^0_{h-1}] + \\
&+ \frac{1}{2}h[2\ell_0P_4'(\ell_0)\alpha^0_{h+1} + 15\ell_2^2(7\ell_0^2 - 1)\alpha^2_{h-1} - 2P_3'(\ell_0)\alpha^0_{h-1} + \\
&+15\ell_0\ell_2(7\ell_0^2 - 2)\alpha^1_h + 35\ell_0\ell_2^3\alpha^3_{h-2} - 15\ell_0\ell_2\alpha^1_{h-2}]\Big\} w_1^{(4)} + \ldots, \qquad (3.41)
\end{aligned}
$$

where $\Omega_j^{(h)}$ continues to be given by Equation (2.12).

A similar integration of $2\Delta y$ as given by Equation (3.23) shows that

$$
\begin{aligned}
f_1^{(h)} = & \frac{1}{3}\Big\{3n_0^2\Im^0_{-1,h+1} + 3n_1^2\Im^0_{1,h-1} + 3n_2^2\Im^2_{-1,h-1} + \\
&+6n_0n_2\Im^1_{-1,h} - \Im^0_{-1,h-1}\Big\} v_1^{(2)} - \\
&- \Big\{3\ell_0^2\Im^0_{-1,h+1} + 6\ell_0\ell_2\Im^1_{-1,h} + 3\ell_2^2\Im^2_{-1,h-1} - \Im^0_{-1,h-1}\Big\} w_1^{(2)} - \\
&- \Big\{5\ell_0^3\Im^0_{-1,h+2} + 15\ell_0^2\ell_2\Im^1_{-1,h+1} + 15\ell_0\ell_2^2\Im^2_{-1,h} + \\
&+5\ell_2^3\Im^3_{-1,h-1} - 3\ell_0\Im^0_{-1,h} - 3\ell_2\Im^1_{-1,h-1}\Big\} w_1^{(3)} - \\
&- \frac{1}{4}\Big\{35\ell_0^4\Im^0_{-1,h+3} + 140\ell_0^3\ell_2\Im^1_{-1,h+2} + 210\ell_0^2\ell_2^2\Im^2_{-1,h+1} + \\
&+ 140\ell_0\ell_2^3\Im^3_{-1,h} + 35\ell_2^4\Im^4_{-1,h-1} - 30\ell_0^2\Im^0_{-1,h+1} - \\
&-60\ell_0\ell_2\Im^1_{-1,h} - 30\ell_2^2\Im^2_{-1,h-1} + 3\Im^0_{-1,h-1}\Big\} w_1^{(4)} + \ldots, \qquad (3.42)
\end{aligned}
$$

and

$$
\begin{aligned}
(r_1/r_2)^{h+1}f_2^{(h)} = & -\frac{1}{3}\Big\{3n_1^2I^0_{1,h-1} + 3n_2^2I^2_{-1,h-1} - I^0_{-1,h-1}\Big\} v_2^2 + \\
&+ \{3\ell_2^2I^2_{-1,h-1} - I^0_{-1,h-1}\}w_2^{(2)} + \\
&+ \{5\ell_2^3I^3_{-1,h-1} - 3\ell_2I^1_{-1,h-1}\}w_2^{(3)} + \qquad (3.43) \\
&+ \frac{1}{4}\Big\{35\ell_2^4I^4_{-1,h-1} - 30\ell_2^2I^2_{-1,h-1} + 3I^0_{-1,h-1}\Big\} w_2^{(4)} + \ldots.
\end{aligned}
$$

The foregoing Equations (3.42)–(3.43) specify the changes of light within minima exhibited by close eclipsing systems and arising from the first-order rotational as well as tidal distortion. In order to describe these in concise form we found it expedient to introduce three new families of auxiliary functions—namely, the associated alpha-functions α_n^m as defined by Equations (3.25)–(3.26), and the "boundary integrals" $\Im^m_{\beta,\gamma}$ and $I^m_{\beta,\gamma}$ defined by (3.35) and (3.38), respectively.

Not all these functions are, to be sure, fully independent of each other. In particular, the $\Im^m_{\beta,\gamma}$-functions can be expressed as simple linear combinations of the α^m_n's for $n > 0$—a fact which will enable us to combine Equations (3.25)–(3.26) and (3.35) largely into one.

In recent years, these (and other) special functions needed for an analytical treatment of theoretical light curves arising from mutual eclipses of the Roche lobes have been developed far beyond the degree indicated in this section. We do not consider it necessary to reproduce our present state of their knowledge in this book, because this aspect of our subject is already available in other and readily accessible sources. The reader is, in this connection, referred in particular to Chapter III of a recent monograph by the present writer (cf. Kopal, 1979a), as well as to many individual papers quoted in the Bibliographical Notes at the end of this chapter. In perusing all this literature (and, in particular, a representation of the light changes arising from mutual eclipses of Roche lobes in terms of Hankel transforms) the reader should, however, keep in mind that all these developments are still of very recent date, and concern a field signposted " Men at Work, Pass at Your Own Risk."

VII.4 Radial-Velocity Variations in Close Binary Systems

Photometric manifestations of close binary systems, discussed in the preceding sections, do not provide the only channel of information by which such systems can disclose their identity and other physical properties. Another avenue to this end can be provided by observations of radial-velocity changes, as disclosed by Doppler shifts of different lines in their spectra.

Up to almost the middle of this century, the analysis of the radial-velocity changes of spectroscopic binaries for the elements of their orbits had been carried out on the basis of the classical problem of two bodies. The founding fathers of our subject saw no reason not to treat, not only visual, but also spectroscopic binaries as systems consisting of two light points; and tacitly identified the measured radial velocities with radial components of the respective Keplerian motions. For visual binaries such a representation is indeed wholly adequate. However, spectroscopic binaries constitute close pairs; and consist of components whose dimensions are not negligible in comparison with their separation (otherwise the radial-velocity changes due to their orbital motion would be observationally insignificant). Therefore, the question is bound to arise as to *the extent to which the "centre of light" of the components in close binary systems will continue to project itself constantly on their centre of mass* (which moves in an elliptic orbit); thus making a direct application of the theory of Keplerian motion legitimate.

As long as the components of such binaries can be regarded as spherical, and to exhibit a distribution of brightness on their apparent discs which remains symmetrical with respect to their projected centres of mass, the radial velocities spectroscopically observed can indeed be identified with those of the Keplerian orbital motions. However, since the 1930's it gradually became obvious that axial rotation as well as tidal forces prevalent in close binaries must cause their components to deviate from spherical form, and render also the distribution of surface brightness over their apparent discs (due to "gravity-darkening"), not only non-uniform, but variable in the course of the orbital cycle. The same will, moreover, be true of the effects caused by mutual irradiation of the components of close binary systems (the "reflection effect") even if their components could be regarded as spheres. All these effects are bound to render the observed radial velocities of the constituent stars *different* from the Keplerian velocities of their mass centres; and terms arising from axial rotation of the components will superpose upon purely orbital velocity to yield a result which, if analyzed in the classical manner, would furnish elements of spectroscopic orbits that are systematically in error.

A study of the effects, on observed radial velocities, invoked by axial rotation of distorted stars commenced with a note entitled "A source of spurious eccentricity in spectroscopic binaries," in which Sterne (1941) pointed out that such effects produced by second-harmonic tidal distortion would be tantamount to those produced by an orbital eccentricity if the apsidal line of the respective orbit were parallel with the line of sight; and Sterne's work was subsequently extended by the present writer (Kopal, 1945) to include the effects of the third and fourth tidal harmonic distortion. It will be the aim of the present section to outline a systematic study of such effects, and to summarize their results.

A: Radial Motion of Rotating Stellar Discs

In order to do so, let us return to the transformation of coordinates represented by the matrix equation (3.6), and consider now the origin of these coordinates to be identified with the centre of mass of the respective star. If so, we continue to have for the distance of any point on the star's surface from a plane tangent to the line of sight and passing through the centre of mass of the star the expression

$$z''' = a'''_{13}x + a'''_{23}y + a'''_{33}z \,, \tag{4.1}$$

where the a'''_{i3}'s stand for the direction cosines of the line of sight. Since, moreover, the position of the latter is regarded as invariant in space, a time-differentiation of Equation (4.1) yields for the radial velocity V of an arbitrary point the expression

$$V \equiv \dot{z}''' = a'''_{13}\dot{x} + a'''_{23}\dot{y} + a'''_{33}\dot{z} = \dot{x} \sin I + \dot{z} \cos I \,, \tag{4.2}$$

by (3.8), where by a differentiation of the relations (3.2)–(3.5) of Chapter II it follows that

$$\dot{x} \equiv u = z\omega_y - y\omega_z + u'_0 \,, \tag{4.3}$$

$$\dot{y} \equiv v = x\omega_z - z\omega_x + v'_0 \,, \tag{4.4}$$

$$\dot{z} \equiv w = y\omega_x - x\omega_y + w'_0 \tag{4.5}$$

in accordance with Equations (1.52)–(1.54) of Chapter VI, in which the angular velocities $\omega_{x,y,z}$ of axial rotation are related with the time-derivatives of the Eulerian angles θ, ϕ and ψ by Equations (1.37)–(1.39) of that chapter, and

$$\left\{ \begin{array}{c} u_0' \\ v_0' \\ w_0' \end{array} \right\} = \left\{ \begin{array}{ccc} a_{11}' & a_{12}' & a_{13}' \\ a_{21}' & a_{22}' & a_{23}' \\ a_{31}' & a_{32}' & a_{33}' \end{array} \right\} \left\{ \begin{array}{c} u' \\ v' \\ w' \end{array} \right\} , \tag{4.6}$$

where

$$u' \equiv \dot{x}', \quad v' \equiv \dot{y}', \quad w' \equiv \dot{z}' \tag{4.7}$$

are the *body* velocity components in the *rotating* (single-primed) coordinates. Accordingly, the quantities u_0', v_0' and w_0' on the right-hand sides of Equations (4.3)–(4.5) as defined by Equation (4.6) represent the *body* velocity components in the direction of the *fixed* spaces axes X, Y and Z. Should our configuration rotate as a *rigid* body, the u', v', w'—and, therefore u_0', v_0', w_0' on the right-hand sides of Equations (4.3)–(4.5)—would be identically zero, and can become finite only if the body in question is *deformable* (or, for spectroscopic binaries, if the medium in which spectral lines are formed is in motion relative to the centre of mass of the respective configuration).

The coordinates x, y, z factoring the angular velocities ω_x, ω_y, ω_z on the right-hand sides of Equations (4.3)–(4.5) refer to the inertial space axes. In order to facilitate the tasks awaiting us in subsequent sections of this chapter we find it, however, of advantage to rewrite Equations (4.2)–(4.5) in terms of the doubly-primed (revolving) and triply-primed coordinates and velocity components. If we ignore, for the time being, the velocity components u_0', v_0', w_0' of non-rotational origin and convert x, y, z into x'', y'', z'' with the aid of the transformation equations (3.6), the radial velocity component $V_{\rm rot}$ arising from rigid-body rotation can be expressed as

$$V_{\rm rot} = \omega_{x''}(n_0 y'' - m_0 z'') + \omega_{y''}(\ell_0 z'' - n_0 x'') + \omega_{z''}(m_0 x'' - \ell_0 y'') , \tag{4.8}$$

where (cf. Equation (1.219) of Chapter VI)

$$\left\{ \begin{array}{c} \omega_{x''} \\ \omega_{y''} \\ \omega_{z''} \end{array} \right\} = \left\{ \begin{array}{ccc} a_{11}'' & a_{21}'' & a_{31}'' \\ a_{12}'' & a_{22}'' & a_{32}'' \\ a_{13}'' & a_{23}'' & a_{33}'' \end{array} \right\} \left\{ \begin{array}{c} \omega_x \\ \omega_y \\ \omega_z \end{array} \right\} \tag{4.9}$$

are the angular velocity components about the doubly-primed axes, and the direction cosines ℓ_0, m_0, n_0 of the line of sight in the same coordinates continue to be given by Equations (3.10)–(3.12).

If, lastly, we wish to rewrite the expression (4.8) for $V_{\rm rot}$ in terms of the triply-primed coordinate system x''', y''', z''', a resort to the transformation equation (3.1) discloses that, in rectangular coordinates of the latter type,

$$n_0 y'' - m_0 z'' = \ell_1 x''' - \ell_2 y''' , \tag{4.10}$$

$$\ell_0 z'' - n_0 x'' = m_1 x''' - m_2 y''' , \tag{4.11}$$

$$m_0 x'' - \ell_0 y'' = n_1 x''' - n_2 y''' , \tag{4.12}$$

where the direction cosines $\ell_{1,2}$, $m_{1,2}$ and $n_{1,2}$ continue to be given by Equations (3.2) and (3.4)–(3.5), so that

$$V_{\rm rot} = \omega_{x''}(-\ell_2 y''') + \omega_{y''}(m_1 x''' - m_2 y''') + \omega_{z''}(n_1 x''' - n_2 y''') , \tag{4.13}$$

since $\ell_1 = 0$.

The foregoing expressions (4.8) or (4.13) for $V_{\rm rot}$ are exact for any (not necessarily constant) values of the angular velocities $\omega_{x''}$, $\omega_{y''}$, $\omega_{z''}$ of our body about the revolving axes. Should, however, the equators of the respective components coincide with the plane of their orbit, then the Eulerian angle θ in Equations (1.37)–(1.39) of Chapter VI vanishes—a fact which renders $\omega_x = \omega_y = 0$. Moreover, since $i = 0$, it also follows that $\omega_{x''} = \omega_{y''} = 0$ and $\omega_{z''} = \omega_z$ —as a result of which Equations (4.8) and (4.13) reduce to

$$V_{\rm rot} = \omega_z(m_0 x'' - \ell_0 y'') = \omega_z(n_1 x''' - n_2 y''') , \tag{4.14}$$

of which extensive use will be made in the sequel.

The tasks set forth in the present section are, however, not yet complete. In order to construct a complete expression for the radial velocity V at any time and any surface point of the components of close binary systems we must adjoin to $V_{\rm rot}$ the contribution $V^{(+)}$ arising from possible deformability or atmospheric motions of the respective star, given by

$$V^{(+)} = a'''_{13} u'_0 + a'''_{23} v'_0 + a'''_{33} w'_0 , \tag{4.15}$$

which, in view of Equations (3.8) and (4.6), can be rewritten as

$$\begin{aligned} V^{(+)} &= u'_0 \sin I + w'_0 \cos I = \\ &= (a'_{11}u' + a;_{12}\, v' + a'_{13}w') \sin I + \\ &\quad + (a'_{31}u' + a'_{32}v' + a'_{33}w') \cos I , \end{aligned} \tag{4.16}$$

where the singly-primed direction cosines a'_{ij} continue to be given by Equations (3.3)–(3.5) of Chapter II.

For $\theta = 0$, these direction cosines reduce to

$$a'_{11} = \cos(\psi + \phi), \quad a'_{12} = -\sin(\psi + \phi), \quad a'_{13} = 0 . \tag{4.17}$$

while

$$a'_{31} = 0, \quad a'_{32} = 0, \quad a'_{33} = 1. . \tag{4.18}$$

As a result, Equation (4.16) simplifies to

$$V^{(+)} = \{u' \cos(\psi + \phi) - v' \sin(\psi + \phi)\} \sin I + w' \cos I , \tag{4.19}$$

where u', v', w' are body-velocity of non-rotational origin. To specify these constitutes a separate problem which will be taken up in the closing section of this chapter; while in what follows we propose to investigate the contributions to the observed radial velocities of close binary systems arising from $V_{\rm rot}$.

B: Effects of Distortion on Radial Velocity

In more specific terms, if V_{rot} stands for the magnitude of the radial component of the velocity-vector at any surface point of a distorted star, the effect ΔV on the observed radial velocity V of the star as a whole should be represented by the ratio

$$V \equiv \frac{\int V_{\text{rot}} d\ell}{\int d\ell} \tag{4.20}$$

of the local value of V_{rot} averaged over the star's apparent disc (or, in eclipsing systems, for a visible fraction thereof) with respect to the light element $d\ell \equiv J \cos\gamma\, d\sigma$ at that point, of bright J and the surface element $d\sigma$ inclined at an angle γ to the line of sight.

The denominator

$$\int d\ell \equiv \mathcal{L} \tag{4.21}$$

on the right-hand side of this equation is obviously identical with the total luminosity $\mathcal{L}_i$ of the respective component—constant if the star were spherical, and varying with the phase during eclipses (if any), or as a result of distortion. The variation of $\mathcal{L}_1$ has already been investigated (to the first order in small quantities) in the preceding section of this chapter.

On insertion from (4.14) for V_{rot}, the numerator in (4.20) can be evaluated by exactly the same method which led us before to (2.29) to establish the light variations of distorted stars; and if—similarly to (2.16)—we set

$$\Delta V = \sum_{h=1}^{n+1} \frac{C^{(n)}}{\nu} \Delta V_1^{(h)} , \tag{4.22}$$

it follows that

$$\begin{aligned} \Delta V_1^{(h)} = & -\frac{\beta_2 + 1 - h}{(h+2)(h+4)} w_1^{(2)} m_0 P_2'(\ell_0) - \\ & - \frac{h(\beta_2 + 7 - h)}{(h+1)(h+3)(h+5)} w_1^{(3)} m_0 P_3'(\ell_0) - \\ & - \frac{(h-1)(\beta_4 + 15 - h)}{(h+2)(h+4)(h+6)} w_1^{(4)} m_0 P_4'(\ell_0) + \\ & + \dots , \end{aligned} \tag{4.23}$$

the constants β_j continue to be given by Equation (2.30); and $w_1^{(j)}$, by (2.22).

Since, moreover,,

$$m_0 P_j'(\ell_0) = -n_1 P_j^1(\ell_0) , \tag{4.24}$$

Equation (4.23) can be (correctly to j = 2, 3, 4) alternatively rewritten as

$$\Delta V = 2\omega_1 R_1 n_1 \sum_{j=2}^{4} w_1^{(j)} f_j^{(n)} P_j^1(\ell_0) , \tag{4.25}$$

where

$$f_2^{(n)} = \sum_{h=1}^{n+1} \frac{\beta_2 + 1 - h}{(h+2)(h+4)} C^{(h)} , \tag{4.26}$$

$$f_3^{(n)} = \sum_{h=1}^{n+1} \frac{h(\beta_3 + 7 - h)}{(h+1)(h+3)(h+5)} C^{(h)} , \tag{4.27}$$

$$f_4^{(n)} = \sum_{h=1}^{n+1} \frac{(h-1)(\beta_4 + 15 - h)}{(h+2)(h+4)(h+6)} C^{(h)} . \tag{4.28}$$

From Equation (4.22) it readily transpires that *whereas the light changes arising from tidal distortion of a rotating star are known to be expressible in terms of the zonal harmonics* $P_j^0(\ell_0)$ *of the phase, the radial velocity changes are found to be expansible in terms of tesseral harmonics of the type* $P_j^1(\ell_0)$ *and of the same argument.*

In the case of uniformly bright stars (no limb-darkening)

$$f_2^{(0)} = \frac{1}{15}\beta_2 = \frac{4}{15}\tau , \tag{4.29}$$

$$f_3^{(0)} = \frac{1}{48}(\beta_3 + 6) = \frac{1}{48}(5\tau + 6) , \tag{4.30}$$

$$f_4^{(0)} = 0 ; \tag{4.31}$$

for the linear law of limb-darkening ($n = 1$),

$$f_2^{(1)} = \frac{8\beta_2 - (3\beta_2 + 5)u_1}{40(3 - u_1)} , \tag{4.32}$$

$$f_3^{(1)} = \frac{35(\beta_3 + 6) - (3\beta_3 + 50)u_1}{560(3 - u_1)} , \tag{4.33}$$

$$f_4^{(1)} = \frac{(\beta_4 + 13)u_1}{64(3 - u_1)} ; \tag{4.34}$$

for $n = 2$ (quadratic limb-darkening),

$$f_2^{(2)} = \frac{56\beta_2 - 7(3\beta_2 + 5)u_1 - 16(2\beta_2 + 3)u_2}{140(6 - 2u_1 - 3u_2)} , \tag{4.35}$$

$$f_3^{(2)} = \frac{140(\beta_3 + 6) - 4(3\beta_3 + 50)u_1 - 35(\beta_3 + 12)u_2}{1120(6 - 2u_1 - 3u_2)} , \tag{4.36}$$

$$f_4^{(2)} = \frac{(\beta_4 + 13)u_1}{32(6 - 2u_1 - 3u_2)} + \frac{4(\beta_4 + 12)u_2}{105(6 - 2u_1 - 3u_2)} ; \tag{4.37}$$

etc.

An inspection of the foregoing results discloses that non-orbital contributions δV to the radial velocity of rotating components in close binary systems vanish not only as regards terms arising from polar flattening, but also from second-harmonic tides if the apparent disc of prolate spheroid were uniformly bright; for if $u_1 = u_2 = \ldots = u_n = 0$ and $\tau = 0$, all $f_2^{(n)}$'s vanish identically. For distortion arising from such tides to produce a significant contribution to δV it is necessary that there be some limb- and (or) gravity-darkening. Third-harmonic distortion of tidal origin will, however, contribute to δV even in the absence of limb- or gravity-darkening; while fourth-harmonic distortion contributes nothing unless there is some darkening at the limb.

The coefficient $f_2^{(n)}$ of the contribution due to the second harmonic may be positive or negative—depending on whether limb- or gravity-darkening preponderates—and vanishes when

$$\beta_2 = \frac{35u_1 + 48u_2 + \ldots}{56 - 21u_1 - 32u_2 - \ldots}; \tag{4.38}$$

or, for the linear law of limb-darkening ($n = 1$), for a value of u_1 which satisfies the condition

$$\beta_2 = \frac{5u_1}{8 - 3u_1} \text{ or, conversely, } u_1 = \frac{8\beta_2}{5 + 3\beta_2}. \tag{4.39}$$

In such a case, δV vanishes for spheroidal configurations, and this can happen if—and only if—the isophotes on the apparent disc of such a star remain symmetrical with respect to the projected centre of mass. Therefore, *to every degree of limb-darkening of a rotating elllipsoid there corresponds a certain amount of gravity-darkening for which* $f_2^{(n)} = 0$ *and which, accordingly, make the "centre of light" of its apparent disc project itself constantly on the centre of mass of the respective star*.

One trivial case when this is true arises if the disc of the star in question is uniformly bright, since if all the u_i's vanish and $\tau = 0$, Equations (4.38) or (4.39) are obviously satisfied. A second, less trivial, case arises if $u_1 = 1$ (i.e., complete linear darkening at limb) and $\beta_2 = 1$ (corresponding to a coefficient of gravity-darkening $\tau = \frac{1}{4}$). For $u_1 = 1$, the limb itself has become an isophote corresponding to $J = 0$, and if, in addition, $\tau = \frac{1}{4}$, all isophotes interior to the limb must represent a family of ellipses similar to the limb and concentric with it. This particular condition corresponding to $u_1 = 1$ and $\tau = 0.25$ was noticed first by Russell (1942), but the more general one represented by Equation (4.39) was subsequently discovered by Kopal (1945).

All this is true, of course, only if we restrict the form of the distorted configuration to remain ellipsoidal. When we come to consider the terms in δV associated with tidal distortion of symmetry described by harmonics higher than the second, we note that—unlike $f_2^{(n)}$—the coefficients $f_4^{(n)}$ associated with the fourth harmonic vanish only if the star is undarkened at limb; and the coefficients $f^{(n)}$ factoring the effects arising from the third-harmonic distortion would fail to vanish even then. This fact indicates that in no circumstances can the isophotes on apparent discs of distorted ellipsoids constitute a family of curves which remain symmetrical with respect to the projected centre of mass—whatever the amount of limb- or gravity-darkening.

In order to investigate the significance of the foregoing results, and their bearing on the determination of the orbital elements of close spectroscopic binaries from observations of their radial velocities, let us compare Equations (4.23) or (4.25) expressing the radial-velocity changes arising from the tidal distortion of their components with the expressions for the radial velocity of the centre of mass of such stars in their orbit. In doing so we shall assume that their actual orbit is circular, so that the true anomaly u becomes identical with the mean longitude $L \equiv \omega + M$; and the phase angle $\psi \equiv u - 90°$ in the direction cosines ℓ_0 and m_0 becomes equal to $L - 90°$. Accordingly,

$$\begin{aligned} \ell_0 &= \sin L \sin I , \\ m_0 &= - \cos L \sin I , \\ n_0 &= \cos I ; \end{aligned} \qquad (4.40)$$

and, as the reader may easily verify,

$$m_0 P_2'(\ell_0) = -\frac{1}{2} P_2^2(n_0) \sin 2L , \qquad (4.41)$$

$$m_0 P_3'(\ell_0) = -\frac{1}{4} P_3^1(n_0) \cos L - \frac{1}{8} P_3^3(n_0) \cos 3L , \qquad (4.42)$$

$$m_0 P_4'(\ell_0) = \frac{1}{12} P_4^2(n_0) \sin 2L + \frac{1}{48} P_4^4(n_0) \sin 4L ; \qquad (4.43)$$

which on insertion in (4.22)–(4.28) yield

$$\begin{aligned} \Delta V = & \frac{1}{24}\{12 f_3^{(n)} w_1^{(3)} P_3^1(n_0) \cos L + 4[6 f_2^{(n)} w_1^{(2)} - f_4^{(n)} w_1^{(4)} P_4^2(n_0)] \sin 2L + \\ & + 6 f_3^{(n)} w_1^{(3)} P_3^3(n_0) \cos 3L - f_4^{(n)} w_1^{(4)} P_4^4(n_0) \sin 4L + \ldots\} \omega_1 R_1 . \end{aligned} \qquad (4.44)$$

In reality, the terms on the right-hand side of the preceding equation will superpose on those arising from Keplerian motion; and would falsify its elements obtained by interpreting the observations in the conventional way. For instance, the term in (4.44) varying as $\cos L$ would superimpose on the amplitude K_1 of the orbital motion, from which the masses and absolute dimensions of the system can be deduced (or estimated). Or the terms in (4.44) varying as $\sin 2L$ would obviously simulate the effects or orbital eccentricity in an orbit characterized by a spurious longitude of periastron 90° or 270°. The existence of such phenomena was noted many years ago (cf. Sterne, 1941b); for their latest discussion cf. Kopal (1980a,c).

All results discussed so far pertain to the radial velocities of rotating components of close binary systems observed in "full light"—i.e., at such phases when the apparent discs of the stars are fully exposed to the observer. Should, however, the binary in question happen to be an eclipsing variable, an analysis of the type developed earlier in this section continues to furnish the non-orbital contributions ΔV to the observed radial velocity of the star undergoing eclipse as well—provided only that the limits of integration in the numerator as well as denominator on the r.h.s. of Equation (4.20) for ΔV are extended over the fraction of the eclipsed star visible at a particular phase.

This has indeed been done by the present writer to the same degree of accuracy to which the corresponding light changes have been so investigated (cf. Kopal, 1945); and the results expressible again in terms of the special functions (3.25)–(3.26), (3.36) and (3.39) introduced in the preceding section are well known, though too long to be quoted in full in this place (for their latest version, cf. Kopal, 1980a).

Only in the case of spherical stars undergoing eclipse the situation becomes simpler, and yet sufficiently illustrative to be rendered here in full (cf. Kopal, 1942b). In such a case, the loss of light becomes given by

$$1 - \ell = \sum_{h=1}^{\infty} C^{(h)} \alpha_{h-1}^{0} , \tag{4.45}$$

where the limb-darkening coefficients $C^{(h)}$ continue to be given by Equations (2.17)–(2.18); and, as a result, Equation (4.20) can be reduced to the form

$$\Delta V = -\frac{\omega_1 R_1 n_1 \sum_{h=1}^{\infty} C^{(h)} \alpha_{h-1}^{1}}{\sum_{h=1}^{\infty} C^{(h)} \left\{ \frac{2}{h+1} - \alpha_{h-1}^{o} \right\}} , \tag{4.46}$$

where the α_n^m's are the "associated α-functions" as defined by Equations (3.25)–(3.26) as functions of the phase, and the direction cosine n_1 is given by

$$n_1 \equiv \frac{m_0}{\ell_2} = \frac{\sin\psi \sin j}{\sqrt{1 - \cos^2\psi \sin^2 j}} \tag{4.47}$$

in accordance with Equations (3.2), (3.4) and (2.14).

The quantity $\Delta V/\omega_1 R_1$ as defined by Equation (4.46) is commonly referred to as the "rotation factor", signifying a deviation of the radial velocity of the remaining crescent of the star undergoing eclipse from that of its centre of mass, equating the observed value V with its theoretical value as given by (4.46) should enable us to extract from each value of ΔV measured at any phase the corresponding absolute value of $\omega_1 R_1$. Moreover, the algebraic sign of the radial velocities ΔV observed before and after the moments of conjunction can disclose the algebraic sign of ω_1—to tell us if the rotation is direct or retrograde. In every case observed so far the radial-velocity deviation ΔV turned out to be *positive* (i.e., spectral lines red-shifted) *before* maximum eclipse, and negative afterwards—facts disclosing that the sense of axial rotation is *direct*. That no case of retrograde rotation has been detected in any system is a fact of undoubted cosmogonic significance, which would seem to testify in favour of *fissional* origin of close binary systems; but all the implications of such a conjecture are not yet fully clear.

But let us return to the absolute values of the products $\omega_1 R_1$ determined spectroscopically in the manner outlined in this section. Should it be legitimate to identify ω_1 with the Keplerian angular velocity $\omega_K = 2\pi/P$ of the star's orbital motion in a period P, a known value of $\omega_1 R_1$ would serve for a determination of the absolute dimensions of so rotating a star. This is often possible in systems characterized by circular orbits (Algol being a typical example), but dubious if the eccentricity is large—in such cases, the components are apt to rotate *faster* than they revolve. The only opposite known case when $\omega_1 << \omega_K$ appears to be the principal (cB8) component of the system of β Lyrae; and the slowness of its axial rotation may be connected with the conservation of its angular momentum in the course of the star's expansion in the relatively recent past.

Lastly, the structure of our equations defining the rotation factors makes it evident that if the orbit of the system is circular, and axis of rotation perpendicular to the orbital plane, the rotational effect before the moment of conjunction should be the mirror-image of the one after conjunction (i.e., symmetrical, but differing in sign). Should, however, the axis of rotation be inclined to the plane of the orbit (and, consequently, its projection on the $Z''' = 0$ plane inclined to the Y'''-axis), the rotational effect within minima would become *asymmetric*—a fact which could again be detected from the observations. Moreover, once the axis of rotation becomes inclined to the plane of the orbit, its direction in space (or its projection on the celestial sphere) can no longer remain fixed, but will be bound to precess—causing its projection to oscillate slowly in the period of precessional motion (cf. Section VI.3A). In such a case, the results of this section based on the use of Equation (4.14) for V_{rot} to compute ΔV would still be incomplete; and recourse would have to be made to the full-dress Equation (4.13). To do so would not, however, confront us with any fundamental obstacles; only the algebra would be more complicated; and can be left as an exercise for the interested reader.

Lastly, it should be stressed that—apart from tidal distortion— one additional physical mechanism is operative in close binary systems which is bound to render the radial-velocity curves of their components asymmetric in full light as well as during eclipses: and that is the so-called "reflection effect", produced by mutual irradiation of both components, capable of warming up the illuminating component by several hundred degrees.

A physical theory of such a phenomenon—discovered first by Dugan (1908) and Stebbins (1910) in the eclipsing systems RT Per and Algol (and since in many others)—was developed by Eddington (1926b) or Milne (1926); and the extent to which it can affect the observed radial velocity of such objects was established by the present writer (cf. Kopal, 1943). We shall, however, not give any more detailed account of it in this place, because its underlying process is intimately connected with another effect—triggered by reflection—which can exert a much more profound influence on the observed radial velocity of close binary systems; and to this we shall turn our attention in the closing section of this book.

C: Irradiation of the Components and its Effect on the Observed Radial Velocity

Throughout the preceding parts of this section we have been concerned mainly with the effects of axial rotation on the radial velocity of distorted components of close binary systems—three-axial if we employ Equation (4.13), or monoaxial if we adopt (4.14)—which cause the "centre of light" of their projected discs to deviate from their projected centres of mass. The actual situation in real binary systems may, of course, become still more complicated; for their components not only attract, but also irradiate each other; and this phenomenon (commonly referred to as the "reflection effect"; although its physical reason is absorption-reemission and (or) scattering) provides an additional reason rendering the geometry of isophotes over their apparent discs to vary with the phase. Moreover, an essential feature of the reflection effect is the fact that it influences these isophotes—and, therefore, the radial velocity of the respective star—even when the latter is spherical, or departs from spherical form to only a negligible extent.

The principal reason of our concern with the effects of reflection on the radial velocity of the components of binary systems goes, however, much deeper: it is not only its effect on the geometry of isophotes which renders the average motion of their surfaces a function of the phase, but (if the physical mechanism of reflection is absorption-reemission) the absorption of incident heat will produce convective motions within the atmosphere, which may be symmetrical with respect to the radius-vector of the illuminating source, but *not* with respect to the line of sight! As is well known, the radial velocity of the components at any phase can be deduced from the Doppler shifts of spectral lines formed in the atmosphere of the respective star; and the classical way to interpret these velocities has been to ascribe them to the Keplerian motions of the components of the system about its centre of mass.

However, such an identification entails a tacit assumption that *the gaseous layers in which the measured spectral lines originate are at rest with respect to the centre of mass of the respective star*. As long as the systems of isophotes over the apparent disc of such a star is indeed symmetrical with respect to the projected centre of mass, the foregoing requirement may well be fulfilled—at least to a good degree of approximation—in the average. Thermal effects of incident radiation are, however, bound to give rise to a global system of gas currents which are not symmetrical with respect to the line of sight, and not limited to the "daylight" hemisphere when the illuminating source is above the horizon.

Should axial rotation of the irradiated star be synchronized with the orbit, its surface would be permanently divided into "day" and "night" hemispheres; with warm gas rising in daytime, and sinking wherever the illuminating source is below the horizon. However, if rotation and revolution are asynchronous, each point on the illuminated surface (apart from the poles) would experience alternation of days and nights—a phenomenon giving rise to atmospheric streaming which is bound to persist even in regions where the intake of heat from the companion star is cut off by the onset of night.

A "dynamical meteorology" in the atmospheres of the components of close binary systems—as distinct from those of single stars whose external energy input is negligible—represents a problem, the foundations of which were laid down already in the earlier chapters of this book. It is also one of great importance; for systems are indeed well known—such as U Cephei (cf. Carpenter, 1930; Struve, 1944) or RZ Scuti (Neubauer and Struve, 1945)—whose actual orbits (as evidenced by the light changes of these eclipsing variables) are essentially circular, but whose radial-velocity curves are asymmetric to such an extent as to simulate an orbital eccentricity in excess of 0.4! The distortion of their components is much too small to be responsible for the anomalies exhibited by them; and gas streams offer the only reasonable approach to their explanation.

In such a case, the deformation ΔV of the radial-velocity curve due to orbital motion should be expressible as

$$\Delta V = \frac{\int V^{(+)} d\ell}{\int d\ell}, \tag{4.48}$$

where the local radial velocity $V^{(+)}$ of non-orbital origin should be expressible by Equation (4.16) in terms of the *body* velocity components u', v', w' of gas motion.

In order to investigate such motions, let us transform first these body velocities from singly-primed to doubly-primed coordinates by means of the matrix equation

$$\begin{Bmatrix} u' \\ v' \\ w' \end{Bmatrix} = \begin{Bmatrix} \lambda_1'' & \lambda_2'' & \lambda_3'' \\ \mu_1'' & \mu_2' & \mu_3' \\ \nu_1'' & \nu_2' & \nu_3' \end{Bmatrix} \begin{Bmatrix} u'' \\ v'' \\ w'' \end{Bmatrix}, \tag{4.49}$$

where the direction cosines λ_j'', μ_j'', ν_j'' are given by Equations (3.11) of Chapter II; in doing so we find that

$$a_{11}'u' + a_{12}'v' + a_{13}'w' = a_{11}''u'' + a_{12}''v'' + a_{13}''w'', \tag{4.50}$$

and

$$a_{31}'u' + a_{32}'v' + a_{33}'w' = a_{31}''u'' + a_{32}''v'' + a_{33}''w'', \tag{4.51}$$

where the direction cosines a_{ij}'' continue to be given by Equations (3.7)–(3.9) of that chapter.

Next, let us change over from the rectangular velocity components u'', v'', w'' to the polar velocity components U, V, W by means of the matrix equation

$$\begin{Bmatrix} U \\ V \\ W \end{Bmatrix} = \{L\} \begin{Bmatrix} u'' \\ v'' \\ w'' \end{Bmatrix}, \tag{4.52}$$

where $\{L\}$ stands for the square matrix (3.18) of Chapter V. If so, it follows that

$$\begin{aligned} a''_{11}u'' + a''_{12}v'' + a''_{13}w'' = \\ = U\{[\cos\Omega\cos(\phi+u) - \sin\Omega\sin(\phi+u)\cos i]\sin\theta + \\ + \sin\Omega\sin i\cos\theta\} + \\ + V\{[\cos\Omega\cos(\phi+u) - \sin\Omega\sin(\phi+u)\cos i]\cos\theta - \\ - \sin\Omega\sin i\sin\theta -\} - \\ - W\{\cos\Omega\sin(\phi+u) + \sin\Omega\cos(\phi+u)\cos i\} \end{aligned} \tag{4.53}$$

and

$$\begin{aligned} a''_{31}u'' + a''_{32}v'' + a''_{33}w'' = \\ = U\{\sin(\phi+u)\sin i\sin\theta + \cos i\cos\theta\} + \\ + V\{\sin(\phi+u)\sin i\cos\theta - \cos i\sin\theta\} + \\ + W\{\cos(\phi+u)\sin i\}\,, \end{aligned} \tag{4.54}$$

where the angles Ω and i specify the position of the orbital plane in space (cf. Section II.3A) and u, the true anomaly of the revolving body measured from the nodal passage.

If, moreover, the angle i between the (instantaneous) orbital and invariable plane of the system is small enough to be ignored, an insertion from Equations (3.7) and (3.8) of Chapter II discloses that

$$\begin{aligned} a''_{11}u'' + a''_{12}v'' + a''_{13}w'' &= U\cos(\Omega+u+\phi)\sin\theta + \\ &+ V\cos(\Omega+u+\phi)\cos\theta - W\sin(\Omega+u+\phi) \end{aligned} \tag{4.55}$$

and

$$a''_{31}u'' + a''_{32}v'' + a''_{33}w'' = U\cos\theta - V\sin\theta\,. \tag{4.56}$$

Within this scheme of our approximation, the non-rotational contribution to non-orbital radial velocity $V^{(+)}$ arising from bodily motion will, in accordance with Equation (4.15), be of the form

$$\begin{aligned} V^{(+)} &= U\{\cos(\Omega+u+\phi)\sin I\sin\theta + \cos I\cos\theta\} + \\ &\quad \{\cos(\Omega+u+\phi)\sin I\cos\theta - \cos I\sin\theta\} - \\ &\quad - W\{\sin(\Omega+u+\phi)\sin I\}\,. \end{aligned} \tag{4.57}$$

In order to determine the velocity-components U, V, W describing the gas motions in rotating stellar atmospheres in Clairaut coordinates, let us return to Equations (3.25)–(3.27) of Section V.C, to which we adjoin the equation (3.8) of continuity of the same section, safeguarding the conservation of mass. However, in order to safeguard also the conservation of energy in irradiated stellar atmospheres, we must adjoin also the equation of radiative transfer represented by

Equation (1.19) of Section VII.1, together with the Equation (1.49) of radiative equilibrium, in which the irradiating flux πS is given by

$$\pi S = L_2/R^2 \tag{4.58}$$

to a plane-parallel approximation; where L_2 stands for the luminosity of the illuminating source at a distance R; while the angle α between the direction of incident flux and surface normal on the r.h.s. of (1.49) is defined by the equation

$$\cos\alpha = \ell_0 L + \ell_2 N , \tag{4.59}$$

where ℓ_0, 0, ℓ_2 are (cf. Equation (3.2)) the direction cosines of the X'''-axis in the doubly-primed systems of coordinates; and L, M, N are those of an arbitrary radius vector in the triply-primed system of coordinates: if

$$\left.\begin{array}{rcl} L & \equiv & \cos\gamma , \\ M & = & \sin\gamma \sin\zeta , \\ N & = & \sin\gamma \cos\zeta , \end{array}\right\} \tag{4.60}$$

where γ continues to denote the angle of reflection,

$$\cos\alpha = \cos\gamma\cos\epsilon + \sin\gamma\sin\epsilon\cos\zeta \tag{4.61}$$

in which ϵ stands for the angle between the radius-vector of the binary orbit (i.e., the X'''-axis) and the line of sight.

In order to establish the contribution ΔV to the radial velocity of the illuminated star arising from atmospheric convection within the layer of optical depth $0 < \tau < \infty$, we have to evaluate the ratio of the integrals of $d\ell$ and $V^{(+)}d\ell$ on the right-hand side of Equation (4.48), taken (cf. Kopal, 1943) between the limits

$$\int \equiv \int_{-1}^{1}\int_{0}^{\sqrt{1-M^2}} - \int_{-1}^{1}\int_{0}^{-\ell_0\sqrt{1-M^2}} \tag{4.62}$$

defining the geometry of the irradiated crescent of the illuminated star, with respect to the light element

$$d\ell \equiv J\cos\gamma\, d\sigma , \tag{4.63}$$

where

$$\cos\gamma \equiv L \quad \text{and} \quad d\sigma = \frac{dM\,dN}{L} . \tag{4.64}$$

The solution of this problem has so far been obtained (see Kopal, 1988) under the simplest possible conditions (parallel incident beam, spherical stars, Lambert's law of reflection, Boussinesq approximation to the equation of state), and much remains yet to be done before the whole subject can be placed on a more solid footing. However, preliminary results already on hand disclose that physical conditions may easily exist in the atmospheres of many known binaries in which a reduction of their observed radial velocities to the mass centres of their components becomes a matter of some complexity; and that the two cannot be automatically identified. The velocities furnished directly by the observations can be related with the orbital motion of the components only through the "filter" of the outer semi-transparent fringe in which the spectral lines whose Doppler shifts we can measure originate. It is only if gas in this fringe is at rest with respect to the centre of mass of the star as a whole that the observed radial velocity can be identified without any ado with those arising from the Keplerian orbital motion. In the present section we have, however, shown that the "reflection effect" in close binary systems—making their components irradiate each other from close proximity—makes this unlikely, if not impossible, to be the case. If we do not heed this message, its neglect may cause the results obtained by short-cutting the process of velocity transfer between observations and the centre of mass to be systematically in error—not only as far as the shape of the orbit is concerned, but also its size—and, above all, the absolute masses of the constituent stars.

VII.5 Bibliographical Notes

An investigation of the effects of radiative transfer in plane-parallel atmospheres on the distribution of brightness over the apparent discs of the stars, treated in the first section of this chapter, goes back to Schuster (1905) and Schwarzschild (1906); in extended atmospheres, to Kozyrev (1934) and Chandrasekhar (1934); for a comprehensive summary of the entire subject cf. Chandrasekhar (1950). For a treatment of the radiative transfer in distorted outer layers of the Roche model, followed in this section, cf. Kopal and Zafiropoulos (1983).

The material of Section 2 follows largely an earlier presentation of this subject in Chapter IV of the author's previous book on *Close Binary Systems* (Kopal, 1959) and Chapter II of his *Language of the Stars* (Kopal, 1979a), which contain extensive references to previous literature.

Up to the early 1930's, the form of the components in close binary systems was regarded as arbitrary ellipsoids, unrelated to other physical elements (fractional dimensions, mass-ratio) of the systems. The first investigator who considered their shape to be governed by the prevalent field of force in hydrostatic equilibrium was Takeda (1934), who was the first to consider the effects of gravity-darkening on the light changes of eclipsing variables—a work developed further to its essentially present form by Kopal (1942a). Takeda considered, to be sure, the photometric effects of gravity-darkening only in total (integrated) light; an extension of his work to monochromatic light (under the assumption of black-body radiation) which led to Equation (2.8) is due to Kopal (1941b). The same author is also responsible for an investigation of the photometric effects of limb-darkening of the form (2.3) to an arbitrary degree of cos γ (cf. Kopal, 1949).

A theory of the light changes due to mutual eclipses of distorted stars goes back to the pioneer work of Takeda (1937); and was developed to its current state by the present writer (cf. Kopal, 1946, 1959, 1979a). For a fuller account of the theory of "special functions" of the form (3.25)–(3.26), (3.35) or (3.38) necessary to this end, cf. Kopal (1947), followed by Sections IV.5 and Chapter V of the source (Kopal, 1959 and 1979a) above quoted.

The effects of distortion on the radial velocities of close binary systems were predicted first by Sterne, and investigated for the ellipsoidal shape of distorted configurations. His work was extended (by a more general method) for all higher harmonics by the present wrtier (1945) to the degree presented in this chapter (cf. also Chapter V of Kopal, 1959); and followed by Kitamura (1953, 1954) or Batten (1957).

The "rotational effect" within eclipses—detected almost simultaneously by McLaughlin (1924) in Algol, and by Rossiter (1924) in β Lyrae—was theoretically analyzed by Petrie (1938); and (more fully) by Kopal (1942b), to whom Equation (4.46) of the present chapter is due.

That effects of similar nature can be invoked also in systems consisting of spherical stars, and regardless of eclipses, by mutual irradiation of their components was foreseen (qualitatively) by Eddington (1926b); but not qualitatively investigated till by the present writer (Kopal, 1943). The physical theory underlying such phenomena goes back to Eddington (1926b) and Milne (1926) who was the first to obtain an approximate solution of radiative transfer of "reflected" light in plane-parallel atmospheres; its exact solution was subsequently obtained by Hopf (1934). For indirect effects of reflection on radial velocity causing thermal convection in irradiated stellar atmospheres cf. Kirbiyik and Smith (1976) or Kopal (1980c, 1988); the latter of which was largely followed in Section VII.3.

REFERENCES

Batten, A. H.: 1957, Mon. Not. R. astr. Soc., **117**, 521.

Bessel, F. W.: 1829, "Untersuchung des Theils der Planetarischen Störungen, welcher aus der Bewegung der Sonne entsteht", Berlin Abhandl.

Bialobrzewski, I.: 1913, Bull. Acad. Sci. Cracoviae, p.64.

Bouguer, P.: 1729, in "Essai d'optique sur la gradation de la lumière", Paris.

Brooker, R. A. and Olle, T. W.: 1955, Mon. Not. R. astr. Soc., **115**, 101.

Brouwer, D.: 1946, Astron. J., **52**, 57.

Bruggencate, P. ten: 1934, Zeit. f. Astrophys., **8**, 344.

Caimmi, R.: 1980, Astrophys. Space Sci., **71**, 415.

Callandreau, P. J. O.: 1888, Bull. Astr., **5**, 474.

Callandreau, P. J. O.: 1889, Ann. de l'Obs. de Paris, **19**E.

Carpenter, F. M.: 1930, Astrophys. J., **72**, 105.

Chanan, G. A., Middleditch, J. and Nelson, J. E.: 1976, Astrophys. J., **208**, 512.

Chandrasekhar, S.: 1933, Mon. Not. R. astr. Soc., **93**, 539.

Chandrasekhar, S.: 1934, Mon. Not. R. astr. Soc., **94**, 444.

Chandrasekhar, S.: 1950, *Radiative Transfer* (pp. XIV + 393), Clarendon Press, Oxford.

Chandrasekhar, S.: 1963, Astrophys. J., **138**, 1182.

Chapman, S.: 1954, Astrophys. J., **120**, 151.

Clairaut, A. C.: 1743, *Théorie de la Figure de la Terre, tirée des Principes de l'Hydrostatique*, Paris.

Courant, R. and Hilbert, D.: 1931, *Methoden der Mathematischen Physik* (2nd ed.), vol. **1**, p.153.

Darboux, G.: 1910 *Leçons sur les Systèmes Orthogonaux et les Coordonnées Curvilignes* (2nd ed.), pp.VI + 567, Gauthier-Villars, Paris; Chapitre II.

Darwin, G. H.: 1879, Phil. Trans. Roy. Soc. London, **170**, 447.

Darwin, G. H.: 1900, Mon. Not. R. astr. Soc., **60**, 82.

Darwin, G. H.: 1906, Phil. Trans. Roy. Soc. London, **206**, 161.

Darwin, G. H.: 1911, *The Tides* (3rd ed.), John Murray, London, p.340.

De Sitter, W.: 1924, Bull. Astr. Inst. Netherlands, **2**, 97.

Dirichlet, G. P. L.: 1850, J. de Crelle, **32**, 183.

Dugan, R. S.: 1908, Publ. Amer. Astr. Soc., **1**, 311.

Eddington, A. S.: 1916, Mon. Not. R. astr. Soc., **77**, 16.

Eddington, A. S.: 1926a, *Internal Constitution of the Stars* (pp.VI + 407), Cambridge Univ. Press.

Eddington, A. S.: 1926b, Mon. Not. R. astr. Soc., **86**, 320.

El-Shaarawy, M. B.: 1974, Ph.D. Thesis, Univ. of Manchester (unpublished).

Euler, L.: 1772, "Theoria Motuum Lunae", in *Impér.* Acad.de Sci.de St. Pétersbourg (reprinted in *Opera Omnia* (Ser. 2), **22**, ed. L. Courvoisier, Lausanne, 1958).

Galilei, G.: 1613, in *Opera di Galileo Galilei*, Edizione Nazionale (ed. by A. Favaro, Firenze), vol.**6**, p.198.

Galilei, G.: 1638, in *Discourses on the Two New Sciences*, Leiden.

Geroyannis, V. S.: 1980, Astrophys. Space Sci., **73**, 453.

Geroyannis, V. S. and Tokis, J. N.: 1977, Astrophys. Space Sci., **51**, 409.

Gyldén, H.: 1871, Nova Acta Reg. Soc. Sci. Uppsalliensis, **8**(3).

Hadrava, P.: 1987, Astrophys. Space Sci., **138**, 61.

Hagihara, Y.: 1972, *Celestial Mechanics* (vol. II.1), pp. IX + 504, MIT Press, Cambridge, Mass.

Hazlehurst, J. and Sargent, W. L. W.: 1959, Astrophys. J., **130**, 276.

Higgins, T. P. and Kopal, Z.: 1968, Astrophys. Space Sci., **2**, 352.

Hopf, E.: 1934, *Mathematical Problems of Radiative Equilibrium* (Cambr. Tracts in Mathematics and Math. Phys.), No. 31, pp.VI + 146, Cambridge Univ. Press.

Jacobi, C. J. G.: 1836, Compt. Rend. Acad. Paris, **3**, 59.

Jeans, J. H.: 1919, *Problems of Cosmogony and Stellar Dynamics* (pp. VII + 393), Cambridge Univ Press.

Jeans, J. H.: 1925, Mon. Not. R. astr. Soc., **85**, 526, 917, 933.

Jeans, J. H.: 1926, Mon. Not. R. astr. Soc., **86**, 328, 444.

Jeans, J. H.: 1928 *Astronomy and Cosmogony* (pp. X + 428), Cambridge Univ. Press.

Kirbiyik, H. and Smith, R. C.: 1976, Mon. Not. R. astr. Soc., **176**, 103.

Kitamura, M.: 1953, Publ. Astron. Soc. Japan, **5**, 114.

Kitamura, M.: 1954, Publ. Astron. Soc. Japan, **5**, 217.

Kitamura, M.: 1970, Astrophys. Space Sci., **7**, 272.

Kopal, Z.: 1941a, Ann. New York Acad. Sci., **41**, 13.

Kopal, Z.: 1941b, Proc. U. S. Nat. Acad. Sci., **27**, 359.

Kopal, Z.: 1942a, Proc. Amer. Phil. Soc., **85**, 399.

Kopal, Z.: 1942b, Proc. U. S. Nat. Acad. Sci., **28**, 133.

Kopal, Z.: 1943, Proc. Amer. Phil. Soc., **86**, 351.

Kopal, Z.: 1945, Proc. Amer. Phil. Soc., **89**, 517.

Kopal, Z.: 1946, *An Introduction to the Study of Eclipsing Variables* (Harvard Obs. Mono. No. 6), pp. X + 220, Harvard University Press, Cambridge, Mass.

Kopal, Z.: 1947, Harvard Obs. Circ., No. 450.

Kopal, Z.: 1949, Harvard Obs. Circ., No. 454.

Kopal, Z.: 1953, Mon. Not. R. astr. Soc., **113**, 769.

Kopal, Z.: 1954, Jodrell Bank Ann., **1**, 37.

Kopal, Z.: 1955, Ann. d'Astrophys., **18**, 379.

Kopal, Z.: 1959, *Close Binary Systems* (pp. XIV + 558), Chapman-Hall and John Wiley, London and New York.

Kopal, Z.: 1960, *Figures of Equilibrium of Celestial Bodies* (pp.VI + 135), Univ. of Wisconsin Press, Madison, Wis.
Kopal, Z.: 1968a, Astrophys. Space Sci., **1**, 179.
Kopal, Z.: 1968b, Astrophys. Space Sci., **2**, 23, 166.
Kopal, Z.: 1969, Astrophys. Space Sci., **5**, 360.
Kopal, Z.: 1970, Astrophys. Space Sci., **8**, 149.
Kopal, Z.: 1971, Astrophys. Space Sci., **10**, 328.
Kopal, Z.: 1972, "The Roche Model" in *Advances in Astronomy and Astrophysics*, Vol. **9**, pp. 1-65, Academic Press, New York.
Kopal, Z.: 1973, Astrophys. Space Sci., **24**, 145.
Kopal, Z.: 1974, Astrophys. Space Sci., **27**, 389.
Kopal, Z.: 1978, *Dynamics of Close Binary Systems* (pp. XIII + 510), D. Reidel Publ. Co., Dordrecht and Boston.
Kopal, Z.: 1979a, *Language of the Stars* (pp. VII + 280), D. Reidel Publ. Co., Dordecht and Boston.
Kopal, Z.: 1979b, Astrophys. Space Sci., **60**, 441.
Kopal, Z.: 1980a, Astrophys. Space Sci., **70**, 329.
Kopal, Z.: 1980b, Astrophys. Space Sci., **70**, 407.
Kopal, Z.: 1980c, Astrophys. Space Sci., **71**, 65.
Kopal, Z.: 1983, Astrophys. Space Sci., **93**, 149.
Kopal, Z.: 1987a, Astrophys. Space Sci., **133**, 157.
Kopal, Z.: 1987b, Astrophys. Space Sci., **134**, 55.
Kopal, Z.: 1988, Astrophys. Space Sci., **144**, 557.
Kopal, Z. and Ali, A. K. M. S.: 1971, Astrophys. Space Sci., **11**, 423.
Kopal, Z.: and Kitamura, M.: 1968, in *Advances in Astronomy and Astrophysics*, Vol. **6**, pp. 125-172, Academic Press, New York.
Kopal, Z. and Lanzano, P.: 1973, Astrophys. Space Sci., **23**, 425.
Kopal, Z. and Lyttleton, A. R.: 1963, Icarus, **1**, 455.
Kopal, Z. and Kamala Mahanta, M.: 1974, Astrophys. Space Sci., **30**, 347.
Kopal, Z. and Song, G.-X.: 1983, Astrophys. Space Sci., **92**, 3.
Kopal, Z. and Zafiropoulos, F.: 1983, Astrophys. Space Sci., **91**, 299.
Kourganoff, V. and Busbridge, I. W.: 1952, *Basic Methods of Transfer Problems* (pp.XI + 281), Oxford Univ. Press.
Kozyrev, N. A.: 1934, Mon. Not. R. astr. Soc., **94**, 430.
Kruszewski, A.: 1966, in *Advances in Astronomy and Astrophysics*, Vol. **4**, pp. 233-299, Academic Press, New York.
Kuiper, G. P. and Johnson, J. R.: 1956, Astrophys. J., **123**, 90.
Lagrange, J. L.: 1770, in Hist. Acad. Roy. Sci. Berlin, **25**, 204.
Lagrange, J. L.: 1772, "Essai d'une nouvelle méthode pour resoudre le problème des trois corps", in Recueil des pièces qui ont remparté les prix de l'Academie Royal des Sciences. Reprinted in *Oeuvres* (M. J. A. Serret, ed.), vol. 6, Gauthier-Villars, Paris 1873.
Lamb, H.: 1932, *Hydrodynamics* (6th ed.), pp. XV + 738, Cambridge Univ. Press.

Langebartel, R. G.: 1979, Astrophys. Space Sci., **60**, 431.
Lanzano, P.: 1962, Icarus, **1**, 121.
Lanzano, P.: 1973, Astrophys. Space Sci., **20**, 71.
Laplace, P. S.: 1825, *Mécanique Céleste*, Paris; vol. **5**.
Lebovitz, N.: 1970, Astrophys. Space Sci., **9**, 398.
Ledoux, P.: 1958, in Handb. d. Phys. (ed. S. Flügge), **51**, 432-455.
Legendre, A. M.: 1793, "Recherches sur la Figure des Planètes", in Mémoire de Mathématique par divers savants pour 1789 (though not actually printed in Paris until 1793).
Liapounov, A. M.: 1904, "Sur l'equation de Clairaut et les equations plus générales" in Mém. de l'Acad. Impérial de St. Petersbourg, (8)**15**, No.10.
Limber, D.N.: 1963, Astrophys. J., **138**, 1112.
Liouville, J.: 1858, Journ. de Math., (2)**3**, 1-25.
McLaughlin, D. B.: 1924, Astrophys. J., **60**, 222.
Milne, E. A.: 1926, Mon. Not. R. astr. Soc., **87**, 43.
Milne, E. A.: 1930, "Thermodynamics of the Stars", in *Handb. der Astrophys.*, Vol. 3 (first half), Springer Verlag, Berlin.
Nechvíle, V.: 1926, Compt. Rend. Acad. Paris, **182**, 310.
Neubauer, F. J. and Struve, O.: 1945, Astrophys. J., **101**, 240.
Newton, I.: 1687, *Philosophiae Naturalis Principia Mathematica*, vol. 3, London.
Nishimura, H. and Mori, H.: 1961, in Progr. Theor. Phys. Japan, **26**, 967.
Oster, L.: 1957, Zeit. f. Astrophys., **42**, 228.
Pannekoek, A.: 1902, "Untersuchungen über den Lichtwechsel Algols", Diss. Leiden.
Petrie, R. M.: 1938, Publ. Dominion Astrophys. Obs., **7**, 133.
Petty, A. F.: 1973, Astrophys. Space Sci., **21**, 189.
Plavec, M.: 1958, Mem. Soc. Roy. Sci. de Liège, (4)**20**, 411.
Plavec, M. and Kratochvíl, P.: 1964, Bull. Astr. Inst. Czechoslovakia, **15**, 165.
Poincaré, H.: 1885, Acta Math., **7**, 259.
Poincaré, H.: 1892, *Méthodes Nouvelles de la Mécanique céleste*, vol. 1, Chap. V, Gauthier-Villars, Paris.
Poincaré, H.: 1899, *Théorie du Potential Newtonien* (lectures at the Sorbonne, edited by E. Leroy and G. Vincent), Carré et Naud, Paris (Chapters IV-V and VII-VIII).
Poincaré, H.: 1902, in *Leçons sur les Figures d'Equilibre d'une masse Fluide* (lectures at the Sorbonne, edited by L. Dreyfus), Carré et Naud, Paris; p.9.
Poincaré, H.: 1910, Bull. Astr., **27**, 321.
Poincaré, H.: 1911, in *Leçons sur les Hypothèses Cosmogoniques* (lectures at the Sorbonne, edited by H. Vergue), A. Hermann et Fils, Paris.
Prendergast, K. H.: 1960, Astrophys. J., **132**, 162.
Radau, R.: 1885, Compt. Rend. Acad. Paris, **100**, 972; Bull Astr., **2**, 157.
Rahimi-Ardabili, M. Y.: 1979, Astrophys. Space Sci., **66**, 325.
Rein, N.: 1940, Trudy Sternberg State Astron. Inst., **14**, 85.
Roach, G. F.: 1975, Astrophys. Space Sci., **36**, 159.

Roach, G. F.: 1981, Astrophys. Space Sci., **80**, 237.
Roche, E. A.: 1849, Mém. de l'Acad. des Sci. de Montpellier, **1**, 243, 333.
Roche, E. A.: 1850, Mém. de l'Acad. des Sci. de Montpellier, **2**, 21.
Roche, E. A.: 1873, Mém. de l'Acad. des Sci. de Montpellier, **8**, 235.
Rödiger, C.: 1902, "Unterschuchungen über das Doppelsternsystem Algol", Diss. Königsberg.
Rosenthal, J. E.: 1931, Astron. Nachr., **244**, 169.
Rossiter, R. A.: 1924, Astrophys. J., **60**, 15.
Roy, A. E. and Ovenden,M. W.: 1961, Mon. Not. R. astr. Soc., **123**, 1.
Russell, H.N.: 1942, Astrophys. J., **95**, 345.
Scheibner, N.: 1866, Crelle J. Reine Angew. Math., **65**, 291.
Scheiner, Chr.: 1626, *Rosa Ursina etc.*, Bracciano; vol. 4, p.618.
Schuster, A.: 1905, Astrophys. J., **21**, 1.
Schwarzschild, K.: 1906, Göttingen Nachr., Nr. 41.
Stebbins, J.: 1910, Astrophys. J., **32**, 213.
Sterne, T. E.: 1941a, Proc. U. S. Nat. Acad. Sci., **27**, 99.
Sterne, T. E.: 1941b, Proc. U. S. Nat. Acad. Sci., **27**, 168.
Struve, O.: 1944, Astrophys. J., **99**, 222.
Szebehely, V. G.: 1963, Astron. J., **68**, 147.
Szebehely, V. G.: 1967, *Theory of Orbits: the Restricted Problem of Three Bodies*, Acad. Press, New York; Chapters 4 and 7.
Takeda, S.-I.: 1934, Mem. Coll. Sci. Kyoto (Ser. A), **17**, 197.
Takeda, S.-I.: 1937, Mem. Coll. Sci. Kyoto (Ser. A), **20**, 47.
Thomas, L. H.: 1930, Quart. J. Math. (Oxford), **1**, 239.
Tisserand, F.: 1891, *Traité de la Mécanique Céleste*, vol. II, Gauthier-Villars, Paris.
Tokis, J. N.: 1974a, Astrophys. Space Sci., **26**, 447, 477.
Tokis, J. N.: 1974b, Astrophys. Space Sci., **31**, 349.
Tokis, J. N.: 1974c, The Moon, **10**, 337.
Tokis, J. N.: 1975, Astrophys. Space Sci., **36**, 427.
Unsöld, A.: 1938, *Physik der Sternatmosphären*, Julius Springer Verlag, Berlin.
Véronnet, A.: 1912, Journ. de Math., (6)**8**, 331.
Wintner, A.: 1941, *The Analytical Foundations of Celestial Mechanics*, Princeton Univ. Press; Chapter VI.
Zeipel, H. von: 1924, Mon. Not. R. astr. Soc., **84**, 702.
Zeipel, H. von: 1925, Mon. Not. R. astr. Soc., **85**, 678.

INDEX